Impact of Plastic Waste on the Marine Biota

Mohd. Shahnawaz • Manisha K. Sangale •
Zhu Daochen • Avinash B. Ade
Editors

Impact of Plastic Waste on the Marine Biota

Springer

Editors
Mohd. Shahnawaz
Biofuels Institute
School of Environment and Safety Engineering
Jiangsu University
Zhenjiang, China

Manisha K. Sangale
Cyfoeth Naturiol Cymru
Natural Resource Wales
Swansea, UK

Zhu Daochen
Biofuels Institute
School of Environment and Safety Engineering
Jiangsu University
Zhenjiang, China

Avinash B. Ade
Department of Botany
Savitribai Phule Pune University
Pune, India

ISBN 978-981-16-5402-2 ISBN 978-981-16-5403-9 (eBook)
https://doi.org/10.1007/978-981-16-5403-9

This Springer imprint is published by the registered company Springer Nature Singapore Pte Ltd.
The registered company address is: 152 Beach Road, #21-01/04 Gateway East, Singapore 189721, Singapore

The editors would like to dedicate this book to their respective parents, spouses, and kids

Foreword

Plastic is one of the highly used polymers around the globe due to its enormous properties. Plastic is ubiquitous, highly inert, and requires several centuries to get mineralized in the environment. Every year, a huge amount of plastic waste is getting accumulated in the environment. Due to the lack of an efficient method to manage plastic waste, most of the plastic waste ends up in landfill (65%), around 15% is incinerated, and only 10% of plastic waste is recycled. Other methods also manage a fraction of plastic waste. However, the plastic waste at the dumping site is being eaten by the animals along with foodstuff and leads to their death. It is also reported that plastic waste blocks drainage system in various metropolitan cities and creates flood-like situations. Using different pathways, most of the plastic waste manages to enter the marine environment. In the oceans, plastic waste is accumulating at an alarming rate and leads to the deaths of billions of marine animals, birds, fishes, etc. annually at the global level.

During the past decade, Dr. Mohd. Shahnawaz, Dr. Manisha K. Sangale, and Prof. Dr. Avinash B. Ade have undertaken a demanding challenge to discover the elite microbial isolates (bacteria and fungi) from 12 ecogeographical locations along the West Coast of India to degrade the polythene. They have reported promising results and had presented their results in their previous book with Springer *Bioremediation Technology for Plastic Waste*.

I know Dr. Mohd. Shahnawaz virtually since his Ph.D. programme at the Department of Botany, Savitribai Phule Pune University, Maharashtra, India.

Recently, he got selected as an Associate Professor at Biofuels Institute, School of the Environment and Safety Engineering, Jiangsu University, Zhenjiang, China. He along with Prof. Dr. Daochen Zhu are involved in the mitigation of marine microplastic using bioremediation technology.

This book *Impact of Plastic Waste on the Marine Biota* is the joint effort of these potential editors. They have selected an important topic to spread awareness among the public to understand the impact of plastic waste on marine biota. I believe this book could be helpful for various governmental agencies around the globe, especially in developing countries to frame some regulations/guidelines to minimize the generation and entry of plastic waste into the oceans to save marine biota. This book is divided into two parts, part A discusses the plastic waste cycle (generation, transportation, accumulation, management, mitigation measures) and part B highlights the impact of plastic waste on the marine ecosystem. It contains a total of 16 chapters contributed by the experts of the international repute from different parts of the world. I am convinced and believe that this book will be proven as a potential awareness companion tool to make the readers aware of the deleterious effect of plastic waste on the marine biota to minimize the usage of plastic-based products to reduce the generation and emission of plastic waste into the marine environment.

Department of Biological Sciences
King Abdulaziz University
Jeddah, Saudi Arabia

Khalid Rehman Hakeem

Preface

Plastics have revolutionized our living standards, and we are all dependent on plastics in one way or the other to meet our daily needs. Plastic has replaced all the traditional polymers used at a domestic and commercial level due to its enormous properties. Every year, the demand of plastic-based products is elevated, which led to an increase in total global plastic production. As per an estimate, in 2018 the global plastic production was 359 million tonnes with the most production from Asia (51%). Plastics of various types are used at domestic and industrial level, viz. polyethylene (PE), polypropylene (PP), polyvinyl chloride (PVC), polyethylene terephthalate (PET), and others. However, PE and PP together represent around 50% of the total plastic production at the global level. Among all types of plastics, one-third of the plastic is used in the production of single used plastic products, viz. packaging material, garbage bags, and domestic items (foodstuff, snacks, and plastic cutlery). The usage of cosmetics or personal care products, synthetic textiles, packaging materials, medical resources, artificial turf materials, wastewater treatment plants, paints, additives, plastic pellets, and domestic items such as foodstuff, snacks, and plastic cutlery leads to the generation of primary microplastic. The fragmentation of plastic due to weathering, degradation, and alterations in the physical and chemical properties leads to the generation of secondary microplastic.

In the current scenario of the Coronavirus disease-19 (COVID-19) outbreak, the usage of plastics for the manufacturing of food packaging, waste bags, hand gloves, mask, protective shields, hand sanitizer bottles, and personal protective equipment (PPE) kits has increased many folds throughout the globe. This led to the generation of huge biomedical waste. The obtained biomedical waste resulted to elevate the overall production of plastic waste at an alarming rate. As per an estimate, around 60–99 million metric tons of plastic waste gets accumulated in the environment annually and if the usage of plastic increases with the current pace, it is predicted that by 2025 the world's non-rural population will generate around 6 million metric tons of solid waste on a daily basis. Due to the lack of a proper efficient waste management system, most of the plastic wastes from all the land sources include leakage from landfills and industrial activities, direct dumping, and sewage discharge finally manages to enter the marine environment through various routes, i.e. wind, river, water canals, and floods. Abandoned fishing nets are important sea-based sources of plastic that contribute to the high abundance of plastic waste in marine waters. As per an estimate, around 4.8–12.7 million metric tons of plastic waste managed to enter in the marine environment annually. In the oceans, due to various environmental and biogeochemcial forces, the plastic waste started deteriorating by following varied pathways and leads to the fragmentation of plastic into different sizes >25 mm, 5–25 mm, <5 mm, and <0.1 μm and are referred to as macroplastic, mesoplastic, microplastic, and nanoplastic, respectively. Plastic waste pollution has attracted huge attention and has been widely reported in the marine environment, including in water bodies, on beaches and shorelines, and in the deep sea. Every year, deaths of billions of marine animals, fishes, birds, benthos, coral reefs, and reptiles are reported due to ingestion, entanglement, disease, and absorption of plastic additives of the plastic waste.

In literature, there is a need to understand the insights of mismanagement of plastic waste, mitigation practices of the plastic litter, transportation of the plastic waste from land and marine sources to the marine environment, fragmentation of the plastic waste in the oceans, entry of plastic particles into the food chain, followed by the impact of plastic waste on the inhabitants of the marine environment.

So, in this book *Impact of Plastic Waste on the Marine Biota* an attempt was made to overview the fate of the plastic waste cycle (generation, transportation, accumulation, management, mitigation measures) and its impact on the inhabitants of the marine biota. The book is generally divided into two main sections: **Part I. Plastic waste cycle** (generation, transportation, accumulation, management, mitigation measures) and **Part II. Impact of plastic waste on the inhabitants of the marine biota**. Accordingly, quality chapters were invited from the experts of the field working in different parts of the world, a total of 16 chapters (Part I. Chaps. 1–10; Part II. Chaps. 11–16) were selected to be considered for this book.

The chapterwise synopsis of the book are as follows:

Chapter 1 overviews the sources, occurrence, and degradation pathways of the marine plastic waste, followed by the recommendations/suggestion for future practices to sustainably bioremediate plastic waste.

Chapters 2 and 3 enlist sources, generation, management of macro- and microplastics and discuss the problems associated with their mitigation measures.

Chapter 4 reports the accumulation, biodegradation mechanisms, and roles of biotic and abiotic factors in plastic biodegradation.

Chapters 5 and 6 brief about the strategies used to mitigate plastic waste in the marine environment.

Chapter 7 enlists various important policies/regulations/guidelines to be followed at global level to minimize the entry of plastic waste from land to marine environment.

Chapter 8 updates about the recent trends around the globe to address the plastic waste pollution.

Chapter 9 emphasizes the role of algae to mitigate the micro- and nanoplastic using bioremediation technology.

Chapter 10 highlights the impact of Coronavirus disease 19 (COVID-19) outbreak to enhance the generation and accumulation of littered plastic-based products used as a precautionary measures to stop the spread of COVID-19.

Chapter 11 describes the meachism of plastic particles (meso-, micro-, and nanoplastic) entry into the food chain across the trophic levels.

Chapters 12 and 13 document the impact of all types of plastic waste on the ecology, distribution, and health of benthos and coral reefs in the marine enviornment.

Chapters 14 and 15 overview the ecotoxic effect of plastic waste on marine flora and fauna followed by suggesting the strategies to be followed to save the marine inhabitants from the menace of plastic pollution.

Chapter 16 provides detailed information of different types of diseases reported in marine inhabitants as a result of plastic waste as a carrier of pathogens.

Suggestions are welcome, from the subject experts, your inputs will enable us in the improvement of ideas and proficiency in the implementation of the regulations to minimize the generation and emission of plastic waste from land into marine environment to save marine inhabitants.

Zhenjiang, China — Mohd. Shahnawaz
Swansea, UK — Manisha K. Sangale
Zhenjiang, China — Zhu Daochen
Pune, India — Avinash B. Ade

Acknowledgements

To begin with the name of Almighty God for his endless blessings on all of us to make this dream come true. We are thankful to anonymous reviewers for their valuable suggestions to improvize the quality of the book.

MS is thankful to the authorities of the Jiangsu University for providing an opportunity to work as an Associate Professor under Invited Talent Programme. MKS is the thankful to the authorities of the Natural Resource Wales, Swansea, United Kingdom for given an opportunity to work as a Scientist. MS and ZD are highly indebted to the financial support of the National Key R&D Program of China (Grant No.2018YFE0107100) and the Primary Research & Development Plan of Jiangsu Province.

It is of immense pleasure to acknowledge the help and support rendered by the editorial and production team (especially Dr. Bhavik Sawhney, Vaishnavi Venkatesh, Ashok Kumar, and Aakanksha Tyagi) of Springer Nature.

Last but not least, we would like to acknowledge our respective parents and spouses for their trust, love, and support throughout this work.

Contents

Editors and Contributors

About the Editors

Mohd. Shahnawaz recently got selected as an Associate Professor at Biofuels Institute, School of the Environment and Safety Engineering, Jiangsu University, Zhenjiang, China. Previously, he worked as a Botany lecturer (on academic arrangement basis) at Govt. Degree College Paloura, Mishriwala, Jammu and Kashmir. Last year, he got selected as a postdoc fellow at Yeungnam University, Gyeongsan, North Gyeongsang, South Korea; however, due to the COVID-19 scenario, he was not able to join the same. He has also completed the National Post-Doctoral Fellowship, funded by DST-SERB, GoI, at Plant Biotechnology Division, CSIR-Indian Institute of Integrative Medicine, Canal Road Jammu, Jammu and Kashmir, India. Under the guidance of Prof. (Dr.) Altafhussain B. Nadaf and Prof. (Dr.) Avinash B. Ade, he earned his M.Phil. (2010) and Ph.D. (2016) in Botany from the Department of Botany, Savitribai Phule Pune University, Pune, Maharashtra, India. Dr. Shahnawaz is recipient of various fellowships awarded by the Savitribai Phule Pune University and University Grants Commission (UGC), India. His research interest is in ecology, microbiology, bioremediation, and plant biotechnology. Currently, he is focused on the mitigation of microplastic using bioremediation technology. He has served as a referee and editor for several of international journals. In the year 2019, he received the Young Scientist Award from VDGood Professional Association at Visakhapatnam, India. His contributed work has been published in *Scientific Reports* (Nature Publishing Group), and his book on bioremediation technology for plastic waste (Springer) has attracted huge media coverage. He has published more than 20 research articles in the peer-reviewed journals and authored/ edited 8 books of international repute.

Manisha K. Sangale is currently working as a Scientist at Cyfoeth Naturiol Cymru, Natural Resource Wales, Swansea, United Kingdom. Previously, she worked as an Assistant Professor of Botany in the Department of Botany, Radhabai Kale Mahila Mahavidyalaya, Ahmednagar, Maharashtra, India. She had also worked as an Assistant Professor on CHB basis in the Department of Botany, S. M. Joshi College, Hadapsar, Pune, and Yashwantrao Chavan Institute of Science, Satara, Maharashtra, India. Under the guidance of Prof. Dr. B. B. Chaugule and Prof. Dr. Avinash B. Ade, she earned her M.Phil. and Ph.D. in Botany from the Department of Botany, Savitribai Phule Pune University, Pune, Maharashtra, India. Dr. Sangale is recipient of various fellowships and awards conferred by Savitribai Phule Pune University and University Grants Commission (UGC), India. Her research interest is in phycology, mycology, ecology, microbiology, and bioremediation. She has served as referee for a number of international journals. Dr. Sangale has more than 5 years teaching experience in phycology, mycology, and microbiology and plant biotechnology at UG and PG level. Her work on plastic degradation using fungi has been published in *Scientific Reports* (Nature Publishing Group) and her co-authored book on bioremediation technology for plastic waste (Springer) has attracted huge media coverage. She has published more than 17 research articles in the peer-reviewed international journals and authored/co-authored 2 books of international repute. Recently, she received the Young Scientist Award by VDGood Professional Associations.

Daochen Zhu is currently working as a Professor at Biofuels Institute, School of the Environment and Safety Engineering, Jiangsu University, Zhenjiang, China. After completing his Ph.D. in environmental biochemistry from Kobe University, Japan, in 2008, he joined Biofuels Institute, School of the Environment and Safety Engineering, Jiangsu University, Zhenjiang, China, in 2009 as an Assistant Professor. Later he was selected as an Associate Professor in the same Institution. He has earned his B.S. in fermentation engineering from Dalian Polytechnic University, China, in 2004. During the tenure of his career, he also worked at Washington State University, Pullman, WA, USA, as a visiting scholar (2015–2016). He is recipient of numerous awards and funding to support his research, by Jiangsu University, Zhenjiang, China. Dr. Zhu is a recognized guide for M.S. students. He is the reviewer and editorial board member of various peer-reviewed journals of international repute. His research group mainly focuses on microorganism resource and diversity, lignin biodegradation and chemicals product, and biodegradation of organic pollutants. He is also interested in the mitigation of microplastic through bioremediation technology. Dr. Zhu has 50 publications to his credit published in peer-reviewed international journals of repute with good impact factor. He also has 12 patents, 6 of those authorized to his credit. Dr. Zhu has also published two book chapters in the books indexed in Book Citation Index—Clarivate.

Avinash B. Ade is currently working as a Professor and Head at the Department of Botany, Savitribai Phule Pune University, Pune-411007, Maharashtra, India. He has earlier served as an Associate Professor (2005–2009) and Assistant Professor (1997–2005) at the Department of Botany, Dr. Babasaheb Ambedkar Marathwada University, Aurangabad, Maharashtra, India. His research interest is in ecology and cytogenetics with specialization in plant–microbes interaction and bioremediation. He has been conferred with various prestigious awards notably, Dr. M. A. Dhore Gold Medal for standing first in the order of merit in M.Sc. (botany) examination conducted by Amravati University, Amravati (1996). He has served as a referee for a number of national and international journals. Dr. Ade has more than 21 years of teaching experience in plant ecology, cytogenetic, genetics, plant pathology, and bioremediation.

He has also published more than 65 research articles in peer-reviewed international journals and authored/edited more than 10 books and numerous book chapters. He is a member of many international scientific societies and organizations, in particular life member of Marathwada Botanical Society, life member of Maharashtra Society of Genetics and Plant Breeding, life member of the Indian Society of Genetics and Plant Breeding, life member of the Indian Ecological Society, and life member of the Indian Society of Plant Pathologists.

Contributors

Ogundolie Frank Abimbola Department of Biotechnology, Baze University, Abuja, Nigeria

Avinash B. Ade Department of Botany, Savitribai Phule Pune University, Pune, Maharashtra, India

Charles Oluwaseun Adetunji Applied Microbiology, Biotechnology and Nanotechnology Laboratory, Department of Microbiology, Edo State University Uzairue, PMB 04, Auchi, Edo State, Nigeria

Kiran Aftab Department of Zoology, University of Gujrat, Gujrat, Pakistan

Syed Abrar Ahmad Department of Zoology, Savitribai Phule Pune University, Pune, Maharashtra, India

Romana Akhtar Department of Zoology, Govt. Degree College, Kishtwar, Jammu and Kashmir, India

Qurratu Aini Mat Ali Institute of Islam Hadhari, The National University of Malaysia, Selangor, Malaysia

Gloria Anwuli Anani Department of Animal and Environmental Biology, Faculty of Life Science, University of Benin, Benin City, Nigeria

Osikemekha Anani Laboratory for Ecotoxicology and Forensic Biology, Department of Biological Science, Edo State University Uzairue, PMB 04, Auchi, Edo State, Nigeria

Osikemekha Anthony Anani Laboratory for Ecotoxicology and Forensic Biology, Department of Biological Science, Faculty of Science, Edo State University Uzairue, Auchi, Edo State, Nigeria

Jobil J. Arackal Department of Food Technology, Saintgits College of Engineering, Kottukulam Hills, Pathamuttam, Kerala, India

Cesar David Benavides Fernández Universidad Santiago de Cali, Department of Microbiology, Cali, Colombia

Bhushan P. Bhusare R&D Department, Excellence Teaching and Research Academy, Ahmednagar, Maharashtra, India

Lida Vivian Carvajal Rodríguez Universidad Santiago de Cali, Department of Microbiology, Cali, Colombia

Doongar R. Chaudhary Plant Omics Division, CSIR-Central Salt and Marine Chemicals Research Institute, Bhavnagar, India

Elsa Cherian Department of Food Technology, Saintgits College of Engineering, Kottukulam Hills, Pathamuttam, Kerala, India

Zhu Daochen Biofuels Institute, School of the Environment and Safety Engineering, Jiangsu University, Zhenjiang, China

Mudasir A. Dar Biofuels Institute, School of the Environment and Safety Engineering, Jiangsu University, Zhenjiang, China
Department of Zoology, Savitribai Phule Pune University, Pune, Maharashtra, India

Neeraja P. Dhole Department of Zoology, Savitribai Phule Pune University, Pune, Maharashtra, India

Alex Ajeh Enenuku Department of Environmental Management and Toxicology, Faculty of Life Science, University of Benin, Benin City, Nigeria

Thabo Falayi Department of Engineering, Malawi University of Science and Technology, Limbe, Malawi

Farah Ayuni Mohd Hatta Institute of Islam Hadhari, The National University of Malaysia, Bangi, Selangor, Malaysia

Khalid Hussain Center for Biodiversity and Taxonomy, Department of Botany, University of Kashmir, , Hazratbal, Srinagar, Jammu and Kashmir, India

Mubashar Hussain Department of Zoology, University of Gujrat, Gujrat, Pakistan

Tunde O. T. Imoobe Department of Animal and Environmental Biology, Faculty of Life Science, University of Benin, Benin City, Nigeria

Abel Inobeme Department of Chemistry, Edo State University Uzairue, PMB 04, Auchi, Edo State, Nigeria

Tasadoq Hussain Jaweed Govt. Higher Secondary School, Kharpora Kokernag, Jammu and Kashmir UT, India

T. Jayasree Joshi Department of Food Technology, Saintgits College of Engineering, Kottukulam Hills, Pathamuttam, Kerala, India

Nor Hafizana Mat Jusoh Herbarium Unit, Department of Landscape Architecture, Kulliyyah of Architecture and Environmental Design, International Islamic University Malaysia, Kuala Lumpur, Malaysia

Sarika N. Kanade Centre for Science Education and Communication, Savitribai Phule Pune University, Pune, Maharashtra, India

Rahul Kapoore Department of Biosciences, College of Science, Swansea University, Swansea, UK

Alka Kumari Plant Omics Division, CSIR-Central Salt and Marine Chemicals Research Institute, Bhavnagar, India

Nur Hanie Mohd Latif International Institute for Halal Research and Training, International Islamic University Malaysia, Kuala Lumpur, Malaysia

Ho Man Leung Department of Biology, Hong Kong Baptist University, Hong Kong SAR, China

Wai Chin Li Department of Science and Environmental Studies, The Education University of Hong Kong, Hong Kong SAR, China

Somia Liaqat Department of Zoology, University of Gujrat, Gujrat, Pakistan

Lakshmi Mohan Department of Food Technology, Saintgits College of Engineering, , Kottukulam Hills, Pathamuttam, Kerala, India

Mufaro Moyo Department of Fibre and Polymer Materials Engineering, National University of Science and Technology, Bulawayo, Zimbabwe
Department of Engineering, Malawi University of Science and Technology, Limbe, Malawi

Olugbemi T. Olaniyan Laboratory for Reproductive Biology and Developmental Programming, Department of Physiology, Edo State University Uzairue, PMB 04, Auchi, Edo State, Nigeria

John Ovie Olomukoro Department of Animal and Environmental Biology, Faculty of Life Science, University of Benin, Benin City, Nigeria

Rashidi Othman Herbarium Unit, Department of Landscape Architecture, Kulliyyah of Architecture and Environmental Design, International Islamic University Malaysia, Kuala Lumpur, Malaysia

Radhakrishna S. Pandit Department of Zoology, Savitribai Phule Pune University, Ganeshkhind, Pune, Maharashtra, India

Kiran D. Pawar School of Nanoscience and Biotechnology, Shivaji University, Vidyanagar, Kolhapur, Maharashtra, India

Razanah Ramya Institute of the Malay World and Civilization, The National University of Malaysia, Bangi, Selangor, Malaysia

Manisha K. Sangale Cyfoeth Naturiol Cymru, Natural Resource Wales, Swansea, UK

Sarawut Sangkham Department of Environmental Health, University of Phayao, Phayao, Thailand

Mohd. Shahnawaz Biofuels Institute, School of the Environment and Safety Engineering, Jiangsu University, Zhenjiang, China

Mohd. Yaseen Sirwal Department of Chemistry, Govt. Degree College Kishtwar, Kishtwar, Jammu and Kashmir, India

Wan Syibrah Hanisah Wan Sulaiman International Institute for Halal Research and Training, International Islamic University Malaysia, Kuala Lumpur, Malaysia

Jianzhong Sun Biofuel's Institute, School of the Environment and Safety Engineering, Jiangsu University, Zhenjiang, China

John Tadso Department of Chemistry, Ibrahim Badamasi Babangida University Lapai, Lapai, Niger State, Nigeria

Isioma Tongo Department of Animal and Environmental Biology, Faculty of Life Science, University of Benin, Benin City, Nigeria

Hin Fung Tse Department of Science and Environmental Studies, The Education University of Hong Kong, Hong Kong SAR, China

Varsha Wankhade Department of Zoology, Savitribai Phule Pune University, Pune, Maharashtra, India

Rongrong Xie Biofuel's Institute, School of the Environment and Safety Engineering, Jiangsu University, Zhenjiang, China

Ying Kit Yue Department of Biology, Hong Kong Baptist University, Hong Kong SAR, China

Vasudeo P. Zambare R&D Department, Om Biotechnologies, Nashik, Maharashtra, India
Aesthetika Eco Research Pvt Ltd, Nashik, Maharashtra, India

Part I

Plastic Waste Cycle (Generation, Transportation, Accumulation, Management, Mitigation Measures)

Generation and Management of Microplastic Waste

1

Mufaro Moyo and Thabo Falayi

Abstract

Microplastics (MPs), which are tiny plastic materials with size below 5 mm, are ubiquitous in both terrestrial and aquatic environments. They are an emerging pollutant posing potential threats to the biosphere. Once they get into the environment, microplastic wastes are difficult to eliminate and hence are continually accumulating in the environment resulting in pollution. Eventually, they end up in the food web, and due to their tiny size, they can easily enter bodies of the biosphere. They also can act as conduits for the proliferation of microbes and fungi. Undoubtedly, the MPs waste needs to be handled safely. Understanding the MPs cycle from the point of generation to disposal can help in the safe use of MPs and handling of MPs waste. This chapter, therefore, discusses the MPs cycle by focusing on the generation of MPs, characterisation of MPs and review of the current challenges associated with MPs waste. The current research trends in the area of MPs pollution will be reviewed together with recommendations on future mitigation measures.

Keywords

Microplastics · Microplastic waste · Biosphere · Microplastics cycle · Pollutant · Microbes · Fungi · Mitigation measures

M. Moyo (✉)
Department of Fibre and Polymer Materials Engineering, National University of Science and Technology, Bulawayo, Zimbabwe

Department of Engineering, Malawi University of Science and Technology, Limbe, Malawi
e-mail: mufaro.moyo@nust.ac.zw

T. Falayi
Department of Engineering, Malawi University of Science and Technology, Limbe, Malawi

M. Shahnawaz et al. (eds.), *Impact of Plastic Waste on the Marine Biota*,
https://doi.org/10.1007/978-981-16-5403-9_1

1.1 Introduction

The agreed definition of plastic is any non-natural polymeric material including additives (Hartmann et al. 2019). One expanded definition of MPs defines them as water-insoluble regular or irregularly shaped solid synthetic particles or polymeric matrices with size ranging from 1 μm to 5 mm and originating from either primary or secondary manufacturing (Frias and Nash 2019). The general definition of MPs describes them as plastics of diameter less than 5 mm (Fu et al. 2020). This includes materials such as rubber, composite material and paints with polymeric binders. MPs are found in various shapes and sizes. The shape and size are determined or influenced by the type of degradation of the plastic, the type of force and the way the plastics are handled. There are five categories of MPs which include fragments, films, foams/sponges, fibres and pellets. Fragments are defined as plastic particles having sharp edges, films are thin transparent plastics that are usually used as plastic bags and foams/sponges are lightweight plastics used for packaging or as cups. Fibres/lines are thin or fibrous plastic and are usually from textile materials or fishing nets. Pellets/beads are spherical plastic which is usually incorporated in the formulation of personal hygiene products or used in the preproduction of larger plastic. MPs are an environmental problem in that they may contain toxic additives or are capable of adsorbing and concentrating organic pollutants from the environment.

Microplastics are ubiquitous in that they are found in all environments. Microplastics distribution in the environment is influenced by the obtaining environment and human/industrial activities (Imhof et al. 2016). The environmental factors include weather patterns, tides, wind direction and currents of waves. Floods have been shown to increase the concentration of MPs by about tenfold (Roebroek et al. 2021).

Two general types of microplastics are primary microplastics and secondary microplastics. The primary MPs are from plastics used as additives for packaging of hygiene products, whilst the secondary source is from the breaking down of larger plastic debris (this also includes fragments of tyres as cars move on the roads). The wearing of tyres contributes about 0.81 kg/year/person to the MPs embedded in the environment (Jan Kole et al. 2017). Primary microplastics are engineered and are present in products like (the list is not exhaustive) toothpaste, sunscreen, shower gels, textile and clothing, drug delivery systems and nail polish. The primary MPs usually flow into the wastewater system from home sinks and the bathroom drainage system. Primary microplastics are deliberately produced for different applications in cosmetics and other industries (Yudi et al. 2021). Primary microfibres include microbeads, microfibres and plastic pellets (nurdles). Secondary microplastics are types of microplastics that come as due to the degradation of macroplastic waste (Wang et al. 2019). These secondary microplastics exist in various shapes and have diversified chemical and physical properties (Helmberger et al. 2020; Rillig et al. 2019). Plastic fragments and foam (Styrofoam) have a large contribution to the formation of secondary plastics. Microplastic particles are predominantly of the following polymer types, HDPE (high-density polyethylene), polyethylene terephthalate, LDPE (low-density polyethylene), polypropylene, linear low-density polyethylene, polystyrene, polyvinylchloride, polymethylacrylate, polyurethane and polyesters (Wu et al. 2016; Yudi et al. 2021).

1.2 Sources of Microplastics

1.2.1 Sources of the Primary Microplastic

The sources of primary microplastics are listed in Fig. 1.1. The major sources of primary microplastics either contain microbeads (for instance, cosmetic products) or are made using micro-sized plastics (for instance, microfibres found in synthetic textiles).

1.2.2 Sources of the Secondary Microplastic

The sources of the secondary microplastics are listed in Fig. 1.2. Alterations, weathering and degradation processes disintegrate macroplastics into micro-sized plastic fragments. Discarded plastic products are exposed to different types of stresses which also fragment them into microplastics.

1.3 Generation of Microplastics Waste

The main source of MPs is the plastic waste dumped into the environment (solid waste) whose particle size is then reduced through UV radiation and physical, chemical and biological conditions (Andrady 2017). It has also been shown that plastics react with oxygen in the air creating free radicals which break down the polymeric chain leading to the creation of MPs from larger chunks of plastic. The

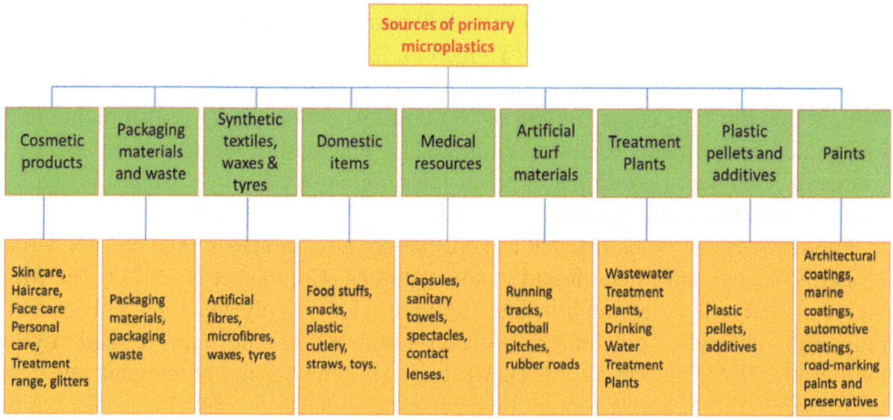

Fig. 1.1 Sources of primary microplastics

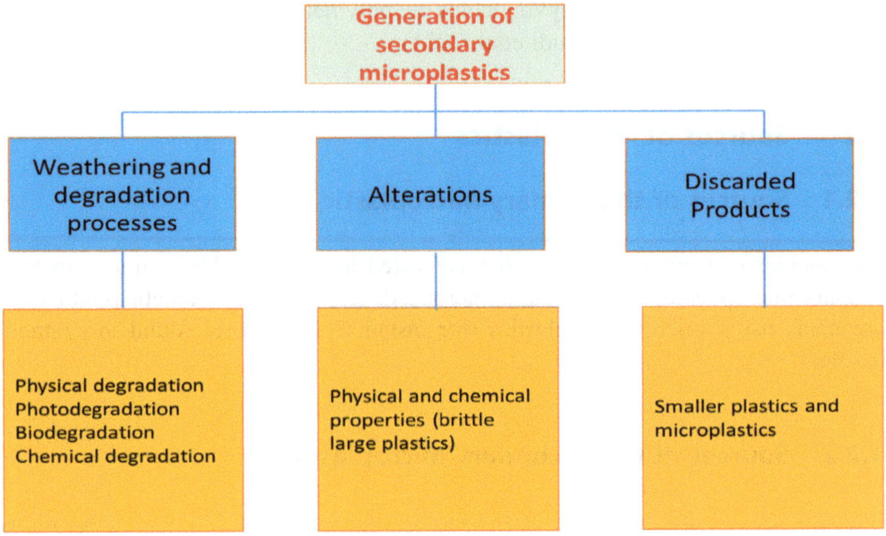

Fig. 1.2 Sources of secondary microplastics

world production level of plastics in 2018 was around 359 million tons (Plastics Europe and Conversion Market and Strategy GmbH 2019); with such production levels, cases of MPs pollution are likely to increase also due to poor disposal practices by consumers. The disposal of plastics which ultimately creates the MPs is most likely to increase due to the emergence of one-use plastics and PPE (personal protective equipment) use with coronavirus (Patrício Silva et al. 2021). One report has estimated that about 79% of the plastics produced between 1950 and 2015 have been disposed of in landfills and due to weathering; therefore, the landfill becomes the source of MPs (Ostle et al. 2019). The use of landfills as a solid waste management system makes them a sink and source of MPs (He et al. 2019). Landfills have been shown to have a 20,000 to 91,000 items/kg concentration of MPs. This is higher than in sediment or topsoil (Yang et al. 2015). The rain and wind help in the transportation of MPs from the landfills to other parts of the environment. It is thought that the MPs concentration of 20,000 to 91,000 items/kg is also higher than in aquatic environments (Golwala et al. 2021). The landfill leachates can also be a source of MPs but at a minimal level as the number of MPs in leachates is significantly lower than in the landfill.

Food processing industries that use plastics are also another source of MPs. Up to 15 items per fish have been found in some studies (Golwala et al. 2021). The MPs included polyethylene-terephthalate, polypropylene, polyethylene, polystyrene, polyvinyl chloride, polyamide, polyacrylonitrile, acrylic, rayon and nylon (Wu et al. 2016; Yudi et al. 2021). Up to 1–10 items/kg in commercial salt have been found; the MPs were polyethylene, polyethylene-terephthalate, polystyrene and polyacrylonitrile (Golwala et al. 2021).

Treatment plants for sewage (WWTPs) are also considered as one of the most significant sources of MPs (Zhang and Chen 2020; Lin et al. 2021). WWTPs are sources in that the traditional operation is not capable of removal of all the MPs. WWTPs on average remove about 98% of MPs meaning that the discharge of WWTPs contains MPs (Ren et al. 2020a). WWTPs are a source of MPs through the use of sludge as a bio-fertiliser. Eighty percent of the influent microplastics are concentrated in the primary sludge. Therefore, the use of primary sludge may contribute to the surface concentration of MPs. The MPs removal rate is dependent on the number of stages for each Sewage Treatment Plant (STP). A two-stage STP with primary and secondary filtration can remove 67% of the influent MPs, whilst a three stage STP comprising primary, secondary and tertiary filtration can remove at least 87% of influent MPs, and the addition of a reverse osmosis stage can reduce MPs to over 90%. MPs can be found in aquatic environments through surface runoff and wind. The source of these MPs is from facial products which contain microbeads. The use of toothpaste has been shown to produce over 4000 microbeads annually (Carr et al. 2016). It has been estimated that 1,125,500,000 particles/day of polyethylene microbeads have been recovered from WWTPs (Kalčíková et al. 2017).

Microfibres are also MPs which result from the washing of garments that are made from synthetic fibres (Gago et al. 2018). The microfibres usually include acrylic, nylon and polyester. The release of microfibres from washing is dependent on the type of fibre, age of cloth and type of detergent and conditioner used. The cleaning of contact lenses and jewellery as well as air blasting, production of Styrofoam products and the textile industry are other sources of MPs (Prata 2018). It has been estimated that mechanical washing of clothing weighing about 6 kg can release more than 700 thousand of fibres into the environment per wash (Napper and Thompson 2016). The type of washing or washing machine can affect the release of MPs which are then deposited into the environment. A study (Hartline et al. 2016) showed that a top-loading machine releases about six times more MPs as compared to a front-loading machine. This is influenced by the amount of friction generated within the washing cycle. Comparison of different experiments conducted in literature is a challenge in that the results are reported in different units (Tiffin et al. 2021). The units include mg/garment, particles per wash and fibres/g/wash. This, therefore, calls for the validation and standardisation of analytical methods. This would allow ease of comparison and reproducibility.

It has been reported that plastic polymers such as polyvinyl chloride (PVC) and polyethylene terephthalate which have greater density than seawater are generally concentrated in sediments, whilst less dense particles are found in aquatic environments (Zhu et al. 2020). Their concentration in the sediment is directly proportional to the proximity of human settlement (de Bhowmick et al. 2021). This, therefore, means indiscriminate disposal of plastics by human beings as a source of generation of MPs. It has been reported that China disposes about 306.9 tons of microbeads into aquatic environments (Cheung and Fok 2017) through WWTPs. The use of sludge in agriculture as a result of a move to avoid incineration or landfilling of sludge culminated in large volumes of dewatered sludge getting into the soil together with large amounts of MPs getting into the environment (Xu et al.

2021). Dris et al. (2016) showed that MPs can be uplifted into the air and remain in the air for some time. This, therefore, makes landfills particularly a dangerous source of MPs.

Literature surveys revealed that drinking water treatment plants (DWTP) were reported as a source of MPs. It has been reported that the removal efficiency of MPs from DWTP ranges from 40% to greater than 80%. Pivokonský et al. (2020) showed that the removal rate of MPs is higher in complex DWTP. Complex DWTP are those whose unit operations include flocculation, filtration (granular activated carbon and sand filtration) and sedimentation, whilst the less complex involve flocculation and sand filtration. A group of researchers (Sarkar et al. 2021) reported a cumulative MPs removal of 63% and 85% at clarification and sand filtration, respectively. In the same study, the screening effect showed higher abundance of microplastic on the sand filter bed. The most abundant MPs were fibres and fragments or films with polyethylene and polyethylene terephthalate as major chemical types.

There have been reports of MPs being present in tap water. One such study was done in Qingdao, China (Zhang et al. 2021). The investigation revealed that MPs concentration was 0.2–1.6 items/L and ranging from 10 to 5000 μm with black, blue, grey and transparent colours being dominant. Fibres were the dominant shape of MPs, whilst the most common polymer was rayon. Research by Bonthuys (2018) showed the availability of MPs in municipality water with one municipality reporting MPs concentration of 15 items/L.

MPS are hydrophobic and, because of their small size, possess a large external surface area (Gatidou et al. 2019). Ren et al. (Ren et al. 2020a) revealed in their study that MPs found in WWTPs could be distinguished due to the following criteria: colour with white being the most abundant and shape with fragment being the most abundant. The chemical nature of MPs varies and includes polyvinyl chloride, polytetrafluoroethylene, polyethylene and polyacrylonitrile (Ren et al. 2020a). However, it is of paramount importance to state that the chemical nature of MPs differs from place to place.

One of the less-studied areas in MPs occurrence is the presence of MPs in indoor environments involving the use of additives in insulation material like polystyrene. The volatilization, polymer degradation and fragment dispersal of additives account for the presence of MPs in indoor environments (Hale et al. 2020). The burning of plastic in poorly controlled places is also a source of MPs. This is further worsened by wildfires which burn households that themselves have plastics (Hale et al. 2020). The wearing of paint is also another source of MPs. MPs from paints were found to be in the region of 1–50 μm due to the brittleness of paint particles (Imhof et al. 2016). It has been shown that in urban areas, the deposition of fibres/m^2/day is about 1.6 and 11 tons, and this is entirely linked to anthropogenic activities of the environment (Dris et al. 2016).

The use of plastics in beverages has not received much attention with regard to the generation of MPs or the presence of eMPs in the beverages. Schymanski and coworkers (Schymanski et al. 2018) showed the existence of MPs in bottled natural mineral water. The concentration was about 118 particles/L for multi-use use bottles, whilst single-use bottles have a concentration of 14 particles/L. This study showed that the reuse of returnable PET bottles increases the quantity of MPs in the mineral water. In another study (Oßmann et al. 2018), the respective concentration of MPs was 4890 particles/L and 2650 particles/L for bottles meant for single-use and multi-use. The vast difference between the two studies was in that the study by (Schymanski et al. 2018) was limited to particles bigger or equal to 5 μm yet the study by (Oßmann et al. 2018) studied also the pigment and additives in addition to MPs which were as small as 1 μm.

1.4 Challenges Associated with Microplastics

Plastics, both in the macro size and micro size, are increasingly threatening the lives of humans and the natural ecosystem. The macro-, micro- and nano-sized plastic debris have been predicted to double by the year 2030 (Patrício Silva et al. 2021). The implication is that there will be increased challenges associated with plastics since microplastics have been identified as the emerging category of plastics posing potential threats to the environment. Challenges associated with microplastics waste are listed in Fig. 1.3.

Microplastics have toxic effects on living organisms. Ingestion of microplastics by organisms is currently the greatest concern of the risks of microplastics (Peng et al. 2017). Studies showed that some living organisms, for instance, filter-feeding shrimp, fish and crabs, ingest microplastics (Eriksen et al. 2018; Westphalen and Abdelrasoul 2018; Zhong and Li 2020). This affects the entire food web by causing various forms of toxicity. This is because waste from microplastics can act as a vector or opening wedge for toxicants into the food chain. Human beings are exposed to microplastics mostly through seafood. The effects of microplastics span across the entire food chain, adversely affecting reproduction capability, growth and survival rate (Usman et al. 2020; Westphalen and Abdelrasoul 2018). Entanglements and ingestion cause injuries of various forms as well as mortality. Also, there are potential risks to human health from microplastics-contaminated drinking water. The potential risks come from particles, chemicals and microorganisms (World Health Organization 2019). The particles present physical hazards, and microorganisms can attach and colonize on microplastics (World Health Organization 2019).

Accumulation of microplastics in mammalian bodies is another challenge associated with microplastics. Mammalian bodies have been found to have an accumulation of microplastics. In mice fed with microplastics-contaminated fresh drinking water and fish, accumulation was in the liver, and effects were neurobehavioural changes, slow locomotion and higher anxiety (da Costa Araújo and Malafaia 2021). In Wister rats, polystyrene microplastics caused cardiac toxicity and other negative effects and myocardial damage and death (Usman et al. 2020).

Microplastics serve as vehicles for the absorption of metals, contaminants and pathogens. Microplastics dumped in the environment form new homes for organisms and microbial colonization (Peng et al. 2017). They also act as vehicles for absorbing and adsorbing other organic pollutants, metals, contaminants and

Fig. 1.3 Challenges of microplastic waste

pathogens from the environment to the living organisms causing more toxic effects on the living organisms (Eriksen et al. 2018; Usman et al. 2020; Westphalen and Abdelrasoul 2018). Microplastics can accumulate POPs (persistent organic pollutants) including organochlorine pesticides (OPs), PCBs and metallic pollutants from the environment (Peng et al. 2017).

Abundant quantities of microplastics waste can change beach sediments' physical properties, thereby affecting some biotas like sea turtles, crustacean and eggs (Peng et al. 2017). Also, the interference of microplastics with the light rays affects the light absorption capabilities of aquatic plants such as algae. Microplastics floating on water bodies interfere with light absorption by blocking and reflecting the sun's rays (Peng et al. 2017; Zhong and Li 2020). This has the effect of affecting the plant's natural processes such as photosynthesis and their normal life which eventually affects the ecosystem.

Microplastics have the potential to cause hormonal disruptions. This is attributed to POPs and other PBT (persistent bioaccumulative toxic chemicals) and toxic chemicals used as additives in plastic manufacture (CIWEM 2017). Kleinteich et al. (2018) revealed that polyethylene microplastic particles can potentially alter the composition of sediment bacterial community and in the long run affect aquatic ecosystem's basic biogeochemical processes (Kleinteich et al. 2018). In addition, microplastics can inhibit gene expression. Some microplastics capable of crossing the cell membrane end up inhibiting growth gene expression of *Chlamydomonas* (Zhong and Li 2020).

Microplastics are precursor materials for the generation of nanoplastics. Microplastics can be degraded further into nanoplastics which have also been identified to pose challenges to the environment. Some studies have shown that when microplastics are reduced to nanometres in size, they can enter the liver, brain and testicles of the *Chlamydomonas* (Zhong and Li 2020). This is because microplastics reduced to nanometre size are capable of passing across the blood-brain barrier of medakas (Zhong and Li 2020). Experiments conducted by Rossi et al. (2014) revealed that insertion of nanoparticles into cross membranes causes structural changes of the membranes, reduces membrane protein activity and hence alters the cell's function (Rossi et al. 2014).

1.5 Management of Microplastic Waste

Literature shows that policies and strategies for managing microplastics should include two key areas, namely, managing the origin of pollution and redressing and clean-up of microplastic waste (Usman et al. 2020). This can be done either upstream (prevention strategies) in the value chain for specific plastic types or end-of-pipe solutions such as clean-ups (Programme 2018). Brennholt et al. (2018) categorized the main environmental policy on managing microplastics into:

- Regulative instruments.
- Economic instruments.
- Persuasive instruments.

These instruments are rated on a scale from soft to hard based on strictness of the measures against the application of certain plastic materials (Brennholt et al. 2018). The soft instruments consist of voluntary approaches, financial incentives and education, and the hard instruments consist of commanding and controlling, imposition of financial burdens and pooling funds for research (Brennholt et al. 2018). Some of the methods that can be used to manage microplastics waste are shown in Fig. 1.4.

Some countries have formulated such policies and legislations as a drive to manage microplastic waste. However, only a few countries have taken this route. Notable examples include the US government's 2005 Microbeads-Free Water Acts (Usman et al. 2020). Canada, China, Ireland, the Netherlands and New Zealand are other countries that have state laws and bans on some plastic materials and microbeads-containing products (Zhang et al. 2018; Zhong and Li 2020). Malaysia crafted a road map for nil single use of plastics in 2018 (Usman et al. 2020). The phased road map includes the setting of strong institutions on zero single-plastic use,

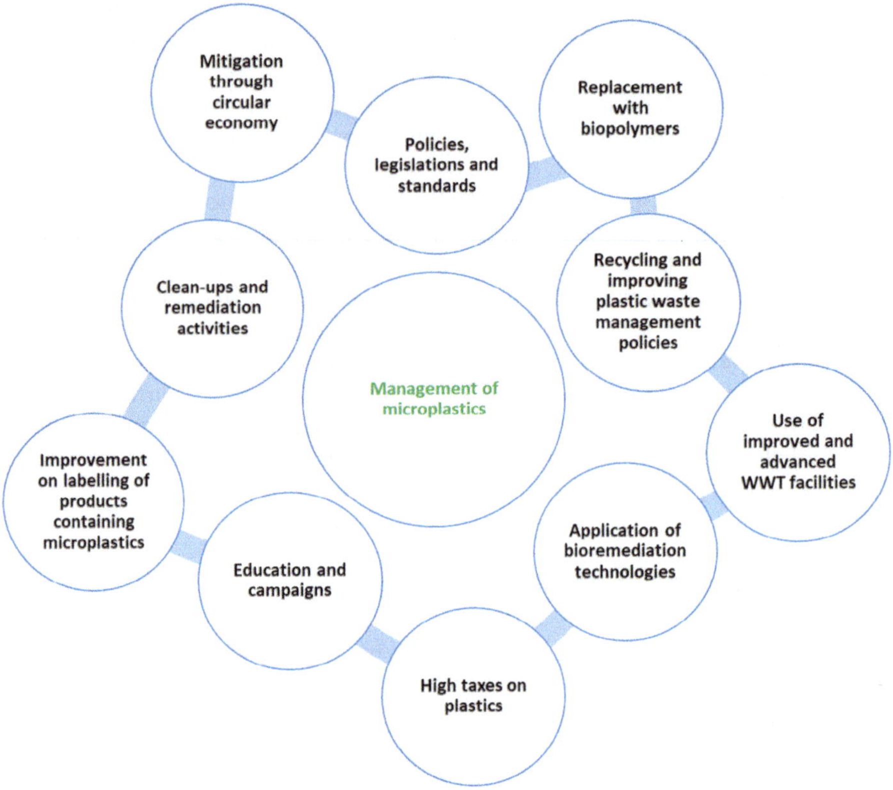

Fig. 1.4 Ways for managing microplastic waste

replacement of plastic bags with biogas, pollution charge on single-use plastics, imposing of a plastic levy to manufacturers of plastics, regional cooperation on marine plastic waste, the Communication, Education and Public Awareness (CEPA) Plan and research and development for alternative environmentally friendly products (Ministry of Energy 2018; Usman et al. 2020). Australia introduced a bill on Recycling and Waste Reduction (Usman et al. 2020). Governments can also provide financial and non-financial incentives for the use of less damaging alternatives to traditional plastics and tackle issues associated with single-use packaging products.

Microplastic pollution can be addressed at the source. The problem of microplastics can be effectively and cheaply addressed at the source since it is practically difficult to remove microplastics once they get released into the environment. The focus should be on manufactures of primary microplastics such as microbeads, the textile industry and white goods manufacturers (CIWEM 2017).

Improving relevant standards for plastic waste control helps in managing microplastics waste. Microplastics monitoring standards should be formulated and strengthened. These include standards on plastics production and recycling; standards on plastic packing, plastic films, plastic bags, and plastic enterprise wastewater; and air emission standards and standards of biodegradable plastics

(Zhong and Li 2020). This should be strengthened with well-formulated monitoring standards for microplastics and the timely publishing of industrial and urban WWTPs and tailwater microplastics discharge monitoring standards (Zhong and Li 2020). Standards set by governments on recovery and recycling of plastics should be high so as to minimize the number of secondary microplastics that end up in the environment (CIWEM 2017). Use of biopolymers is also a measure for managing microplastics waste. An aggressive promotion and awareness campaign for the use of biopolymers such as polylactic acid (PLA) and polyhydroxyalkanoates (PHA) could help in reducing the appetite for non-biodegradable plastics. Biopolymers are suitable alternatives to traditional plastic polymers (Usman et al. 2020). Recycling of plastic materials should be enhanced to reduce the generation of more plastics from virgin materials. Presently, in Europe and Germany, plastic recycling rate is on average above 45% and 60%, respectively (Zhong and Li 2020). The EU has projected plans of recycling all plastic-made packaging materials by the year 2030 (Zhong and Li 2020). Use of improved and advanced wastewater treatment facilities can also help in managing microplastics waste. Such facilities should be able to effectively filter or block microplastics from getting deposited into flowing waters and other water bodies. Some of the advanced water treatment technologies include rapid sand filters (RSF), disc filters (DF), dissolved air flotation (DAF) and membrane bioreactors (MBR) (Talvitie et al. 2017). These systems have been found in removing more than 95% of microplastics (>20 mm) from effluents (Talvitie et al. 2017). Application of bioremediation technologies helps in managing microplastics waste. Research and development of bioremediation technologies could significantly help in the management of plastic waste. Technology like microbes-assisted biodegradation of microplastic materials can be used (Wu et al. 2016).

Imposing a high tax on single-use plastic bags also helps in managing microplastics waste. This has the potential of restricting plastic bags user at both personal, family, community, and industrial levels. Education and campaigns can positively contribute to the management of microplastics waste. People should be educated on the dangers of plastics to the environment, how they help in the generation of microplastics and how these microplastics will end up in human beings with the potential of causing some health problems. Researchers and the public domain should also contribute to educating people and carrying out campaigns on plastics and microplastics. Some companies are contributing to education and campaigns by communicating with consumers to stop the use of microbeads via different media platforms such as emails and mobile applications (Chang 2015). Research should also focus on microplastics transportation mechanisms and their fate within the environment (CIWEM 2017). Proper labelling of products containing microplastics helps in making users aware of the contents of the products and how they should be disposed of. Use of microorganisms capable of degrading the microplastics is also a strategy that can help in managing microplastics waste. Microorganisms with the capability to degrade plastics from the Antarctic cold regions have been explored (Urbanek et al. 2018). Clean-ups and remediation activities such as cleaning of beaches and application of technology-assisted collection of plastic waste from oceans enable the removal of plastic materials which are already spread in the environment (OECD 2018; Programme 2018).

Today's linear economy model of "take, make, dispose of" to mitigate MPs is relying on huge amounts of easily accessible, cheap materials, and energy is reaching its physical limits (Programme 2018). There is a need to rethink the system along with the value chain approach. Leakages of plastic materials at all their life cycle phases can be potentially addressed by a circular economic model (Eriksen et al. 2018). Modifying the plastic value chain throughout all stages of their functional life is one key issue to the implementation of the circular economic model.

1.6 Challenges in Managing Microplastic Waste

The initiatives on management of microplastics waste face some challenges which include

- Weak enforcement of laws and poor coordination of international agreements. It has been reported that the countries that have formulated restrictive laws and legislation governing the use of plastics and products with microplastics are failing to make them effective and are finding it tedious to enforce (Usman et al. 2020). Poor coordination of agreements at an international level poses challenges in managing microplastics waste. Usman et al. (2020) reported that the agreements to prevent the hazards of plastic pollution are poorly coordinated at the international level resulting in disjointed governance systems (Usman et al. 2020). Involvement of proactive intersectoral teams buttressed with a globalised mindset and local actions are necessary to combat microplastic pollution and hence reduce the hazardous effects of microplastic waste.
- Insignificant change in consumer behaviour on plastics. The imposition of tariffs on single-plastic usage is not significantly changing consumer behaviour (Usman et al. 2020). This has been exacerbated by the plastic industries' resistance against plastic use or taxation (Usman et al. 2020).
- Poor awareness, high cost of alternative non-plastic products and low recycling rate. The case of Malaysia showed that plastic waste is habit-driven, chief contributors being the habits of littering and unsustainable consumption habits (Ministry of Energy 2018). Recycling industries focus on plastics which are easy to collect and have a huge value. Consequently, other plastic materials are not being recycled. The case of Malaysia showed that plastic waste materials such as PET-made bottles are the ones undergoing mass recycling whilst other plastic wastes such as packaging materials for food, polystyrene-made products and straws are hardly recycled (Ministry of Energy 2018). Another challenge is that the biodegradable alternatives are currently more expensive than traditional plastics. This discourages or in a way hinders businesses to embrace the alternatives.

1.7 Current Research Trends on Microplastics

The review of current research or state of the science was done focusing on research in 2020 and 2021. The review may not be exhaustive, but it gives a picture of the current trends. The research can be broken into several themes which include monitoring, environmental effects, removal of microplastics and quantification of microplastics as shown in Tables 1.1, 1.2, 1.3, 1.4

Table 1.1 Monitoring

Environment	Area of monitoring	Summary of results	Reference
Sewage sludge	Yangling, Shaanxi province, China	Size: 8–400.00 μm Type: 41% PVC	(Ren et al. 2020b)
Water	Danjiangkou reservoir, China	The microplastic accumulated more in the middle layer of the reservoir	(Lin et al. 2021)
Beaches	Puerto Rico	The most polluted beach had MPs concentration of 17 items/kg dw Type: 40% fibres, 28% fragments	(Pérez-Alvelo et al. 2021)
Coastal	Cochin coastal, India	Estuarine-751.7 ± 452.21 particles/m^3 Sediment 1340 ± 575.22 particles/kg Type: Cellophane, polystyrene, polyethylene, polypropylene and nylon	(Suresh et al. 2020)
Water	Ganges River, India	Identification of 140 microplastic waste particles with higher concentration during the monsoon Type: 91% fibres and 9% fragments	(Napper and Thompson 2016)
Water	Netravathi River, India	Water: 288 pieces/m3 Sediment: 96 pieces/kg Soil: 84.45 pieces/kg Type: Fibres, fragments and films	(Amrutha and Warrier 2020)
Water	Urban tap water in urban areas and Qingdao (China) water sources	Tap water: 0.3–1.6 items/L. Water sources: 0.2–0.7 items/L Type: Fibre (99.2%) Polymers: 48.9% rayon and 29.6% PET	(Zhang and Chen 2020)

Table 1.2 Environmental effects

Environmental effects	Type of MP	Summary of results	Reference
UV-C	Antifouling paint	The accelerated weathering increased the release and formation of volatile components and hydrophilic groups, respectively	(Simon et al. 2021)
Mechanical abrasion	Plastic film	Decreased light transmittance by the plastic film	(Ren et al. 2020a)

Table 1.3 Removal of plastics

Source	Location	Summary of results	Reference
Drinking water	India	Sand filtration had 85% removal Pulse clarification had 63% removal from a source with 17.88 items/L MPs	(Sarkar et al. 2021)

Table 1.4 Quantification

MPs	Method	Reliability/results	Source
Textile fibre	GyroWash followed by vacuum-assisted single filtration	90%	(Tiffin et al. 2021)
Others	Microscopy: μ-FTIR (Fourier-transform infrared microspectroscopy), μ-Raman (Raman microspectroscopy), Σ-GC/MS (thermoextraction and desorption or pyrolysis combined with gas chromatography coupled to mass spectrometry), SEM and particle counter	Best performance from Σ-GC/MS and μ-Raman Σ-GC/MS alone gave questionable results	(Müller et al. 2020)

1.8 Conclusion

This chapter provided an overview of the generation and management of microplastic waste. The generation of microplastics depends on the type, that is, primary microplastic type or secondary microplastic type. Primary microplastics are generated from sources such as cosmetics or personal care products, synthetic textiles, packaging materials, medical resources, artificial turf materials, wastewater treatment plants, paints, additives, plastic pellets and domestic items such as foodstuff, snacks and plastic cutlery. Conversely, secondary microplastic type is created through processes such as weathering, degradation and alterations of the chemical and physical properties of large plastics. The microplastic waste ends up in the environment through different pathways. They get deposited in freshwater systems, contaminate drinking water, and also become part of the food web, especially seafood. The existence of microplastic materials in the environment has activated the safety alarms since they are an emerging threat to the environment. They pose potential risks to living organisms and the food chain. Consequently, microplastic waste needs to be managed well to eradicate or reduce its projected detrimental effects on the environment. Some of the ways of managing microplastics include formulation of strong policies, legislation and standards for governing the generation of microplastics and regulating their use and disposal, replacement with suitable alternatives such as biopolymers, strong educational programmes and campaigns on the potential dangers of microplastics and how they can be managed and mitigations through circular economy. There is also a need to fund research on microplastics.

The current research focuses on areas such as monitoring, environmental effects, removal of plastics and quantification of microplastics.

Acknowledgement The authors are grateful to the Department of Fibre and Polymer Materials Engineering at the National University of Science and Technology, Zimbabwe, and the Engineering Department at the Malawi University of Science and Technology, Malawi, for infrastructural support.

References

(CIWEM) CIoWaEM (2017) Addicted to plastic: microplastic pollution and prevention. London

(OECD) OfEC-oaD (2018) Improving plastics management: trends, policy responses, and the role of international co-operation and trade

Amrutha K, Warrier AK (2020) The first report on the source-to-sink characterization of microplastic pollution from a riverine environment in tropical India. Sci Total Environ 739:140377. https://doi.org/10.1016/j.scitotenv.2020.140377

Andrady AL (2017) The plastic in microplastics: a review. Mar Pollut Bull 119:12–22. https://doi.org/10.1016/j.marpolbul.2017.01.082

de Bhowmick G, Sarmah AK, Dubey B (2021) Microplastics in the NZ environment: current status and future directions. Case Stud Chem Environ Eng 3:100,076. https://doi.org/10.1016/j.cscee.2020.100076

Brennholt N, Heß M, Reifferscheid G (2018) Freshwater microplastics: challenges for regulation and management. In: Wagner M, Lambert S (eds) Freshwater microplastics: emerging environmental contaminants. Springer International Publishing, Cham, pp 239–272. https://doi.org/10.1007/978-3-319-61615-5_12

Carr SA, Liu J, Tesoro AG (2016) Transport and fate of microplastic particles in wastewater treatment plants. Water Res 91:174–182. https://doi.org/10.1016/j.watres.2016.01.002

Chang M (2015) Reducing microplastics from facial exfoliating cleansers in wastewater through treatment versus consumer product decisions. Mar Pollut Bull 101:330–333. https://doi.org/10.1016/j.marpolbul.2015.10.074

Cheung PK, Fok L (2017) Characterisation of plastic microbeads in facial scrubs and their estimated emissions in mainland China. Water Res 122:53–61. https://doi.org/10.1016/j.watres.2017.05.053

da Costa Araújo AP, Malafaia G (2021) Microplastic ingestion induces behavioral disorders in mice: a preliminary study on the trophic transfer effects via tadpoles and fish. J Hazard Mater 401:123,263. https://doi.org/10.1016/j.jhazmat.2020.123263

Dris R, Gasperi J, Saad M et al (2016) Synthetic fibres in atmospheric fallout: a source of microplastics in the environment. Mar Pollut Bull 104:290–293. https://doi.org/10.1016/j.marpolbul.2016.01.006

Eriksen M, Thiel M, Prindiville M, Kiessling T (2018) Microplastic: what are the solutions? In: Wagner M, Lambert S (eds) Freshwater microplastics: emerging environmental contaminants. Springer International Publishing, Cham, pp 273–298. https://doi.org/10.1007/978-3-319-61615-5_13

Frias JPGL, Nash R (2019) Microplastics: finding a consensus on the definition. Mar Pollut Bull 138:145–147. https://doi.org/10.1016/j.marpolbul.2018.11.022

Fu Z, Chen G, Wang W, Wang J (2020) Microplastic pollution research methodologies, abundance, characteristics and risk assessments for aquatic biota in China. Environ Pollut 266:115098. https://doi.org/10.1016/j.envpol.2020.115098

Gago J, Carretero O, Filgueiras AV, Viñas L (2018) Synthetic microfibres in the marine environment: a review on their occurrence in seawater and sediments. Mar Pollut Bull 127:365–376. https://doi.org/10.1016/j.marpolbul.2017.11.070

Gatidou G, Arvaniti OS, Stasinakis AS (2019) Review on the occurrence and fate of microplastics in sewage treatment plants. J Hazard Mater 367:504–512. https://doi.org/10.1016/j.jhazmat.2018.12.081

Golwala H, Zhang X, Iskander SM, Smith AL (2021) Solid waste: an overlooked source of microplastics to the environment. Sci Total Environ 769:144581. https://doi.org/10.1016/j.scitotenv.2020.144581

Hale RC, Seeley ME, la Guardia MJ et al (2020) A global perspective on microplastics. J Geophys Res Oceans 125:1–40. https://doi.org/10.1029/2018JC014719

Hartline NL, Bruce NJ, Karba SN et al (2016) Microfiber masses recovered from conventional machine washing of new or aged garments. Environ Sci Technol 50:11532–11538. https://doi.org/10.1021/acs.est.6b03045

Hartmann NB, Hüffer T, Thompson RC et al (2019) Are we speaking the same language? Recommendations for a definition and categorization framework for plastic debris. Environ Sci Technol 53:1039–1047. https://doi.org/10.1021/acs.est.8b05297

He P, Chen L, Shao L et al (2019) Municipal solid waste (MSW)landfill: a source of microplastics. Evidence of microplastics in landfill leachate. Water Res 159:38–45. https://doi.org/10.1016/j.watres.2019.04.060

Helmberger MS, Tiemann LK, Grieshop MJ (2020) Towards an ecology of soil microplastics. Funct Ecol 34:550–560. https://doi.org/10.1111/1365-2435.13495

Imhof HK, Laforsch C, Wiesheu AC et al (2016) Pigments and plastic in limnetic ecosystems: a qualitative and quantitative study on microparticles of different size classes. Water Res 98:64–74. https://doi.org/10.1016/j.watres.2016.03.015

Jan Kole P, Löhr AJ, van Belleghem FGAJ, Ragas AMJ (2017) Wear and tear of tyres: a stealthy source of microplastics in the environment. Int J Environ Res Public Health 14. https://doi.org/10.3390/ijerph14101265

Kalčíková G, Alič B, Skalar T et al (2017) Wastewater treatment plant effluents as source of cosmetic polyethylene microbeads to freshwater. Chemosphere 188:25–31. https://doi.org/10.1016/j.chemosphere.2017.08.131

Kleinteich J, Seidensticker S, Marggrander N, Zarfl C (2018) Microplastics reduce short-term effects of environmental contaminants. Part II: polyethylene particles decrease the effect of polycyclic aromatic hydrocarbons on microorganisms. Int J Environ Res Public Health 15:287

Lin L, Pan X, Zhang S et al (2021) Distribution and source of microplastics in China's second largest reservoir—Danjiangkou reservoir. J Environ Sci (China) 102:74–84. https://doi.org/10.1016/j.jes.2020.09.018

Ministry of Energy, Science, Technology, Environment & Climate Change (MESTECC), Malaysia (2018) Malaysia's roadmap towards zero single-use plastics 2018–2030 towards a sustainable future. Ministry of Energy, Science, Technology, Environment & Climate Change (MESTECC), Malaysia

Müller YK, Wernicke T, Pittroff M et al (2020) Microplastic analysis—are we measuring the same? Results on the first global comparative study for microplastic analysis in a water sample. Anal Bioanal Chem 412:555–560. https://doi.org/10.1007/s00216-019-02311-1

Napper IE, Thompson RC (2016) Release of synthetic microplastic plastic fibres from domestic washing machines: effects of fabric type and washing conditions. Mar Pollut Bull 112:39–45. https://doi.org/10.1016/j.marpolbul.2016.09.025

Oßmann BE, Sarau G, Holtmannspötter H et al (2018) Small-sized microplastics and pigmented particles in bottled mineral water. Water Res 141:307–316. https://doi.org/10.1016/j.watres.2018.05.027

Ostle C, Thompson RC, Broughton D et al (2019) The rise in ocean plastics evidenced from a 60-year time series. Nat Commun 10. https://doi.org/10.1038/s41467-019-09506-1

Patrício Silva AL, Prata JC, Walker TR et al (2021) Increased plastic pollution due to COVID-19 pandemic: challenges and recommendations. Chem Eng J 405:126683. https://doi.org/10.1016/j.cej.2020.126683

Peng J, Wang J, Cai L (2017) Current understanding of microplastics in the environment: occurrence, fate, risks, and what we should do. Integr Environ Assess Manage 13:476–482. https://doi.org/10.1002/ieam.1912

Pérez-Alvelo KM, Llegus EM, Forestier-Babilonia JM et al (2021) Microplastic pollution on sandy beaches of Puerto Rico. Mar Pollut Bull 164. https://doi.org/10.1016/j.marpolbul.2021.112010

Plastics Europe GMR, Conversion Market & Strategy GmbH (2019) Plastics—the FActs 2019. 14, 35

Prata JC (2018) Microplastics in wastewater: state of the knowledge on sources, fate and solutions. Mar Pollut Bull 129:262–265. https://doi.org/10.1016/j.marpolbul.2018.02.046

Programme UNE (2018) Challenges and opportunities of plastics management. Brussels, Belgium

Ren SY, Sun Q, Ni HG, Wang J (2020a) A minimalist approach to quantify emission factor of microplastic by mechanical abrasion. Chemosphere 245:125630. doi: 10.1016/j.chemosphere.2019.125630

Ren X, Sun Y, Wang Z et al (2020b) Abundance and characteristics of microplastic in sewage sludge: a case study of Yangling, Shaanxi province, China. Case Stud Chem Environ Eng 2:100,050. https://doi.org/10.1016/j.cscee.2020.100050

Rillig MC, Lehmann A, Ryo M, Bergmann J (2019) Shaping up: toward considering the shape and form of pollutants. Environ Sci Technol 53:7925–7926. https://doi.org/10.1021/acs.est.9b03520

Roebroek CTJ, Harrigan S, van Emmerik THM et al (2021) Plastic in global rivers: are floods making it worse? Environ Res Lett 16. https://doi.org/10.1088/1748-9326/abd5df

Rossi G, Barnoud J, Monticelli L (2014) Polystyrene nanoparticles perturb lipid membranes. J Phys Chem Lett 5:241–246. https://doi.org/10.1021/jz402234c

Sarkar DJ, Das Sarkar S, Das BK et al (2021) Microplastics removal efficiency of drinking water treatment plant with pulse clarifier. J Hazard Mater 413:125347. https://doi.org/10.1016/j.jhazmat.2021.125347

Schymanski D, Goldbeck C, Humpf HU, Fürst P (2018) Analysis of microplastics in water by micro-Raman spectroscopy: release of plastic particles from different packaging into mineral water. Water Res 129:154–162. https://doi.org/10.1016/j.watres.2017.11.011

Simon M, Vianello A, Shashoua Y, Vollertsen J (2021) Accelerated weathering affects the chemical and physical properties of marine antifouling paint microplastics and their identification by ATR-FTIR spectroscopy. Chemosphere:274. https://doi.org/10.1016/j.chemosphere.2021.129749

Suresh A, Vijayaraghavan G, Saranya KS et al (2020) Microplastics distribution and contamination from the Cochin coastal zone, India. Reg Stud Mar Sci 40:101533. https://doi.org/10.1016/j.rsma.2020.101533

Talvitie J, Mikola A, Koistinen A, Setälä O (2017) Solutions to microplastic pollution—removal of microplastics from wastewater effluent with advanced wastewater treatment technologies. Water Res 123:401–407. https://doi.org/10.1016/j.watres.2017.07.005

Tiffin L, Hazlehurst A, Sumner M, Taylor M (2021) Reliable quantification of microplastic release from the domestic laundry of textile fabrics. J Text Inst 0:1–9. doi: https://doi.org/10.1080/00405000.2021.1892305

Urbanek AK, Rymowicz W, Mirończuk AM (2018) Degradation of plastics and plastic-degrading bacteria in cold marine habitats. Appl Microbiol Biotechnol 102:7669–7678. https://doi.org/10.1007/s00253-018-9195-y

Usman S et al (2020) Microplastics pollution as an invisible potential threat to food safety and security, policy challenges and the way forward. Int J Environ Res Public Health 17:9591

Wang J, Liu X, Li Y, Powell T, Wang X, Wang G et al (2019) Microplastics as contaminants in the soil environment: a mini-review. Sci Total Environ 691:848–857. https://doi.org/10.1016/j.scitotenv.2019.07.209

Westphalen H, Abdelrasoul A (2018) Challenges and treatment of microplastics in water. In: Glavan M (ed) Water challenges of an urbanizing world. IntechOpen, London, pp 71–83. https://doi.org/10.5772/intechopen.71494

World Health Organization (2019) Microplastics in drinking-water. Switzerland
Wu W-M, Yang J, Criddle CS (2016) Microplastics pollution and reduction strategies. Front Environ Sci Eng 11:6. https://doi.org/10.1007/s11783-017-0897-7
Xu Z, Bai X, Ye Z (2021) Removal and generation of microplastics in wastewater treatment plants: a review. J Clean Prod 291:125,982. https://doi.org/10.1016/j.jclepro.2021.125982
Yang G, Zhang G, Wang H (2015) Current state of sludge production, management, treatment and disposal in China. Water Res 78:60–73. https://doi.org/10.1016/j.watres.2015.04.002
Yudi ML, Timon L, Lydia TL, Anika L, Matthias CR (2021) Microplastic shape, polymer type, and concentration affect soil properties and plant biomass. Front Plant Sci 12:1–14. https://doi.org/10.3389/fpls.2021.616645
Zhang Z, Chen Y (2020) Effects of microplastics on wastewater and sewage sludge treatment and their removal: a review. Chem Eng J 382:122955. https://doi.org/10.1016/j.cej.2019.122955
Zhang K, Shi H, Peng J, Wang Y, Xiong X, Wu C, Lam PKS (2018) Microplastic pollution in China's inland water systems: a review of findings, methods, characteristics, effects, and management. Sci Total Environ 630:1641–1653. https://doi.org/10.1016/j.scitotenv.2018.02.300
Zhang K, Hamidian AH, Tubić A et al (2021) Understanding plastic degradation and microplastic formation in the environment: a review. Environ Pollut 274:116,554. https://doi.org/10.1016/j.envpol.2021.116554
Zhong W, Li S (2020) Microplastic pollution control strategy. In: 6th international conference on advances in energy, environment and chemical engineering (AEECE2020), Jinan, China, June 19–21, 2020. IOP Publishing, pp 1–7. doi: https://doi.org/10.1088/1755-1315/546/3/032046
Zhu C, Kanaya Y, Nakajima R et al (2020) Characterization of microplastics on filter substrates based on hyperspectral imaging: laboratory assessments. Environ Pollut 263:114296. https://doi.org/10.1016/j.envpol.2020.114296

Generation and Management of Macroplastic Waste

2

Manisha K. Sangale, Mohd. Shahnawaz, Rahul Kapoore, Zhu Daochen, and Avinash B. Ade

Abstract

Since the decades plastic importance and utilization have increased significantly. Due to the enormous properties of plastic, its increased global demand leads to production of huge quantity of plastic-based products annually. Due to the luxuriant usage of plastic-based products at an elevated rate, billion tons of plastic is getting generated in the environment. Plastic is persistent in nature and needs some centuries to get degraded in the environment. Most of the unmanaged plastic waste gets accumulated in the oceans. In the marine environment, plastic gets fragmented into various particle sizes due to various environmental forces. The resulted plastic particles are reported as macroplastic, mesoplastic, microplastic and nanoplastic. In the current chapter, an attempt was made to overview the generation and management of macroplastic at global level to reduce the plastic pollution in the environment.

The original version of this chapter was revised: has affiliation of the author Mohd. Shahnawaz and Zhu Daochen updated. The correction to this chapter can be found at https://doi.org/10.1007/978-981-16-5403-9_17.

Manisha K Sangale, Mohd. Shahnawaz, Rahul Kapoore, Zhu Daochen, and Avinash B. Ade have contributed equally to this work and are all first authors.

M. K. Sangale (✉)
Cyfoeth Naturiol Cymru, Natural Resource Wales, Swansea, UK

M. Shahnawaz · Z. Daochen
Biofuels Institute, School of Environment and Safety Engineering, Jiangsu University, Zhenjiang, China

R. Kapoore
Department of Biosciences, College of Science, Swansea University, Swansea, UK

A. B. Ade
Department of Botany, Savitribai Phule Pune University, Pune, Maharashtra, India

M. Shahnawaz et al. (eds.), *Impact of Plastic Waste on the Marine Biota*, https://doi.org/10.1007/978-981-16-5403-9_2

Keywords

Plastic waste · Single-use plastic · Fishing gear · Macroplastic · Mitigation measures

2.1 Introduction

Plastics is a group of versatile materials that can be shaped into solid objects of different shape and sizes (Geyer et al. 2017). Plastic polymer is made up of long chain of hydrocarbon molecules. Plastic is having various important properties and had replaced all the previously used traditional polymers to meet our daily based needs (Shah et al. 2008). In the present scenario, plastic is commercially used at domestic and industry level (Sangale et al. 2012). Based on the needs of plastics and the plastic-based products throughout the globe, every year billion tons of plastic-based products are manufactured (Shahnawaz et al. 2019b). Due to the utility of the plastics, a huge quantity of the plastic waste get generated in the environment (Ayeleru et al. 2019). Various methods are available to tackle the plastic waste; however no method is 100% accurate, so huge amount of unmanaged plastic waste travels towards oceans through wind, flood, river and water canals (Lebreton et al. 2017; Meijer et al. 2017; Van Emmerik et al. 2019). In the marine environment, the plastic waste starts degrading due to various environmental forces which leads to the breaking of the big plastic pieces into small plastic particles of different sizes which are called macroplastic, mesoplastic, microplastic and nanoplastic (Lamb et al. 2018; Shahnawaz et al. 2019c). All these plastic particles are reported to pose serious threat to the inhabitants of the marine environment.

In the current chapter, an attempt was made to pool all the information regarding the generation and management of macroplastic waste under the following lines: (i) to enlist different types of plastic and their utility at domestic and industry level, (ii) to discuss different sources of macroplastic generation, (iii) to highlight the plastic waste generation, (iv) to emphasise the impact of macroplastic pollution and (v) to discuss the challenges associated with the mitigation measures of macroplastic.

2.2 Plastic and its Types

Plastics are divided into three categories according to their material use and their size.

2.2.1 Natural Plastics

Polymers that occur from natural resources and can be easily moulded by heat, e.g. amber, albumin in egg, wax from bees, etc. (Kulkarni Vishakha et al. 2012).

2.2.2 Semi-Synthetic Plastics

These are also available from nature, but they are mixed with other materials to enhance their properties, e.g. cellulose acetate, etc. (Chiellini and Solaro 2003).

2.2.3 Synthetic Plastics

Synthetic plastics are derived from modified molecular structure of carbon-based materials by applying heat and pressure, e.g. crude oil, coal, gas, etc. (Matjasic et al. 2020).

Based on the reaction of the plastic on heat, synthetic plastics were grouped into two categories (Singh and Sharma 2008).

2.2.3.1 Thermoplastics

They can be formed into desired shape upon heating and cooling. They can be reformed by heating again and cooling, e.g. polythene, polypropylene acrylic, styrene, etc. (Drobny 2014).

2.2.3.2 Thermoset Plastics

They can be formed into desired shape upon heating and cooling; however, this is a permanent process. They can't be reformed again by applying heat, e.g. polyester, bakelite, melamine formaldehyde, etc. (Dodiuk 2013).

2.2.3.3 Other Types of Plastic

There are various types of plastic like high-density polythene (HDPE), low-density polythene (LDPE), polyvinyl chloride (PVC), polystyrene (PS), polypropylene (PP), polyethylene tetrachloride (PET), etc. (Shah et al. 2008). To enhance their properties, additives are added into the plastic. For example, vinyl chloride is added into plastic to make polyvinyl chloride (PVC) (Akovali 2012). Vinyl chloride is reported as a carcinogenic agent that could cause cancer (Langärd 2013). Even after considering the disadvantages in human being as well as in natural biota, we can't imagine our daily life without plastic (Shahnawaz et al. 2019d). The word plastic is derived from Greek word "plasticos" that means it takes desired shape if heated (Sangale et al. 2019). Heat is the fundamental requirement of plastic to gain desired shape in manufacturing process (Shah et al. 2008). Versatile products of plastic are available today with reliable, robust, beautiful, and other qualities because of application of plastic and additives together. Different types of plastics are used to fulfil the requirements. Advanced manufacturing processes and technologies eased to make plastic to be made of desired shape and size to meet the requirements. For example, papier mache is still in use to make pipe stems and mouthpieces for musical instruments. It was patented in 1772 due to its mouldable properties (Annomyous 2021).

2.3 Applications of the Plastics

Plastic is one of the important polymers with versatile properties and is widely used to make products as per the requirement, e.g. cheap cost, lightweight, long-lasting, stiffness, electric insulation, etc. make it to use for various applications of plastic (Thompson et al. 2009). Common examples of plastic utilization are in automotive applications, construction, packaging, furniture, electronics, households, agriculture, electrical goods, other industrial applications, etc. (Chauhan et al. 2019). Plastics have proven their importance in medical application (McKeen 2014). Eighty-five percent of the medical instruments are made up of plastics due to their biocompatible properties, and they are also commonly used to make disposable devices, e.g. intravenous bags, disposable syringes, etc. (Van Meer et al. 2005; North and Halden 2013; Czuba 2014).

Due to increased demand and popularity, plastic is used in extensive applications. There is evolution in plastic production and process of manufacturing in the last century. Various technologies are also involved to emerge plastic produced from various petrochemicals. Plastics have competitive advantages over their counterparts (metals, wood, etc.), e.g. low cost, lightweight, etc. (Andrady and Neal 2009). Therefore their extensive applications and demand get increased from domestic usage to industrial tools. In modern society plastics satisfy various essential functions, and we would not be able to live without plastic materials today. Plastic has unavoidable part in medical field right from scissors to implementation in surgery.

Out of total production of plastic worldwide, most commonly used polymers are polyethylene terephthalate (PET), polypropylene (PP), polystyrene (PS), polyvinyl chloride (PVC), low-density polyethylene (LDPE) and high-density polyethylene (HDPE), polyamides (PA) (nylons) and polycarbonate (PC) (Shah et al. 2008); (Grigore 2017).

Application of plastic is a vital part of product design and engineering in today's world. It has given us opportunity to develop a versatile nature of products in a number of fields, e.g. medical applications, toys, kitchen utensils, automotive, etc.

2.3.1 Single-Use Plastic

Most of the plastics are designed for single use and are thrown away or disposed of after single use (Chen et al. 2020). Single-use plastic waste is usually reported from beaches in the form of plastic bags, lids, bottles, food wrappers, etc. (Schnurr et al. 2018). At present, mostly the production of these types of plastic is elevated at an alarming rate due to coronavirus disease-19 (COVID-19) outbreak to minimise the spread of the infection (Hale and Song 2020). In European countries around 40% of the total plastics is used for packaging purpose (Europe 2016). Mostly the single-use plastic carry bag and foamed plastic litter are reported in environment, water bodies (river, canals), seacoast and oceans (Syberg et al. 2020). It was estimated in 2015, Asia contribute 49% of plastic production in the world, while China alone reported highest production of plastic (28%), followed by North America (19%) and Europe (19%) (d'Ambrières 2019). Other regions also showed elevated level of plastic consumption which also needs to be taken into consideration. The government of UK banned the supply of plastic straws in April 2018 (Arkesteyn 2020). Single-use plastic shares around 43% of total plastic waste generated at global level (Winton et al. 2020).

2.4 Generation of the Plastic Waste

Plastic is emerging as a potential environmental pollutant at both terrestrial and marine environment (Thushari and Senevirathna 2020). Utility and production of plastic increased globally at high rate over the last eras and are estimated to further increase (Geyer et al. 2017). Plastic has replaced other convention materials, e.g. wood, metal, glass, etc. in the last few decades (Gourmelon 2015). Due to extensive use of plastic, a tremendous growth in mismanaged plastic waste (Jambeck et al. 2015; Lebreton and Andrady 2019) was reported. The percentage of mismanaged plastic waste generated per year in coastal areas in 2010 is as follows: China (76%), Bangladesh (89%), the United States (2%), Pakistan (88%), Indonesia (83%) and India (87%) (Shahnawaz et al. 2019a). Total mismanaged plastic waste is calculated for populations within 50 km of the coast in the countries considered. The main cause of plastic pollution is its very slow rate of degradation process as it needs about 1000 years to degrade in nature (Sangale et al. 2012; Shah et al. 2008) (Hammer et al. 2012). Due to their persistent nature, it leads to negative effects on environment, human life and aquatic animals (Gall and Thompson 2015; Wilcox et al. 2015; Koelmans et al. 2017; Conchubhair et al. 2019). Chalmin (Chalmin 2019) reported an estimated 2 billion metric tons of waste was generated in 2016, and, among the total waste generated, plastic waste alone shares 242 million metric tons (mostly 57 million metric tons from Asia, 45 million metric tons from European countries and 37 million metric tons from North America). Ho and Not (Ho and Not 2019) investigated accumulation of macro- and microplastic near seashore where waves break and backflush occurs. Lee et al. (Lee et al. 2013) suggested higher amount of macroplastic in backshore water. The accumulation time of macroplastic was also studied (Kataoka et al. 2015; Sander 2016; Hinata et al. 2017), varying from 209 to 518 days (Kataoka et al. 2013; Kataoka et al. 2015). Chalmin (Chalmin 2019) reported about 90% of plastic waste in the ocean was contributed by ten German rivers alone. Plastic accumulation in sea or marine region is recognized in 1997 by the discovery of Great pacific garbage pack and recognized globally; however, still macroplastic studies are not much focused as microplastic pollution (Van Emmerik et al. 2019; Al-Zawaidah et al. 2021). Macroplastic is reported to take thousands of years to degrade (Thompson et al. 2009); however it gets degraded into microplastic and nanoplastics during transportation to the oceans or in the marine environment (Wayman and Niemann 2021).

2.4.1 Macroplastic

Plastic litter with diameter ≥5 mm in size is referred to as a macroplastic (Takada and Tanaka 2016). However this classification of plastic litter based on size is not internationally standardised. In literature, plastic litter with diameter >20 mm (Browne et al. 2011), >25 mm (Directive 2013), >5 cm (van Emmerik et al. 2018) or >1 cm (Hartmann et al. 2019) is also referred to as macroplastic. Most of the researchers preferred ≥5 mm as diameter of the macroplastic (Lechthaler and Waldschläger 2020). Macroplastic is also called by various terms such as macrolitter, plastic debris and plastic litter (Andrady 2011; Chin and Fung 2018); however, the most frequently used term is macroplastic.

2.4.2 Sources of Macroplastic

The importance of plastics and their utility increased day by day; hence, the mass production began in 1950, and it was increased to around 8300 million metric tons in the year 2015. Among the total plastic waste generated, only 600 million metric tons were recycled, whereas a total of 800 million metric tons were retrieved and end up in landfilling and around unmanaged 4900 million metric tons of the plastic waste were accumulated in the environment (Geyer et al. 2017). Hoornweg and Bhada-Tata (Hoornweg and Bhada-Tata 2012) reviewed estimation of solid waste per person per year and reported mismanaged solid waste (MSW) increased from 2.9 billion tons, i.e. 0.64–1.2 kg per person per year, to 3 billion tons in just one decade (2002–2012), which led to accumulation of around 1.3 billion of tons per year. The literature survey revealed around 4.8 million metric tons to 12.7 million metric tons plastic waste are leaked from the inland sources into the marine environment (Jambeck et al. 2015). Ryberg et al. (Ryberg et al. 2018) revealed a discharge of 8.28 million metric tons of plastic waste into the ocean per year.

2.4.2.1 River-Based

The largest share of macroplastic is land-based source for plastic pollution. Rivers are the transportation medium for plastic according to Lebreton et al. (Lebreton et al. 2017) and 1.15 million metric tons per annum to 2.41 million metric tons per annum are transported by rivers. Lebreton and Andrady (Lebreton and Andrady 2019) predicted that over 90% of mismanaged plastic waste is carried by rivers to the marine environment.

2.4.2.2 Disposable Plastic Products

Macroplastic found in the environment is mostly comprised of disposable plastic waste due to its short life cycle and the enormous production. Among the disposable plastic litter, the most quantity (8.92%) is shared by the plastic used for food packings, followed by bottles and their lids (7.51%), plastic bags (5.49%), cigarette butts (4.02%), sanitary items (3.72%) and smoking-related packaging (3.40%)

(Winton et al. 2020). As per report, among the total 71% is plastic waste generated, and macroplastic shared around 59% (Winton et al. 2020).

2.4.2.3 Coastal Area

The infrastructure of coastal area is also responsible for direct entry of plastic waste into ocean; unfortunately, natural events, viz. storms and tsunamis, also lead to direct transportation of plastic material into the oceans (Roebroek et al. 2021). Turrell (2020) studied comparative account of land-based and ocean-based macroplastic around Scotland, while in Japan microplastic and macroplastic were studied by Nihei et al. (Nihei et al. 2020), and they reported discharge of around 210–4776 million metric tons of the inland plastic waste into the oceans.

2.4.2.4 Littering

Littering can be defined as inappropriate disposal of plastic into the ecosystem which is a potential source of macroplastic. It is a very difficult and challenging job to estimate the exact amount of waste disposed of into the environment, e.g. cigarette butts. As per an estimate, around 4.5 trillion cigarette butts are disposed of into the ecosystem annually at global level (Slaughter et al. 2011). The yearly amount of disposable plastic entering the ecosystem is estimated as 0.8 million metric tons (Conservancy 2019; Arnaud 2019). A huge portion of the disposed plastic are comprised of plastic bottles made up of polyethylene terephthalate (PET) (Ryberg et al. 2019; Kawecki and Nowack 2019).

New source of plastic waste is generated due to COVID-19 outbreak with increased usage of plastic in preventative measures to stop the spread of virus (Silva et al. 2020).

2.4.2.5 Agriculture

The covers of the greenhouse in Europe, Spain (53,235 ha) and Italy (25,000 ha) are mostly comprised of macroplastics (Lechthaler et al. 2020). Globally agricultural field uses 6.5 million metric tons of plastics every year. The major part of the film covers for mulching is plastics (Scarascia-Mugnozza et al. 2011). For agricultural activities in Europe, the total requirement of plastic is around 1.74 million metric tons per year. By using mulch and silo films, farming plastic disposable was estimated to be 2.297 million metric tons per year (Piehl et al. 2018). Plastic mulch film is a potential source of macroplastic into environment due to its direct contact with soil (Blasing and Amelung 2018).

2.4.2.6 Wastewater Treatment Plants

Macroplastic is discharged into the environment from wastewater treatment plant through excess water flow from tanks (Kawecki and Nowack 2019). Macroplastic also enters the environment through drainage and surface runoff in urban areas. Every year disposable flow is projected at 3000 million tons in Germany (Bertling et al. 2018). Kawecki and Nowack (Kawecki and Nowack 2019) reveled macroplastic discharge from wastewater treatment plant into environment.

2.4.2.7 Landfills

Macroplastic enters ecosystem through transportation of litter from the place of generation to the landfills (Barnes and Mann 2009). In Europe, the amount of plastic waste dumped in landfill was 440,000 tons (2016) as compared to 1370,000 tons (2012). This shows decreasing trend of landfilling of disposed plastic. Dumping of plastic waste near the sea coasts and banks of the rivers also leads to the transportation of plastic waste into marine environment (Jambeck et al. 2015).

2.4.2.8 Ocean-Based Macroplastic

Fishing gear lost or discarded into environment leads to increased macroplastic and microplastic into environment (Macfadyen et al. 2009). Every year, approx. 640,000 tons of fishing gears are accumulated into the environment, which is equivalent to about 10% of the total marine waste (Good et al. 2010). All these marine-based wastes, viz. fishing gear, nets, ropes, hooks, lines and fishing rods, lead to generation of macroplastic (Jéssica et al. 2019). Coastal aquaculture practices also lead to discharge of plastic waste into the marine sediments (Kruger et al. 2020). About 46% buoyant fishing nets were accumulated in the North Pacific garbage zone (Lebreton et al. 2018), whereas 68% of plastic waste associated with fishing material were reported in the Gulf of Alicante (Garcia-Rivera et al. 2017). It is very difficult to determine the exact amount of yearly entry of discarded fishing gear in the marine environment due to lack of standard method (Macfadyen et al. 2009).

2.4.2.9 Lost Cargo

During transportation of cargo through maritime routes, storms may be encountered in oceans which results in spilling of a huge amount of plastics into the marine environment (Walker et al. 2019). A major quantity of bath toys, e.g. rubber material ducks, were released into the Pacific Ocean in 1992 and are still being cleaned up on the coasts of Alaska, Australia and Scotland (Ebbesmeyer and Ingraham Jr 1994). Recently, 342 containers containing microplastic (among others, 22.5 tons PE pellets) and macroplastic (toys, household items and white goods, and packaging materials) from Mediterranean Shipping Company's ultra-large containership (MSC Zoe) across the Wadden Sea Protected Area were lost due to accidents during the maritime transportation (Annomyous 2019).

2.5 Plastic Waste Generation

The regular use of plastic, viz. single-use plastic, littered after single use leads to the generation of huge amount of plastic wastes. Lechthaler et al. (2020) reported various countries, viz. France, Bangladesh, Rwanda and India, had banned plastic bags lower than 20 micron thickness. Alternatives to plastics are highly encouraged; recently, paper straw has replaced the plastic straws significantly. For reducing plastic pollution globally, it is needed to develop an appropriate method for collection, treatment and management of plastic litter (Blettler et al. 2017).

2.6 Method to Tackle Macroplastic

Application of SPR model that consists of source, pathway and receptor depicts the amount of macroplastic that enters into the environment (Lechthaler et al. 2020). Vriend et al. (Vriend et al. 2020) provide new rapid method to estimate the amount of macroplastic debris floating over River Rhine and used simple methods to perform a swift evaluation of macroplastic waste transportation in most European rivers to enhance plastic pollution control and relief strategies. By analysis of the pathway and receptor, we can at least control macroplastic, mesoplastic and microplastic pollution. Again, the awareness among the public is needed to tackle hazardous macroplastic (Waldschlaÿager and Schuttrumpf 2020). Liro et al. (Liro et al. 2020) presented the outline for estimation of the amount of plastics accumulated in rivers. There are some methods used for detection and quantification of macroplastics such as by aerial photograph (Lavers and Bond 2017), visual counting (van Emmerik and Schwarz 2020), trackers (Van Emmerik et al. 2019), riverbank (van Emmerik et al. 2018) or net sampling (Van Emmerik et al. 2019). Emmerik et al. (Van Emmerik et al. 2019) suggested usage of hydrodynamics modelling method for estimation of macroplastic in future.

2.6.1 Bioremediation

There are various methods available to tackle plastic, viz. photodegradation, thermo-oxidative degradation, chemical and biodegradation. Among all these methods, biodegradation is considered as the most accepted eco-friendly method (Shah et al. 2008; Sangale et al. 2012; Wierckx et al. 2018; Anani and Adetunji 2021). Various researchers have reported the important role of microorganism as an eco-friendly method to tackle plastic pollution (Zhang et al. 2008; Sangale et al. 2012; Webb et al. 2013; Muthukumar and Veerappapillai 2015; Krueger et al. 2015; Senthilkumar et al. 2016; Trivedi et al. 2016; Yang et al. 2018; Danso et al. 2015; Oberbeckmann and Labrenz 2020). Microbes collected from various sources such as dumping sites, saline water and mangrove area provide potential solution to tackle plastic pollution (Shahnawaz et al. 2016; Sangale et al. 2019). However, we can accelerate the plastic degradation process if we combine physical, chemical and microbial treatments (Sangale et al. 2019). The attachment of microorganism on the surface of plastic initiates the formation of biofilm, which is followed by the secretion of enzymes by microbes to utilize the plastic backbone as a carbon source, which leads to the generation of plastic fragments. Due to biodegradation process, the initiation of fragmentation leads to the mineralization with the formation of water and carbon dioxide as end products (Mueller 2006). Jacquin et al. (Jacquin et al. 2019) studied the behaviour of microbes after getting attached to plastic surface and their impacts on marine biota.

2.7 Impact of Macroplastic Pollution

Ingestion of macroplastic by terrestrial as well as marine animals, birds, or reptiles leads to various serious health issues, viz. enzyme over-secretion, reproduction system, effect on endocrinal system causes long-term health issues and leading towards death (Li et al. 2016). A huge amount of plasticizers associated with plastic also get released in the marine environments along with the macro- and microplastic particles (Carpenter and Smith 1972). The accumulation of fishing nets on coastal areas is emerging as a potential threat to coral areas (Edward et al. 2020). The ingestion of macroplastic not only affects the corals; however, it affects all susceptible inhabitants in marine and terrestrial environment (Kühn et al. 2015; Schuyler et al. 2014). In various species of sea birds, elevated level of macroplastic accumulation was reported (LI et al. 2016; Rapp et al. 2017).

2.8 Challenges Associated with Macroplastic Mitigation

Macroplastics are occurring in various forms; therefore it is difficult to determine the exact percentage of macroplastic in the oceans (Van Sebille et al. 2020). Similarly, the percentage of solid waste is transported into the landfill which causes difficulty to find out actual percentage of macroplastics, and sometimes they are accidently transferred to the marine environment due to flood-like situations (Jambeck et al. 2015; Van Emmerik et al. 2019). However, there is still scarcity of data regarding the accumulation of macroplastic in estuaries (Vermeiren et al. 2016). Macroplastic is considered as a potential persistent pollutant in the environment (Nakashima et al. 2016). Due to availability of different types of macroplastics, their luxuriant usage and generation of waste at an alarming rate emerged as a challenging issue to manage. Macroplastics share highest percentage (56%) in plastic wastes (Lechthaler et al. 2020) due to their wide usage in production of single-use plastics. Commonly used macroplastics are food wrappers, bottles, bags, cigarette butts and packets (Winton et al. 2020). There are necessities for expansion of entry gate for plastic litter after natural calamities at coastal area (Carlson et al. 2019). If macroplastics will not be managed properly, it will cause negative impact on economy and is reported to spread algal blooms disease (Maso et al. 2003).

2.9 Conclusion

Macroplastic is a ubiquitous in nature and shares most portion of the plastic waste. There are various inland and marine sources of macroplastic. In the marine environment, macroplastic gets fragmented to generate microplastic and nanoplastic and poses detrimental effect on the marine biota. The mitigation of the macroplastic is nowadays a big and challenging issue. There is a need to focus on the source, transportation path and finally deposition of waste area and management. Among all the methods used to mitigate the macroplastic, bioremediation is considered as the most efficient, reliable and eco-friendly method at global level; however, researchers are still discovering the most efficient macroplastic-degrading microbe to commercialise the mitigation process of macroplastic pollution at the global level.

Acknowledgments MKS is indebted to the authorities of Natural Resource Wales, Swansea, United Kingdom for providing an opportunity to work as a Scientist. MS is thankful to the authorities of the Jiangsu University for providing an opportunity to work as an Associate Professor under Invited Talent Programme. MS and ZD are highly indebted to the financial support of the National Key R&D Program of China (Grant No.2018YFE0107100) and the Primary Research and Development Plan of Jiangsu Province.

References

Akovali G (2012) Plastic materials: polyvinyl chloride (PVC), toxicity of building materials. Elsevier, pp. 23–53

Al-Zawaidah H, Ravazzolo D, Friedrich H (2021) Macroplastics in rivers: present knowledge, issues and challenges. Environ Sci: Processes Impacts 23:535–552

Anani OA, Adetunji CO (2021) Bioremediation of polythene and plastics using beneficial microorganisms, microbial rejuvenation of polluted environment. Springer, pp. 281–302

Andrady AL (2011) Microplastics in the marine environment. Mar Pollut Bull 62:1596–1605

Andrady AL, Neal MA (2009) Applications and societal benefits of plastics. Philos Trans R Soc B Biol Sci 364:1977–1984

Annomyous (2019) Press release: loss of containers from MSC Zoe in the Wadden Sea protected area shows that maritime transport regulations are not being enforced. KIMO International Secretariat Shetland Islands Council, Lerwick, Shetland

Annomyous (2021) TCE ENVIS—plastic waste management. Thiagarajar College of Engineering, Environmental Information System

Arkesteyn C 2020 Framing environmental movements: a multidisciplinary analysis of the anti-plastic straw movement, University of Minnesota

Arnaud F (2019) Field actions science reports, Reinventing Plastic Institut Veolia, pp 1–114

Ayeleru OO, Dlova S, Akinribide OJ, Ntuli F, Kupolati WK, Marina PF, Blencowe A, Olubambi PA (2019) Challenges of plastic waste generation and management in sub-Saharan Africa: a review. Waste Manag 110:24–42

Barnes RSK, Mann KH (2009) Fundamentals of aquatic ecology. John Wiley & Sons

Bertling J, Bertling R, Hamann L (2018) Kunststoffe in der Umwelt: Mikro-und Makroplastik. Ursachen, Mengen, Umweltschicksale, Wirkungen, LosungsansÃ¤tze, Empfehlungen, (Frauenhofer-Institut fur Umwelt-, Sicherheits-und Energietechnik UMSICHT, 2018) *Original not seen

Blasing M, Amelung W (2018) Plastics in soil: analytical methods and possible sources. Sci Total Environ 612:422–435

Blettler MC, Ulla MA, Rabuffetti AP, Garello N (2017) Plastic pollution in freshwater ecosystems: macro-, meso-, and microplastic debris in a floodplain lake. Environ Monit Assess 189:1–13

Browne MA, Crump P, Niven SJ, Teuten E, Tonkin A, Galloway T, Thompson R (2011) Accumulation of microplastic on shorelines woldwide: sources and sinks. Environ Sci Technol 45:9175–9179

Carlson RR, Foo SA, Asner GP (2019) Land use impacts on coral reef health: a ridge-to-reef perspective. Front Mar Sci 6:562

Carpenter EJ, Smith KL (1972) Plastics on the sargasso sea surface. Science 175:1240–1241

Chalmin P (2019) The history of plastics: from the capitol to the Tarpeian rock. Field actions science reports. J Field Actions:6–11

Chauhan V, Karki T, Varis J (2019) Review of natural fiber-reinforced engineering plastic composites, their applications in the transportation sector and processing techniques. J Thermoplast Composit Mater:0892705719889095
Chen Y, Awasthi AK, Wei F, Tan Q, Li J (2020) Single-use plastics: production, usage, disposal, and adverse impacts. Sci Total Environ 141772
Chiellini E, Solaro R (2003) Biodegradable polymers and plastics. Springer Science & Business Media
Chin LW, Fung TH (eds) (2018) Plastic in marine litter. In: Plastics and the environment. Royal Society of Chemistry, pp 21–59
Conchubhair DO, Fitzhenry D, Lusher A, King AL, van Emmerik T, Lebreton L, Ricaurte-Villota-C, Espinosa L, O'Rourke E (2019) Joint effort among research infrastructures to quantify the impact of plastic debris in the ocean. Environ Res Lett 14:065001
Conservancy O (2019) McKinsey & Company (2015) Stemming the tide-land-based strategies for a plastic-free ocean
Czuba L (2014) Application of plastics in medical devices and equipment. In: Handbook of polymer applications in medicine and medical devices. Elsevier, pp 9–19
d'Ambrières W (2019) Plastics recycling worldwide: current overview and desirable changes. Field actions science reports. J Field Actions:12–21
Danso D, Chow J, Streit WR (2015) Plastics: microbial degradation, environmental and biotechnological perspectives. Appl Environ Microbiol 31:204–209
Directive SF (2013) Guidance on monitoring of marine litter in European seas. Luxembourg 10: 99475
Dodiuk H (2013) Handbook of thermoset plastics. William Andrew
Drobny JG (2014) Handbook of thermoplastic elastomers. Elsevier
Ebbesmeyer CC, Ingraham WJ Jr (1994) Pacific toy spill fuels ocean current pathways research. EOS Trans Am Geophys Union 75:425–430
Edward JKP, Mathews G, Raj KD, Laju RL, Bharath MS, Kumar PD, Arasamuthu A, Grimsditch G (2020) Marine debris-an emerging threat to the reef areas of gulf of Mannar, India. Mar Pollut Bull 151:110,793
van Emmerik T, Schwarz A (2020) Plastic debris in rivers. Wiley Interdiscip Rev Water 7:e1398
van Emmerik T, Kieu-Le T-C, Loozen M, van Oeveren K, Strady E, Bui X-T, Egger M, Gasperi J, Lebreton L, Nguyen P-D (2018) A methodology to characterize riverine macroplastic emission into the ocean. Front Mar Sci 5:372
Europe P (2016) Plastics-the facts 2016. An analysis of European latest plastics production, demand and waste data
Gall SC, Thompson RC (2015) The impact of debris on marine life. Mar Pollut Bull 92:170–179
Garcia-Rivera S, Lizaso JLS, Millan JMB (2017) Composition, spatial distribution and sources of macro-marine litter on the Gulf of Alicante seafloor (Spanish Mediterranean). Mar Pollut Bull 121:249–259
Geyer R, Jambeck JR, Law KL (2017) Production, use, and fate of all plastics ever made. Sci Adv 3: e1700782
Good TP, June JA, Etnier MA, Broadhurst G (2010) Derelict fishing nets in Puget Sound and the Northwest Straits: patterns and threats to marine fauna. Mar Pollut Bull 60:39–50
Gourmelon G (2015) Global plastic production rises, recycling lags. Vital Signs 22:91–95
Grigore ME (2017) Methods of recycling, properties and applications of recycled thermoplastic polymers. Recycling 2:24
Hale RC, Song B (2020) Single-use plastics and COVID-19: scientific evidence and environmental regulations. ACS Publications
Hammer J, Kraak MHS, Parsons JR (2012) Plastics in the marine environment: the dark side of a modern gift. Rev Environ Contam Toxicol:1–44
Hartmann NB, Huffer T, Thompson RC, Hasselloÿv M, Verschoor A, Daugaard AE, Rist S, Karlsson T, Brennholt N, Cole M (2019) Are we speaking the same language?

Recommendations for a definition and categorization framework for plastic debris. ACS Publications

Hinata H, Mori K, Ohno K, Miyao Y, Kataoka T (2017) An estimation of the average residence times and onshore-offshore diffusivities of beached microplastics based on the population decay of tagged meso-and macrolitter. Mar Pollut Bull 122:17–26

Ho NHE, Not C (2019) Selective accumulation of plastic debris at the breaking wave area of coastal waters. Environ Pollut 245:702–710

Hoornweg D, Bhada-Tata P (2012) What a waste: a global review of solid waste management, … knowledge papers no. 15. World Bank, Washington, DC

Jacquin J, Cheng J, Odobel C, Pandin C, Conan P, Pujo-Pay M, Barbe V, Meistertzheim A-L, Ghiglione J-F (2019) Microbial ecotoxicology of marine plastic debris: a review on colonization and biodegradation by the "plastisphere". Front Microbiol 10:865

Jambeck JR, Geyer R, Wilcox C, Siegler TR, Perryman M, Andrady A, Narayan R, Law KL (2015) Plastic waste inputs from land into the ocean. Science 347:768–771

Jéssica L, Segal B, Casarini LM (2019) Abandoned, lost or otherwise discarded fishing gear in Brazil: a review. Perspect Ecol Conserv 17:1–8

Kataoka T, Hinata H, Kato S (2013) Analysis of a beach as a time-invariant linear input/output system of marine litter. Mar Pollut Bull 77:266–273

Kataoka T, Hinata H, Kato S (2015) Backwash process of marine macroplastics from a beach by nearshore currents around a submerged breakwater. Mar Pollut Bull 101:539–548

Kawecki D, Nowack B (2019) Polymer-specific modeling of the environmental emissions of seven commodity plastics as macro-and microplastics. Environ Sci Technol 53:9664–9676

Koelmans AA, Kooi M, Law KL, Van Sebille E (2017) All is not lost: deriving a top-down mass budget of plastic at sea. Environ Res Lett 12:114028

Krueger MC, Harms H, Schlosser D (2015) Prospects for microbiological solutions to environmental pollution with plastics. Appl Microbiol Biotechnol 99:8857–8874

Kruger L, Casado-Coy N, Valle C, Ramos M, Sanchez-Jerez P, Gago J, Carretero O, Beltran-Sanahuja A, Sanz-Lazaro C (2020) Plastic debris accumulation in the seabed derived from coastal fish farming. Environ Pollut 257:113,336

Kühn S, Rebolledo ELB, Franeker JAV (2015) Deleterious effects of litter on marine life. Mar Anthropogenic Litter:75–116

Kulkarni Vishakha S, Butte Kishor D, Rathod Sudha S (2012) Natural polymers- a comprehensive review. Int J Res Pharmaceut Biomed Sci 3:1597–1613

Lamb JB, Willis BL, Fiorenza EA, Couch CS, Howard R, Rader DN, True JD, Kelly LA, Ahmad A, Jompa J (2018) Plastic waste associated with disease on coral reefs. Science 359:460–462

Langärd S (2013) The use of national cancer registers for causality studies and prevention, Krebsrisiken am Arbeitsplatz: Beiträge und Ergebnisse zum gleichlautenden Workshop im Rahmen des Programms "Arbeit und Technik" am 13. und 14. Dezember 1990 im Bundesministerium für Forschung und Technologie in Bonn. Springer-Verlag, pp. 95

Lavers JL, Bond AL (2017) Exceptional and rapid accumulation of anthropogenic debris on one of the world's most remote and pristine islands. Proc Natl Acad Sci 114:6052–6055

Lebreton L, Andrady A (2019) Future scenarios of global plastic waste generation and disposal. Palgrave Commun 5:1–11

Lebreton LCM, Van Der Zwet J, Damsteeg J-W, Slat B, Andrady A, Reisser J (2017) River plastic emissions to the world's oceans. Nat Commun 8:1–10

Lebreton L, Slat B, Ferrari F, Sainte-Rose B, Aitken J, Marthouse R, Hajbane S, Cunsolo S, Schwarz A, Levivier A (2018) Evidence that the great Pacific garbage patch is rapidly accumulating plastic. Sci Rep 8:1–15

Lechthaler S, Waldschläger K (2020) Macroplastic. Encyclopedia

Lechthaler S, Waldschlager K, Stauch G, Schuttrumpf H (2020) The way of macroplastic through the environment. Environments 7:73

Lee K-W, Shim WJ, Kwon OY, Kang J-H (2013) Size-dependent effects of micro polystyrene particles in the marine copepod Tigriopus japonicus. Environ Sci Technol 47:11278–11283

Li WC, Tse HF, Fok L (2016) Plastic waste in the marine environment: a review of sources, occurrence and effects. Sci Total Environ 566:333–349
Liro M, Tv E, Wyzga B, Liro J, Mikus P (2020) Macroplastic storage and remobilization in rivers. Water 12:2055
Macfadyen G, Huntington T, Cappell R (2009) Abandoned, lost or otherwise discarded fishing gear. United Nations Environment Programme, Food and Agriculture Organization of the United Nations Rome
Maso M, Garces E, Pages F, Camp J (2003) Drifting plastic debris as a potential vector for dispersing harmful algal bloom (HAB) species. Sci Mar 67:107–111
Matjasic T, Simcic T, Medveŝček N, Bajt O, Dreo T, Mori N (2020) Critical evaluation of biodegradation studies on synthetic plastics through a systematic literature review. Sci Total Environ 141959
McKeen LW (2014) Plastics used in medical devices. In: Handbook of polymer applications in medicine and medical devices. Elsevier, pp. 21–53
Meijer LJJ, van Emmerik T, van der Ent R, Schmidt C, Lebreton L (2017) More than 1000 rivers account for 80% of global riverine plastic emissions into the ocean. Sci Adv 7:eaaz5803
Mueller R-J (2006) Biological degradation of synthetic polyesterse enzymes as potential catalysts for polyester recycling. Process Biochem 41:2124–2128
Muthukumar A, Veerappapillai S (2015) Biodegradation of plastics: a brief review. Int J Pharm Sci Rev Res 31:204–209
Nakashima E, Isobe A, Si K, Itai T, Takahashi S, Guo X (2016) The potential of oceanic transport and onshore leaching of additive-derived lead by marine macro-plastic debris. Mar Pollut Bull 107:333–339
Nihei Y, Yoshida T, Kataoka T, Ogata R (2020) High-resolution mapping of Japanese microplastic and macroplastic emissions from the land into the sea. Water 12:951
North EJ, Halden RU (2013) Plastics and environmental health: the road ahead. Rev Environ Health 28:1–8
Oberbeckmann S, Labrenz M (2020) Marine microbial assemblages on microplastics: diversity, adaptation, and role in degradation. Annu Rev Mar Sci 12:209–232
Piehl S, Leibner A, Loder MGJ, Dris R, Bogner C, Laforsch C (2018) Identification and quantification of macro-and microplastics on an agricultural farmland. Sci Rep 8:1–9
Rapp DC, Youngren SM, Hartzell P, Hyrenbach KD (2017) Community-wide patterns of plastic ingestion in seabirds breeding at French frigate shoals, northwestern Hawaiian islands. Mar Pollut Bull 123:269–278
Roebroek CTJ, Harrigan S, Van Emmerik THM, Baugh C, Eilander D, Prudhomme C, Pappenberger F (2021) Plastic in global rivers: are floods making it worse? Environ Res Lett 16:025003
Ryberg MW, Laurent A, Hauschild M 2018 Mapping of global plastics value chain and plastics losses to the environment: with a particular focus on marine environment, Department of Management Engineering, Technical University of Denmark, 2800 Kgs. , Lyngby, Denmark
Ryberg MW, Hauschild MZ, Wang F, Averous-Monnery S, Laurent A (2019) Global environmental losses of plastics across their value chains. Resour Conserv Recycl 151:104,459
Sander L (2016) Date-prints on stranded macroplastics: inferring the timing and extent of overwash deposition on the Skallingen peninsula, Denmark. Mar Pollut Bull 109:373–377
Sangale MK, Shahnawaz M, Ade AB (2012) A review on biodegradation of polythene: the microbial approach. J Bioremed Biodegr 3:1–9
Sangale MK, Shahnawaz M, Ade AB (2019) Potential of fungi isolated from the dumping sites mangrove rhizosphere soil to degrade polythene. Sci Rep 9:1–11
Scarascia-Mugnozza G, Sica C, Russo G (2011) Plastic materials in European agriculture: actual use and perspectives. J Agri Eng 42:15–28
Schnurr REJ, Alboiu V, Chaudhary M, Corbett RA, Quanz ME, Sankar K, Srain HS, Thavarajah V, Xanthos D, Walker TR (2018) Reducing marine pollution from single-use plastics (SUPs): a review. Mar Pollut Bull 137:157–171

Schuyler Q, Hardesty BD, Wilcox C, Townsend K (2014) Global analysis of anthropogenic debris ingestion by sea turtles. Conserv Biol 28:129–139

Senthilkumar P, Jayapriya D, Indumathi A, Kavitha KK, Ganapathy G (2016) Biodegradation of polythene waste through soil bacteria: a review. Int J Sci Eng Res 4:2347–3878

Shah AA, Hasan F, Hameed A, Ahmed S (2008) Biological degradation of plastics: a comprehensive review. Biotechnol Adv 26:246–265

Shahnawaz M, Sangale MK, Ade AB (2016) Rhizosphere of Avicennia marina (Forsk.) Vierh. As a landmark for polythene degrading bacteria. Environ Sci Pollut Res 23:14621–14635

Shahnawaz M, Sangale MK, Ade AB (2019a) Conclusions and future needs, bioremediation Technology for Plastic Waste. Springer, Singapore, pp 127–130

Shahnawaz M, Sangale MK, Ade AB (2019b) Bioremediation technology for plastic waste. Springer Nature Singapore Pte Ltd, Singapore

Shahnawaz M, Sangale MK, Ade AB (2019c) Microplastics, bioremediation technology for plastic waste. Springer, Singapore, pp 11–19

Shahnawaz M, Sangale MK, Ade AB (2019d) Bioremediation technology for plastic waste, 1. Springer, Singapore

Silva ALP, Prata JC, Walker TR, Campos D, Duarte AC, Soares AMVM, Barcelo D, Rocha-Santos T (2020) Rethinking and optimising plastic waste management under COVID-19 pandemic: policy solutions based on redesign and reduction of single-use plastics and personal protective equipment. Sci Total Environ 742:140,565

Singh B, Sharma N (2008) Mechanistic implications of plastic degradation. Polym Degrad Stab 93: 561–584

Slaughter E, Gersberg RM, Watanabe K, Rudolph J, Stransky C, Novotny TE (2011) Toxicity of cigarette butts, and their chemical components, to marine and freshwater fish. Tob Control 20: i25–i29

Syberg K, Palmqvist A, Khan FR, Strand J, Vollertsen J, Clausen LPW, Feld L, Hartmann NB, Oturai N, Moller S (2020) A nationwide assessment of plastic pollution in the Danish realm using citizen science. Sci Rep 10:1–11

Takada H, Tanaka K (2016) Microplastic fragments and microbeads in digestive tracts of planktivorous fish from urban coastal waters. Sci Rep 6:34351

Thompson RC, Moore CJ, Vom Saal FS, Swan SH (2009) Plastics, the environment and human health: current consensus and future trends. Philos Trans Roy Soc B Biol Sci 364:2153–2166

Thushari GGN, Senevirathna JDM (2020) Plastic pollution in the marine environment. Heliyon 6: e04709

Trivedi P, Hasan A, Akhtar S, Siddiqui MH, Sayeed U, Khan MKA (2016) Role of microbes in degradation of synthetic plastics and manufacture of bioplastics. J Chem Pharm Res 8:211–216

Van Emmerik T, Loozen M, Van Oeveren K, Buschman F, Prinsen G (2019) Riverine plastic emission from Jakarta into the ocean. Environ Res Lett 14:084033

Van Meer F, Giraud A, Esteve D, Dollat X (2005) A disposable plastic compact wrist for smart minimally invasive surgical tools, 2005 IEEE/RSJ international conference on intelligent robots and systems. IEEE, pp. 919–924

Van Sebille E, Aliani S, Law KL, Maximenko N, Alsina JM, Bagaev A, Bergmann M, Chapron B, Chubarenko I, Cozar A, P D (2020): The physical oceanography of the transport of floating marine debris. Environ Res Lett 15, 023003

Vermeiren P, Munoz CC, Ikejima K (2016) Sources and sinks of plastic debris in estuaries: a conceptual model integrating biological, physical and chemical distribution mechanisms. Mar Pollut Bull 113:7–16

Vriend P, Van Calcar C, Kooi M, Landman H, Pikaar R, Emmerik TV (2020) Rapid assessment of floating macroplastic transport in the Rhine. Front Mar Sci 7:10

Waldschlaÿager K, Schuttrumpf H (2020) Infiltration behavior of microplastic particles with different densities, sizes, and shapes' from glass spheres to natural sediments. Environ Sci Technol 54:9366–9373

Walker TR, Olubukola Adebambo, , Feijoo MCDA, Elhaimer E, Hossain T, Edwards SJ, Morrison CE, al. e (2019) Environmental effects of marine transportation, World seas: an environmental evaluation . Academic Press, pp. 505–530

Wayman C, Niemann H (2021) The fate of plastic in the ocean environment-a minireview. Environ Sci Processes Impacts 23:198–212

Webb HK, Arnott J, Crawford RJ, Ivanova EP (2013) Plastic degradation and its environmental implications with special reference to poly (ethylene terephthalate). Polymers 5:1–18

Wierckx N, Narancic T, Eberlein C, Wei R, Drzyzga O, Magnin A, Ballerstedt H, Kenny ST, Pollet E, Averous L (2018): Plastic biodegradation: challenges and opportunities. Consequences of microbial interactions with hydrocarbons, oils, and lipids: biodegradation and bioremediation, pp. 1–29

Wilcox C, Van Sebille E, Hardesty BD (2015) Threat of plastic pollution to seabirds is global, pervasive, and increasing. Proc Natl Acad Sci 112:11899–11904

Winton DJ, Anderson LG, Rocliffe S, Loiselle S (2020) Macroplastic pollution in freshwater environments: focusing public and policy action. Sci Total Environ 704:135,242

Yang SS, Brandon AM, Xing DF, Yang J, Pang JW, Criddle CS, Ren NQ, Wu WM (2018) Progresses in polystyrene biodegradation and prospects for solutions to plastic waste pollution, IOP Conference Series: Earth and Environmental Science. IOP Publishing, pp. 012005

Zhang Z-H, Wang H-l, Chen Y-Q, Xu P-j (2008) Preparation of fuel oil from waste polyethylene by delayed coking. J Fuel Chem Technol 36:223

Impact of the COVID-19 Outbreak on the Generation of Plastic Waste

3

Sarawut Sangkham

Abstract

Since the start of the outbreak COVID-19 epidemic by a novel coronavirus 2019 (SARS-CoV-2) early in December 2020, several countries worldwide have sought to shield their citizens by numerous prevention measures which protective gear is often considered high-demand personal protective equipment (PPE). It has promoted important advancement in the environment and public health, particularly during the COVID-19 outbreak. The PPE kits including face masks, gowns, gloves, hand sanitizer bottles, and other PPE packaging materials used at hospitals, quarantine centers, and households to prevent of spreading of SARS-CoV-2 particles are made up of single-use plastic. Hence, the extensive usage of these protective gears at the global level leads to the generation of plastic residues worldwide. Plastics are difficult to degrade and decay in natural environment, and plastic needs several hundred years to get degraded. The weathering of the plastic waste due to various environmental forces leads to the generation of smaller plastic particles, viz., microplastic, mesoplastic, and nanoplastic. So, due to the ultra-usage of protective gears, huge quality of plastic waste including microplastic got generated in the environment. In the present chapter, an attempt was made to overview the impact of the COVID-19 outbreak on the generation of plastic waste under the following lines: (i) to discuss the properties and types of plastic used in the COVID-19 outbreak as a protective measure, (ii) to enlist the types of plastic waste generated in the current epidemic, (iii) to highlight the methods employed for plastic waste treatment (PWTs), and (iv) to recommend the mitigation measures to manage the plastic waste.

S. Sangkham (✉)
Department of Environmental Health, School of Public Health, University of Phayao, Phayao, Thailand
e-mail: sarawut.sa@up.ac.th

M. Shahnawaz et al. (eds.), *Impact of Plastic Waste on the Marine Biota*,
https://doi.org/10.1007/978-981-16-5403-9_3

Keywords

Face masks · Plastic waste · Medical waste · Waste-to-energy recovery · COVID-19 pandemic

3.1 Introduction

In early December 2019, the outbreak of a novel coronavirus disease (COVID-19) caused by severe acute respiratory syndrome coronavirus 2 (SARS-CoV-2) has led to a dramatic loss of human lives worldwide. The disease first emerged in Wuhan, China (Patrício Silva et al. 2021; Yang et al. 2021). With its rampant spread across the globe, the World Health Organization (WHO) declares COVID-19 a concerning international public health emergency. Consequently, several measures and prevention have been taken by the severely affected countries (Hantoko et al. 2021). The COVID-19 situation has also led to numerous use of personal protective equipment (PPE) and single-use plastics (SUPs) (Haque et al. 2021), resulting in an increasing number of plastic waste from both households and healthcare sectors. In particular, there has been a sudden rose in the demand for plastic products for personal protective equipment to protect the general population, infected persons, and medical and health professionals (Klemeš et al. 2020). This has raised concerns for environmental pollutants related to single-use plastics and waste disposal during the COVID-19 pandemic. It has been reported that about 97% of plastic residues were obtained from the diagnosis process of the coronavirus with real-time polymerase chain reaction (RT-PCR), generating nearly 15,440 tons of plastic residues worldwide (Celis et al. 2021). Additionally, Sangkham's (2020) study estimated a total usage of 2.2 billion pieces of face masks in Asia. About 584 million tons of plastic wastes were also generated globally during the pandemic (Benson et al. 2021). An unprecedented dramatic rise in the use of PPEs and other use plastics including face masks, gloves, hand sanitizer bottles, take-out plastic bags, polyethylene bags, and medical laboratory diagnostic test kits has become a key concern, in addition to the effects of the disease (Benson et al. 2021). These products are commonly manufactured from slowly degradable plastic materials, for example, polypropylene (PP), polyurethane (PU), and polyethylene terephthalate (PET). Moreover, these masks are made of polymer materials that are difficult to degrade in the environment and it may take you 450 years for the entire degradation process to occur (Fadare and Okoffo 2020; Zhang et al. 2021).

Various reviews in literature (Celis et al. 2021; Parashar and Hait 2021) on plastic waste generation are due to the current COVID-19 epidemic. The management of plastic waste associated with the COVID-19 outbreak **requires** knowing the origin sources, types, volumes, properties, and technology by used to proper sanitation and reduced the impact on the environment. Therefore, researchers have reported several options from recent previous studies for the disposal of plastic waste. However, this current thermal treatment is an efficient way to reduce plastic waste and serve as an alternative energy source in the future. In the present chapter, an attempt was made to overview the impact of the COVID-19 outbreak on plastic waste generation with mitigation measures under the following lines: (i) to discuss the properties and types of plastic used in COVID-19 outbreak as a protective measure, (ii) to enlist the types of plastic waste generated in the current epidemic, (iii) to highlight the methods employed for plastic waste treatment (PWTs), and (iv) to recommend the mitigation measures to manage the plastic waste.

3.2 Properties and Types of Plastic Used in COVID-19 Outbreak

Personal protective equipment (PPE) during the COVID-19 outbreak including surgical masks, clothes mask, face shields, gloves, hand sanitizer bottles, SUPs, etc. are mostly made of plastic. Different types of plastic-based PPE used during the COVID-19 outbreak are enlisted in Tables 3.1 and 3.2.

3.3 Types of Plastic Waste Generated in the Current Epidemic

Most personal protective equipment is made from polymer materials such as polypropylene (PP), polyurethane (PU), low-density polyethylene (LDPE), polyvinyl chloride (PVC), and polycarbonate (PC), whereas polyethylene terephthalate (PET) is the most used plastic material used in packaging as well as LDPE, high-density polyethylene (HDPE), polystyrene (PS), etc. (Parashar and Hait 2021). Mainly sources of plastic waste generation related with the outbreak from diagnosis, treatment, and self-protect from the COVID-19 infection in general populations as well as human behavior changes of food consumption during the lockdown. Moreover, surgical masks, gloves, and disposable gowns are often made from polymers that slowly degradable such as PP, PU, and PET (Zhang et al. 2021), for instance,

Table 3.1 Classification of plastics usage during the COVID-19 outbreak

Type of PPE discarded	Polymers	Weight (g)	References
– Respirators	Polypropylene (PP)	18.1	Selvaranjan et al. (2021)
– Surgical mask	Polypropylene (PP)	3.5	Haque et al. (2021), Selvaranjan et al. (2021)
– Surgical gloves	Polyethylene (PE)	15.0	Haque et al. (2021), Selvaranjan et al. (2021)
– Single-use gloves	Polyethylene (PE)	2.5	Haque et al. (2021), Selvaranjan et al. (2021)
– Hand sanitizer bottles, disinfectant wipe container	Polyethylene terephthalate (PET)	18.5	Haque et al. (2021), Selvaranjan et al. (2021)
– Single-use plastic bags and disposable gloves	Polyethylene (PE)	4.0	Haque et al. (2021), Selvaranjan et al. (2021)

Table 3.2 Specific density and properties of type of polymers

Categories of plastic	Types of polymers	Density (g/mL)	Properties	References
Plastics-thermoplastics	Polyethylene (PE)	0.85–0.90	Float	Coyle et al. (2020), Hall (1981), Hidalgo-Ruz et al. (2012), Anonymous (2021)
	Polypropylene (PP)	0.82–0.90	Float	
	High-density polyethylene (HDPE)	0.95–0.96	Float	
	Low-density polyethylene (LDPE)	0.90–0.92	Float	
	Polystyrene (PS)	1.05–1.06	Sink	
	Polyvinyl chloride (PVC)	1.37–1.40	Sink	
	Polyethylene terephthalate (PET)	1.37–1.45	Sink	
	Polyamide (PA, nylon)	1.02–1.05	Sink	
	Polyester (PES)	1.24–2.30	Sink	
	Acrylic	1.09–1.20	Sink	
	Polycarbonate (PC)	1.53–1.69	Sink	
	Polytetrafluorethylene (PTFE)	1.93–2.23	Sink	
Plastics-thermosets	Polyoxymethylene (POM)	1.14–1.36	Sink	Coyle et al. (2020), Hall (1981), Hidalgo-Ruz et al. (2012), Anonymous, (2021)
	Polyvinyl alcohol (PVOH)	1.17–1.33	Sink	
	Poly(butyl acrylate)	1.06–1.10	Sink	
	Alkyd	1.24–2.10	Sink	
	Epoxy or phenoxy resin (epoxy, phenoxy)	1.16–1.19	Sink	
Elastomers	Polychloroprene (neoprene)	1.25–1.26	Sink	Coyle et al. (2020), Hall (1981), Hidalgo-Ruz et al. (2012), Anonymous 2021
	Poly(1,2-butadiene)	0.89–0.94	Float	
	Polyisoprene (IR)	0.89–0.93	Float	

common plastic material related with COVID-19 outbreak, for example, healthcare sectors, hospitals, clinical laboratories, and quarantine such as surgical masks, SARS-CoV-2 test kits, syringes, respirators, saline bottles, gloves, aprons, shoe boot, goggles, gowns, biohazard bags, net caps, catheter, and oxygen pipeline (Dharmaraj et al. 2021b). These materials main components of types of polymers include PP, PS, PVC, LDPE, HDPE, and PET, which can be divided into four sources of plastic waste produced during the COVID-19 outbreak including household, hospital/hospital filed, quarantine, and commerce and industry as shown in Fig. 3.1.

RT-PCR diagnosis of COVID-19 leads to a 15,440 tons generation of plastic waste (Table 3.3). As per report (Benson et al. 2021) with the usage of daily face

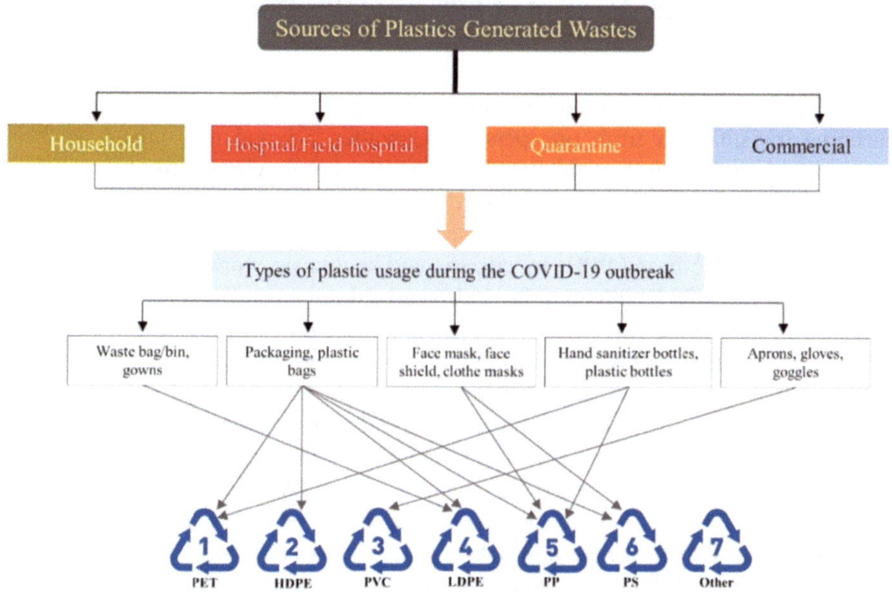

Fig. 3.1 Sources of plastic waste and types of plastics usage with recyclability symbols of the various polymers during the COVID-19 outbreak. PET and HDPE (widely recycled), PS and LDPE (rarely recycled), PVC and PP (often not recycled) based on Klemeš et al. (2020)

Table 3.3 Estimated plastic waste generation related COVID-19 by continent

Region	Plastics residues with SARS-CoV-2 tests (tons)[a]	Plastic discarded with daily face mask usage (tons)[b]
Asia	9617.18	348,079,108
North America	2570.16	27,665,223
Europe	2205.01	56,072,702
South America	563.07	49,046,434
Africa	278.94	100,544,861
Oceania	205.23	3,200,836

[a]Plastic residue derived from RT-PCR coronavirus tests were estimated based on Celis et al.'s study (2021)

[b]Plastic waste generation with face mask was estimated based on Benson et al.'s study (2021)

masks, around 584,609,165 tons of plastic waste gets generated at the global level. In addition, of all the plastics used from confirmatory diagnostic tests for SARS-CoV-2, about 89.99% corresponds to PP, 8.22% (PES), and 1.78% to PE (Celis et al. 2021).

3.3.1 Supply and Demand of PPE and SUPs During the COVID-19 Outbreak

The coronavirus disease generally spread through respiratory droplet aerosol or airborne transmission from one person to another. Several measures have been consequently taken to prevent the spread of the coronavirus. This has led to a suddenly surged in the volume of plastic waste, especially the use of PPEs in both the healthcare sectors and the general (Klemeš et al. 2020). Thus, the increase of demand for plastic-based PPEs by citizens and healthcare workers to protect themselves from COVID-19 infection has exhibited a dramatic rise in plastic production and distribution worldwide (Patrício Silva et al. 2021; Kalina and Tilley 2020). It has estimated that the monthly use of medical masks and gloves reaches approximately 129 and 65 million, respectively. In addition, the World Health Organization forecasts demand for 89 million masks per month, 76 million (gloves), 30 million (robes), and 1.6 million (glasses) with 2.9 million of hand sanitizer, as an important part of the safety measures for frontline health workers (WHO 2020). These products have become common tools for preventing the spread of the coronavirus and being compulsory to wear face masks in public areas. To meet the skyrocketing demand, the production of PPEs and subsequent PPE wastes has increased dramatically in recent times. It is also important to note that disposable masks are made from plastics with a variety of melt-blown properties that cannot be reused. Must be disposed of by employing the incinerators will be appropriate. Moreover, improper disposal of these products may increase the risk of microbial contamination and environmental pollution. With an estimated annual global production of 52 million face masks, it is expected that about 1.56 million masks will enter the ocean in 2020, accounting for 4680–6240 tons of plastic waste (Phelps Bondaroff and Cooke 2020). In addition, estimated demand for plastics is expected to increase 40% and 17% for medical and other uses (Patrício Silva et al. 2020; Patrício Silva et al. 2021).

3.3.2 The Impact of Delivery Food Services and Online Shopping on the Plastic Utilities During the COVID-19 Outbreak (Parashar and Hait 2021)

Nationwide lockdowns to curb the spread of the COVID-19 have resulted in citizens relying on online shopping for groceries and take-out service for food. The lockdown also included the close of markets, and the closure of restaurants and canteens is each country's measure implemented to impose physical distancing among the population and prevent the spread of the SARS-CoV-2 (Dharmaraj et al. 2021b). These activities increase the demand SUP bags and plastic packagings (Parashar and Hait 2021). Additionally, the outbreak has induced a new pattern of consumption and behavioral changes such as hoarding food items and groceries. It eventually resulted in a rise in the use of plastic-based packaging items globally (Parashar and Hait 2021). As a result, plastic becomes increasingly important to everyday life during the COVID-19 epidemic, inevitably deteriorating plastic pollution and contamination. This is a result of the COVID-19 outbreak due to hygiene problems and a growing dependence on take-out food, which has led to an increase in the use of plastics, especially plastic packaging.

3.4 Highlight the Methods Employed for Plastic Waste Treatment (PWTs)

One of the environmental issues that will inevitably arise as a result of the COVID-19 epidemic is the accumulation of plastic waste residues, particularly the medical plastic waste, hygiene face mask used for the general public, and the plastic generated from food packaging during the lockdown of each country worldwide, a lot of this waste. In particular, if this waste is handled inefficiently. Inevitably, there will be problems and impacts on the environment and aquatic life in the future. The author has summarized the method for plastic waste management during the COVID-19 outbreak. Several studies, the result showed that methods for plastic waste disposal during the COVID-19 such as recycling, pyrolysis, landfill, incineration, and composting as shown in Table 3.4.

3.5 Recommend the Mitigation Measures to Manage the Plastic Waste

Plastic wastes, especially PPEs, SUPs, and medical wastes generated during COVID-19, have a high calorific value compared to other conventional fuels. Based on Klemeš et al.'s (2020) study on the calorific value of a plastic material by the types of plastics, the lower heating value (LHV, MJ/Kg) was summarized as follows: PE (42–45 MJ/Kg), PVC (15–25 MJ/Kg), PP (30.90–45 MJ/Kg), PS (38.97–40 MJ/Kg), and PET (21.81 MJ/Kg). Therefore, it has the potential to be used as an alternative source of energy (Miandad et al. 2019). Additionally, the pyrolysis experimental process of face masks and gloves was carried out in a closed reactor at 400 °C for 1 h. It resulted in a 75% rate of liquid or wax fuel, char of 10%, and the remainder is a gas that cannot be condensed (Aragaw and Mekonnen 2021). It has been shown that masks can be converted to oil due to their thermoplastic properties and the possibility of high oil content (Aragaw and Mekonnen 2021). Therefore, the principle of pyrolysis can produce crude oil obtained from many types of plastic waste in an absence of an air process. It is therefore possible to recycle it to replace conventional energy sources. It is worth noting that diesel, wax, or crude oil content depends on the calorific value of the plastic material (Schwarz et al. 2021). The above values showed that a lot of plastic waste was generated during the COVID-19 outbreak. Consequently, pyrolysis combustion can be adapted for sustainable waste disposal (Dharmaraj et al. 2021a). The pyrolysis method is recommended as an effective technology that benefits the environment and shows potential for fuel in waste-to-energy (WTE) recovery.

Table 3.4 Methods for plastic waste disposal during the COVID-19 outbreak

References	Study design	Plastic waste disposal methods				
		Landfill	Incineration	Pyrolysis	Recycling	Composting
Aragaw and Mekonnen (2021)	Experiment			/		
Dharmaraj et al. (2021a)	Review				/	
Dharmaraj et al. (2021b)	Review			/		
Klemeš et al. (2020)	Review		/		/	
Patrício Silva et al. (2020)	Review	/			/	
Sangkham, (2020)	Survey	/	/			
Saberian et al. (2021)	Experiment				/	
Kilmartin-Lynch et al. (2021)	Experiment				/	
Pereira de Albuquerque et al. (2021)	Experiment					/
Yousef et al. (2021)	Experiment			/		

Option of pyrolysis can be an environmentally friendly and efficient process for managing plastic waste from the COVID-19 scenario, simultaneously producing a value-added product (Dharmaraj et al. 2021b). Pyrolysis is thus considered a good choice and a simpler, cleaner, and environmentally friendly method for waste-to-energy production. In addition, pyrolysis can reduce a large amount of plastic waste during the outbreak and an alternative policy for plastic waste disposal in the future, especially from PPE and SUP wastes (Aragaw and Mekonnen 2021). Whereas, in the case of incinerating plastic waste to generate electricity, ineffective control of the released and removed exhaust gas will cause air pollution. It will also affect the health of the population living nearby incineration facilities, especially during the COVID-19 epidemic.

3.6 Conclusion

An increase in the volume of plastic wastes is indeed relevant to the coronavirus outbreak. Plastic wastes particularly from self-defense and livelihood measures during the COVID-19 pandemic, such as face masks, food packaging, waste bags, hand sanitizer bottles, and plastic waste resulting from medical diagnostic and therapeutic procedures, have become a key concern. Most plastics are made of polymers such as PP, PU, LDPE, PVC, and PC. On the other hand, plastic materials used in packaging mainly consist of PET, LDPE, HDPE, and PS. These materials have become both protectors and polluters in case of improper waste management. The recycling of PPE and SUP via fuel recovery through pyrolysis has been suggested, which was sustainable waste management options to mitigate the problem at massive of single-use waste disposal. Thus, governments should raise awareness of the issue of waste with their citizens and adopt a more efficient plastic waste management system to reuse waste, to reduce the accumulation of plastic waste in the country and into the marine ecosystem, and, furthermore, should be adapting legislation used to appropriate plastic waste recycling and WTE recovery before the final disposal.

Acknowledgments The author would like to thank the editors and anonymous reviewers for suggestions and recommendations to improvise the book chapter.

References

Anonymous (2021) Polymer properties database. https://www.polymerdatabase.com/home.html. Accessed 23 February 2021

Aragaw TA, Mekonnen BA (2021) Current plastics pollution threats due to COVID-19 and its possible mitigation techniques: a waste-to-energy conversion via pyrolysis. Environ Syst Res 10 (1):11. https://doi.org/10.1186/s40068-020-00217-x

Benson NU, Bassey DE, Palanisami T (2021) COVID pollution: impact of COVID-19 pandemic on global plastic waste footprint. Heliyon 7(2):9, e06343. https://doi.org/10.1016/j.heliyon.2021.e06343

Celis JE, Espejo W, Paredes-Osses E, Contreras SA, Chiang G, Bahamonde P (2021) Plastic residues produced with confirmatory testing for COVID-19: classification, quantification, fate, and impacts on human health. Sci Total Environ 760:7, 144167. https://doi.org/10.1016/j.scitotenv.2020.144167

Coyle R, Hardiman G, Driscoll KO (2020) Microplastics in the marine environment: a review of their sources, distribution processes and uptake into ecosystems. Case Stud Chem Environ Eng 2:7, 100010. https://doi.org/10.1016/j.cscee.2020.100010

Dharmaraj S, Ashokkumar V, Hariharan S, Manibharathi A, Show PL, Chong CT, Ngamcharussrivichai C (2021a) The COVID-19 pandemic face mask waste: a blooming threat to the marine environment. Chemosphere 272:20, 129601. https://doi.org/10.1016/j.chemosphere.2021.129601

Dharmaraj S, Ashokkumar V, Pandiyan R, Halimatul Munawaroh HS, Chew KW, Chen WH, Ngamcharussrivichai C (2021b) Pyrolysis: an effective technique for degradation of COVID-19 medical wastes. Chemosphere 275:20, 130092. https://doi.org/10.1016/j.chemosphere.2021.130092

Fadare OO, Okoffo ED (2020) Covid-19 face masks: a potential source of microplastic fibersin the environment. Sci Total Environ 737:4, 140279. https://doi.org/10.1016/j.scitotenv.2020.140279

Hall C (1981) Polymers: molecular structure. In: Hall C (ed) Polymer materials: an introduction for technologists and scientists. Macmillan Education, London, pp 1–31

Hantoko D, Li X, Pariatamby A, Yoshikawa K, Horttanainen M, Yan M (2021) Challenges and practices on waste management and disposal during COVID-19 pandemic. J Environ Manage 286:9, 112140. https://doi.org/10.1016/j.jenvman.2021.112140

Haque MS, Sharif S, Masnoon A, Rashid E. 2021; SARS-CoV-2 pandemic-induced PPE and single-use plastic waste generation scenario. Waste Manag Res. Doi: https://doi.org/10.1177/0734242X20980828

Hidalgo-Ruz V, Gutow L, Thompson RC, Thiel M (2012) Microplastics in the marine environment: a review of the methods used for identification and quantification. Environ Sci Technol 46 (6):3060–3075. https://doi.org/10.1021/es2031505

Kalina M, Tilley E (2020) "This is our next problem": cleaning up from the COVID-19 response. Waste Manag 108:202–205. https://doi.org/10.1016/j.wasman.2020.05.006

Kilmartin-Lynch S, Saberian M, Li J, Roychand R, Zhang G (2021) Preliminary evaluation of the feasibility of using polypropylene fibres from COVID-19 single-use face masks to improve the mechanical properties of concrete. J Clean Prod 296:8, 126460. https://doi.org/10.1016/j.jclepro.2021.126460

Klemeš JJ, Fan YV, Tan RR, Jiang P (2020) Minimising the present and future plastic waste, energy and environmental footprints related to COVID-19. Renew Sustain Energy Rev 127:7, 109883. https://doi.org/10.1016/j.rser.2020.109883

Miandad R, Rehan M, Barakat MA, Aburiazaiza AS, Khan H, Ismail IMI, Dhavamani J, Gardy J, Hassanpour A, Nizami AS (2019) Catalytic pyrolysis of plastic waste: moving toward pyrolysis based biorefineries. Front Energy Res 7:17, 27. https://doi.org/10.3389/fenrg.2019.00027

Parashar N, Hait S (2021) Plastics in the time of COVID-19 pandemic: protector or polluter? Sci Total Environ 759:15, 144274. https://doi.org/10.1016/j.scitotenv.2020.144274

Patrício Silva AL, Prata JC, Walker TR, Campos D, Duarte AC, Soares AMVM, Barcelò D, Rocha-Santos T (2020) Rethinking and optimising plastic waste management under COVID-19 pandemic: policy solutions based on redesign and reduction of single-use plastics and personal protective equipment. Sci Total Environ 742:8, 140565. https://doi.org/10.1016/j.scitotenv.2020.140565

Patrício Silva AL, Prata JC, Walker TR, Duarte AC, Ouyang W, Barcelò D, Rocha-Santos T (2021) Increased plastic pollution due to COVID-19 pandemic: challenges and recommendations. Chem Eng J 405:9, 126683. https://doi.org/10.1016/j.cej.2020.126683

Pereira de Albuquerque F, Dhadwal M, Dastyar W, Mirsoleimani Azizi SM, Karidio I, Zaman H, Dhar BR (2021) Fate of disposable face masks in high-solids anaerobic digestion: experimental observations and review of potential environmental implications. Case Stud Chem Environ Eng 3:7, 100082. https://doi.org/10.1016/j.cscee.2021.100082

Phelps Bondaroff T, Cooke S (2020) Masks on the beach: the impact of COVID-19 on marine plastic pollution. OceansAsia. https://oceansasia.org/wp-content/uploads/2020/12/Marine-Plastic-Pollution-FINAL.pdf. Accessed 1 March 2021

Saberian M, Li J, Kilmartin-Lynch S, Boroujeni M (2021) Repurposing of COVID-19 single-use face masks for pavements base/subbase. Sci Total Environ 769:8, 145527. https://doi.org/10.1016/j.scitotenv.2021.145527

Sangkham S (2020) Face mask and medical waste disposal during the novel COVID-19 pandemic in Asia. Case Stud Chem Environ Eng 2:9, 100052. https://doi.org/10.1016/j.cscee.2020.100052

Schwarz AE, Ligthart TN, Godoi Bizarro D, De Wild P, Vreugdenhil B, van Harmelen T (2021) Plastic recycling in a circular economy; determining environmental performance through an LCA matrix model approach. Waste Manag 121:331–342. https://doi.org/10.1016/j.wasman.2020.12.020

Selvaranjan K, Navaratnam S, Rajeev P, Ravintherakumaran N (2021) Environmental challenges induced by extensive use of face masks during COVID-19: a review and potential solutions. Environ Challenges 3:11, 100039. https://doi.org/10.1016/j.envc.2021.100039

WHO (2020) Shortage of personal protective equipment endangering health workers worldwide. https://www.who.int/news/item/03-03-2020-shortage-of-personal-protective-equipment-endangering-health-workers-worldwide. Accessed 16 Febuery 2021

Yang L, Yu X, Wu X, Wang J, Yan X, Jiang S, Chen Z (2021) Emergency response to the explosive growth of health care wastes during COVID-19 pandemic in Wuhan, China. Resour Conserv Recycl 164:8, 105074. https://doi.org/10.1016/j.resconrec.2020.105074

Yousef S, Eimontas J, Striūgas N, Abdelnaby MA (2021) Pyrolysis kinetic behaviour and TG-FTIR-GC–MS analysis of coronavirus face masks. J Anal Appl Pyrolysis 156:10, 105118. https://doi.org/10.1016/j.jaap.2021.105118

Zhang EJ, Aitchison LP, Phillips N, Shaban RZ, Kam AW (2021) Protecting the environment from plastic PPE. BMJ 372:2, n109. https://doi.org/10.1136/bmj.n109

Regulations to Minimize the Entry of Plastic Waste Into the Oceans

4

Mubashar Hussain, Somia Liaqat, and Kiran Aftab

Abstract

Unsustainable consumption and disposal of plastics is posing threats to ecosystem functioning which ultimately endangers biodiversity and threatens human life. The ever-increasing magnitudes of plastic wastes loaded in the ecosystems generate ecological consequences. Unprecedented, indiscriminate, and uncontrolled accumulation of plastic contaminants in the oceans is choking our marine life. This has led to devaluing the services and values of the ecosystem. In this chapter, we highlighted various strategies and regulations to minimize the entry of plastic waste into marine environments. Plastic disposal strategies like wastewater management, recycling, coastal zone improvements, reduced consumption, developing alternatives to plastics, and using biodegradable plastics need to be employed to conserve the environment. Global, regional, and national level initiatives, media campaigns, policies, legislations, and regulations are key to reducing plastic debris in marine ecosystems. The United Nations, International Laws, multinational companies, and educating entrepreneurs and civil society about plastic management are key drivers in mitigating plastic pollution. The regional-scale initiatives may include various political and economic associations like ASEAN, SAARC, Arab League, Asia-Pacific Economic Cooperation, European Union, Three Seas Initiative, Global Covenant of Mayors for Climate & Energy, etc., which provide an important platform for nations to develop and enforce regulations for preventing plastic entry into the oceans. In conclusion, marine reservoirs may be prevented from the entry of plastics by focusing on the 3R rule (reuse, recycle, and reduce), removal by regular

M. Hussain (✉) · S. Liaqat · K. Aftab
Department of Zoology, University of Gujrat, Gujrat, Pakistan
e-mail: dr.mubashar@uog.edu.pk

M. Shahnawaz et al. (eds.), *Impact of Plastic Waste on the Marine Biota*,
https://doi.org/10.1007/978-981-16-5403-9_4

monitoring of debris and introducing cleaning initiatives, regulation of litter disposal, and educational approaches including awareness campaigns.

Keywords

Biodiversity loss · Marine choking · Plastic pollution · Plastic management

4.1 Introduction

Overpopulation over the past few decades has accelerated the demand for plastic products which has resulted in 381 million tonnes (in 2015) equivalent to two-third (2/3) mass of the world population (Hannah and Max 2018). The ever-increasing use of disposable plastic products is resulting in an increased ratio of plastic in municipal solid waste (MSW). Plastic litter has contaminated all ecosystems with far-reaching effects on wildlife and human health (Law 2017; Chae and An 2018; Wagner and Lambert 2018; Bergmann et al. 2019). Poor plastic waste management is the main source of plastic pollution in oceans. Marine sources of plastic pollution contribute about 20% to marine plastic pollution which comes mainly from fishing nets, ropes, and lines (Hannah and Max 2018). Recently, the Mariana Trench that is the deepest part of the Pacific Ocean reported plastic debris (Strategy 2018). Plastic is a large and diverse group of synthetic polymers, and it can be categorized in different ways. Based on thermal resistance, plastic can be categorized as thermoplastic and thermosets.

Thermoplastics like polypropylene, polystyrene, polyamide, polyvinyl chloride, polyethylene, and polyethylene terephthalate are recyclable due to their property of melting upon heating and solidifying on cooling. Thermosets cannot be recycled because they undergo chemical alteration upon heating. Polyurethanes, silicone, unsaturated polyester, epoxy, melamine, acrylic, and phenolic resins are examples of thermosets (Aisbl 2019). Plastics can also be categorized as fossil-based plastic or biobased plastics based on carbon sources which can be either fossils or freshly grown biomass (Álvarez-Chávez et al. 2012; Kühn et al. 2015). Another way to categorize plastics is based on the biodegradability of plastic material (Bioplastics 2016).

Plastic pollution is considered one of the greatest threats to biodiversity due to its accumulation in the environment at an alarming rate. IUCN reports that about eight million tonnes out of 300 million tonnes of plastics enter into the oceans contributing about 80% of all marine debris from surface waters. This demands an understanding of the issue, reviewing corrective measures, and suggesting future roadmaps for dealing with plastics polluting marine environments.

4.1.1 Production, Use, and the Fate of Plastics

The Plastics Europe Market Research Group (PEMRG) is the commonly used source of data for plastic production. Fibers and additives are not included in PEMRG data. PEMRG data suggests an 8.3% annual growth rate from 1950 to 2017 (Geyer 2020b). Data on plastic additives is sparse and is not included in the statistical analysis of data for plastic production. In non-fiber plastics, the ratio of additives is 7%, while that of polymer resin is 93% (Rajaram 2009; Geyer 2020a). Plasticizers, fillers, and flame retarders are the commonly used additives (Geyer et al. 2017). Data obtained in 2017 suggest that the plastic production sector is dominated by China and Asia with 29% production in China, 21% in Asia, 19% in Europe, 18% in NAFTA countries, and 13% in the rest of the world (CPMA 2015; Geyer et al. 2017; Aisbl 2018; Geyer 2020b). The production of biobased plastics (about two million metric tonnes annually) is insignificant and some of which are biodegradable plastics (Bioplastics 2018).

The low cost and diverse properties make plastics suitable for use in every industrial sector. Plastics are commonly used for packaging, transportation, electrical equipment, medical supplies, sports goods, and construction (ACC 2013; Aisbl 2018). PVC is used for the production of pipes, window and door frames, hoses, and cable insulation. PET is usually used to produce bottles and clamshell containers. High-density polyethylene (HDPE) is used in the manufacture of milk bottles, toiletries, and washing and cleaning products (Siskind 1990; Siddique et al. 2008). Foam products like mattresses and pillows are made from polyurethane (PUR), while low-density polyethylene (LDPE) is commonly used for packaging (Emblem 2012; Geyer et al. 2017; Koerner and Koerner 2018; Han 2019; Das and Mahanwar 2020).

Pre-consumer waste is produced at the time of plastic production and can be easily recycled. Plastic pollution is mostly generated in the seas and gulfs that are surrounded by densely populated coasts (Hinojosa and Thiel 2009; Collignon et al. 2012; Reisser et al. 2013; Ryan 2013). Toxic chemicals are adsorbed on the surface of plastics found in all oceans across the globe (Mato et al. 2001; Teuten et al. 2007; Beaman et al. 2016). Non-biodegradable plastics are destined for landfills or water reservoirs. Plastics can also be recycled to produce the secondary plastic product but cannot be recycled over and over (Zink and Geyer 2018). Photodegradation causes the disintegration of plastic into fragments that get dispersed in the oceans (Barnes et al. 2009, 2010).

4.1.2 Types of Plastic

4.1.2.1 Polycarbonates

Polycarbonates are polymers with carbonate linkages. Depending upon the variation in the R-group, polycarbonates have been classified as aliphatic and aromatic polycarbonates. Polycarbonates have high heat resistance, rigidity, and transparency

(Takeuchi 2012). These are used to manufacture eyeglasses, traffic lights, discs, and lenses (Rhodes 2018).

4.1.2.2 Polyester

Polyesters are produced as a result of the condensation of terephthalic acid and ethylene glycol. Polyesters have an ester functional group in the polymer chain (Grishanov 2011). Polyester is used to manufacture textiles and fibers (Rhodes 2018).

4.1.2.3 Polyethylene

Polyethylene is a polyolefin resin having an amorphous crystalline structure (Agboola et al. 2017). Polyethylene is manufactured by the polymerization of ethylene (Ucankus et al. 2018). It is used to manufacture wrapping and packing materials (Rhodes 2018).

4.1.2.4 Polyethylene Terephthalate

Polyethylene terephthalate is produced by transesterification of dimethyl terephthalate and ethylene glycol. PET is chemically more stable and has high mechanical strength. PET is used to manufacture food jars, plastic films, and carbonated drink bottles (Rhodes 2018).

4.1.2.5 Polypropylene

Polypropylene is a thermoplastic which is a polymer of propylene monomers. Polypropylene is hard, chemically resistant, and highly crystalline (Begum et al. 2020). It is used to manufacture yogurt containers, straws, bottle caps, and pipe systems (Rhodes 2018).

4.1.2.6 Polystyrene

Polystyrene is an aromatic hydrocarbon produced by the polymerization of styrene (Begum et al. 2020). Polystyrene is amorphous, transparent, and thermoplastic (Mallakpour and Behranvand 2017). Polystyrene is used to manufacture plates, cutlery, discs, food containers, and disposable cups (Rhodes 2018).

4.1.2.7 Polyurethane

Polyurethane is a diverse class of organic polymers that are synthesized by reacting a polyol with di-isocyanate (Samavedi et al. 2014). It provides heat insulation, to manufacture printing rollers and surface coatings (Rhodes 2018).

4.1.2.8 Polyvinyl Chloride

Polyvinyl chloride is the most common synthetic polymer produced by the polymerization of chloroethylene monomers (Pascault et al. 2012). PVC is chemically inert and it can be flexible or rigid (McKeen 2012). Frames of doors and plumbing pipes are made up of PVC (Rhodes 2018).

4.1.2.9 Innovations in Plastic Materials

Fossil-based sources of carbon are nowadays replaced by biobased raw materials to synthesize various chemicals including plastics. Plastic products obtained by using biobased raw materials are called biobased plastics. Biobased plastics may or may not be biodegradable. Some biobased plastics are produced from naturally occurring polymers, while others are produced by the polymerization of biobased monomers (Tsuji 2014).

4.1.2.10 Cellulose Derivatives

Cellulose carbamate is the biobased plastic derived from the chemical modification of cellulose. Cellulose acetates, butyrates, and propionates are some of the cellulose esters. The reactions proceed in a controlled environment to avoid too much degradation of cellulose chains which can result in low-quality products (Tsuji 2014).

4.1.2.11 Starch Derivatives

Thermoplastic starch is obtained by heat treatment and mechanical modification of starch. This modification leads to the destruction of the original crystalline structure of starch. Various additives are added to avoid the production of crystals upon cooling. Water (below 30%) is the most commonly used additive. Vegetable oils, nucleating agents, glycerol, and some water-soluble polymers are also added to enhance performance (Tsuji 2014).

4.1.2.12 Polyhydroxyalkanoates

Polyhydroxyalkanoates are synthesized by various genera of archaea, eubacteria, and cyanobacteria as intracellular products. Polyhydroxyalkanoates are the only natural polymers that are thermoplastic. Microbes possess numerous substrates for the production of polyhydroxyalkanoates. Various microbial species possess the ability to degrade polyhydroxyalkanoates (Tsuji 2014).

4.1.2.13 Polylactic Acid

Polylactic acid (a thermoplastic polymer) is synthesized by lactic acid polymerization which is a product of carbohydrate fermentation. It exists in two enantiomeric forms, i.e., poly (L-lactic acid) and poly (D-lactic acid). Pure polymers of both these forms are optically active; however, a mixture of both these forms is optically inactive. After cellulose and starch derivatives, PLA is a biobased polymer with the highest production rate globally (Tsuji 2014).

4.1.2.14 Polybutylene Succinate

Polybutylene succinate is a linear polymer manufactured by the polycondensation of 1,4-butanediol and succinic acid. It is 50% biobased because succinic acid is a product of carbohydrate fermentation, while 1,4-butanediol is obtained via a fossil-based production route. To obtain 100% biobased polybutylene succinate, 1,4-butanediol is obtained by the process of succinic acid hydrogenation (Patel et al. 2018).

4.1.2.15 Biobased Polyethylene

Polyethylene is produced by the polymerization reaction of ethylene monomers. Biobased ethylene is manufactured by the dehydration of ethanol which is produced by carbohydrate fermentation (Inaba et al. 2011). Different types of polyethylene are produced by variations in manufacturing conditions, e.g., temperature, pressure, and catalysts (Koopmans 2014).

4.1.2.16 Biobased Polyamides

Polyamides are linear polymers having amide linkages throughout the chain. Polyamides are produced by the polymerization of a single amine consisting of an amino group and a carboxylic acid group. Biobased polyamides are obtained from undecenoic acid or sebacic acid, and both of these are obtained from castor oil (Brehmer 2013).

4.2 Importance of Plastic

Plastics constitute a major part of the widely used materials, and plastic products in numerous forms cover every aspect of everyday life. Most of the plastic products are used for packaging and building applications such as plastic bags, containers, pipes, etc. (Mutha et al. 2006). Common applications of plastics include as follows.

4.2.1 Clothing

Polyethylene terephthalate (PET) is used to manufacture polycotton clothing, while high-quality clothing is entirely made up of nylons, polyesters, and fluoropolymers. Fleece clothing consists of 100% PET, and it can also be manufactured with recycled PET. The soles and footbeds of footwear are made from polyurethane, and the upper part is mainly composed of vinyl or any other polymer (Andrady and Neal 2009).

4.2.2 Healthcare

Plastics also find applications in the public healthcare sector by facilitating the supply of clean water. Plastics are also used to produce medical devices including medical equipment, drips, pill packaging material, etc. Plastic packaging helps to reduce food wastage by increasing the shelf life of meat and vegetable items (Mullan 2002).

4.2.3 Transportation

Vehicles typically contain 20% plastics in the form of steering wheels, door lines, electrics, and electronics. About 30% of the aeroplane's structure is composed of plastic. This results in reduced weight, improved performance, and enhanced fuel efficiency (Andrady and Neal 2009). Plastics have helped to implement various safety measures in vehicles such as shock absorbers, seat belts, airbags, and child safety seats.

4.2.4 Sports and Recreation

Plastics have been widely used for the production of sports shoes, footballs, tennis racquets, helmets, knee protectors, skateboards, and toys (Scherer et al. 2018; Laad 2020). Running shoes are light in weight but provide sufficient strength and support to the athlete (Huang 2019). Nylon, polyether, polypropylene, and rubber polymers have been widely used to manufacture products used in sports (Coleman 2008; Geyer et al. 2017).

4.3 Sources of Marine Plastic Pollution

All ecosystems are interconnected with each other so changes in one ecosystem have effects on the other ecosystem (Adams 2005). Humans have been exploiting natural resources from both terrestrial and aquatic ecosystems. Anthropogenic activities have resulted in litter accumulation in oceans and freshwater bodies.

Among all types of litter including glass, paper, food waste, wood, and metal, plastic litter is the most persistent type of pollutant (Rosevelt et al. 2013). Improper disposal has resulted in the accumulation of more than 260 metric tonnes of plastic waste in oceans (Eriksen et al. 2014). Land-based plastic litter sources are thought to contribute 80% to the total marine plastic pollution (UNEP 2009). Mismanagement of plastic litter is the root cause of all environmental and health issues caused by plastic pollution. Major contributors of plastic waste in the marine ecosystem include as follows.

4.3.1 Residential and Domestic Activities

Coastal areas are highly populated and industrialized. The waste generated by local communities is directly discharged into the coastal zone. Plastic waste gets accumulated in freshwater bodies or leaches down to groundwater deposits which ultimately enter into oceans. Plastic waste carried by lotic water usually accumulates in coastal areas. Debris accumulation in the coastal environment is mainly caused by domestic and residential activities (Thushari et al. 2017).

4.3.2 Tourism and Recreational Activities

Ocean-based sources of plastic waste dominate in areas that have high recreational activities like fishing and boating. Research studies have shown that tourism and recreational activities significantly increase the litter inputs into oceans (Thiel et al. 2003; Hoellein et al. 2015). Plastic waste accumulated on the seashore is subjected to photodegradation and then enters into the ocean as microplastics (Cole et al. 2011).

4.3.3 Maritime and Navigation Activities

Major sources of plastic waste in oceans include ships because of the maritime routine of dumping waste in the sea (Hagen 1989). Discarded fishing gears serve as fishing apparatus and continue to target fish species. The accumulation rate of plastic litter dumped by marine vessels is almost 6.5 million tons per year by early 1990 (Derraik 2002).

4.3.4 Fishing and Aquaculture-Related Activities

Aquaculture-related activities and offshore fishing are also major sources of plastic pollution. Fishing nets disposed at the seashore are carried into the ocean where they float on the water surface. These nets are carried to different locations along with water currents (Cole et al. 2011).

4.4 Effects of Plastic Pollution

Numerous research studies have explored the impact of plastic pollution on various aspects of aquatic wildlife (Azzarello and Van Vleet 1987; Sigler 2014; Vegter et al. 2014; Hardesty et al. 2015; Seltenrich 2015; Wilcox et al. 2016; Carney Almroth and Eggert 2019; Machovsky-Capuska et al. 2019; Bucci et al. 2020; Galgani and Loiselle 2020; Litchfield et al. 2020; Roman et al. 2020; Thushari and Senevirathna 2020; Alfonso et al. 2021). Plastic pollution affects mainly by the ingestion, suffocation, and entanglement of hundreds of marine species. Plastic debris disturbs fishes, seabirds, whales, and turtles by entering into their stomach leading to death due to starvation (Thevenon et al. 2014; Boucher and Friot 2017) and also lacerations, infections, and internal injuries (Cassoff et al. 2011; Moore and Van der Hoop 2012; Abreo et al. 2016). Following are the ways by which plastics affect aquatic wildlife.

4.4.1 Entanglement

Entanglement involves the trapping of marine animals in plastic waste dumped in the sea. Discarded fishing nets, plastic ropes, and packaging materials are the most common entanglements (Laist 1987; Page et al. 2004; Lawson et al. 2015; Wilcox et al. 2016; Jepsen and de Bruyn 2019; Gall and Thompson 2015; Kühn et al. 2015). Abandoned fishing gears continue to capture and entangle fishes (Macfadyen et al. 2009; Wilcox et al. 2015; Adelir-Alves et al. 2016; He and Suuronen 2018). Marine turtles, seals, whales, and seabirds are severely affected (Wehle and Coleman 1983; Laist 1997; Kühn et al. 2015; Gall and Thompson 2015; Li et al. 2016; Pawar et al. 2016; Law 2017; Thiel et al. 2018; Isangedighi et al. 2020; Senko et al. 2020).

4.4.2 Ingestion

Marine animals directly or indirectly ingest plastic waste along with water currents or through food (Derraik 2002; Alimba and Faggio 2019; Kühn and Van Franeker 2020; López-Martínez et al. 2021; Savoca et al. 2021). The size of ingested plastic debris depends upon the size of the organism. Microplastics are ingested by small animals like mussels and oysters, while large plastic films and fibers are taken up by large fish species, marine-associated bird species, and whales (Lazar and Gračan 2011; Possatto et al. 2011; Avery-Gomm et al. 2013; Markic et al. 2020, De Stephanis et al. 2013). Ingested plastic reduces the stomach capacity, can cause ulcers, and may rupture the gastric lining ultimately leading to death (Day et al. 1985; Carson 2013; Gilbert et al. 2016; Avery-Gomm et al. 2018). Biochemical responses to plastic include reduction in enzyme activity, cellular breakdown, and imbalance in the production and accumulation of oxygen (Browne et al. 2013; Oliveira et al. 2013; Rochman et al. 2013).

4.4.3 Interaction

Collision or obstruction of wildlife with plastic can also affect in multiple ways like injury, killing, entanglement, and threatening sensitive habitats (Jones 1995; Bugoni et al. 2001; Donohue and Foley 2007; Allen et al. 2012; Bond et al. 2013; Baulch and Perry 2014; Kühn et al. 2015; Awuchi and Awuchi 2019). About 700 marine species have been reported to interact with marine debris (Gall and Thompson 2015; Wilcox et al. 2016). Substrate interference by plastic can affect the ecosystem structure, and the abrasion of fishing nets affects the coral reef communities (Donohue et al. 2001; Arthur et al. 2014). However, the impact of debris on wildlife is difficult to quantify mainly due to a lack of information and understanding of wildlife–debris interactions under natural environment (Laist 1997; Ryan et al. 2009; Wilcox et al. 2016).

4.4.4 Effects on Humans

Humans ingest microplastics (0.1 μm–1 mm) and nanoplastics (<0.1 μm) through water or seafood such as fishes that have ingested plastic (Revel et al. 2018). It moves in trophic levels leading to bioaccumulation of toxic compounds (Galloway 2015; Güven et al. 2017). Microplastics are mostly present in the gut tract of fishes which is not eaten so any transmission of plastic through fish occurs only if the plastic fibers are present in muscle tissue (Galloway 2015; Bouwmeester et al. 2015). The hydrophobic nature of plastic material provides a large surface area for the adsorption of contaminants which are released when ingested by humans (Iñiguez et al. 2017). Further investigations would provide evidence regarding the bioaccumulation of these pollutants and plastic additives in humans. Biomonitoring of human tissue has indicated that chemicals used in plastic synthesis are found in humans. Research has shown that these chemicals affect experimental organisms adversely when trialed under laboratory conditions (Talsness et al. 2009). Endocrine disorders, cancer, and developmental and reproductive anomalies are the most pronounced effects of chemicals in plastic (Trasande et al. 2015).

4.4.5 Climate Change

Sea surface is a very unique place as it contains the compounds synthesized by marine life (Malve 2016; Pakistan 2016). The quantity of chemical compounds on sea surface increases with the increase in plastic waste as marine microbes, i.e., bacteria, become more active, hence reducing the oxygen content of water (Galgani and Loiselle 2021). The sea surface is connected with oceans as well as the environment and is responsible for exchange between the two systems, i.e., flow of gases like oxygen and carbon dioxide which is a major greenhouse gas (Halmann and Steinberg 1998; Heinze et al. 2015). Anaerobic decomposition of plastic litter in dead zones releases methane, carbon dioxide, and some other trace gases. These are the greenhouse gases that have high global warming potential (Change 2007). Reduced usage and recycling of plastic waste are expected to mitigate the effects of climate change. Innovative packaging and plastic waste recycling would reduce the drastic effects on the environment (Chandegara et al. 2015).

4.5 Regulations to Minimize the Entry of Plastic Waste to Oceans

Both terrestrial and aquatic sources of plastics in various modes pollute the marine environment (Thushari and Senevirathna 2020). Several rules have been developed at national, international, and regional scales to minimize the negative impacts of plastic waste (van Truong and Bei Ping n.d.; Zhang and Chen 2014; Mwanza and Mbohwa 2017; Hasan 2018; Shahnawaz et al. 2019; Abril Ortiz et al. 2020; Jyothsna and Chakradhar 2020). Waste management may prove valuable in restoring the quality of oceans (Palumbi et al. 2009; Aburto et al. 2012). Diverse usage of plastics and poor disposal result in the accumulation of waste in natural habitats causing problems for wildlife and their transference to wildlife and humans (Thompson et al. 2009).

4.5.1 Recycling

Plastic recycling involves a complex process of waste collection, separation of contaminants and recyclable products, grinding and separation based on color and polymer type using Fourier-transformed near-infrared spectroscopy, pellet formation, and selling the pellets to companies (Ren 2003; Bing et al. 2014; Ignatyev et al. 2014; Schneider and Ragossnig 2015; Walker and Xanthos 2018). Primary recycling uses uncontaminated materials to produce high-quality plastic products like the plastic back of televisions, whereas secondary recycling involves the use of consumer waste to produce low-quality plastic products, i.e., textiles, concrete, asphalt, and composites (Selwyn 2007; Peeters et al. 2012; Najafi 2013; Gu and Ozbakkaloglu 2016; Poulikakos et al. 2017; Walker and Xanthos 2018). Granulated plastic waste could improve the properties of asphalt concrete (Gu and Ozbakkaloglu 2016; Poulikakos et al. 2017). Cheap thermoset composites can be produced by the crosslinking or polymerization of plastic and non-plastic mixtures (Najafi 2013; Ignatyev et al. 2014; Walker and Xanthos 2018).

Plastic recycling has various limitations, i.e., high recycling cost relative to the production cost of new plastic products (Al-Salem et al. 2009), reduced recycling cycles due to decomposition and contamination (Braungart et al. 2007; Walker and Xanthos 2018), and less recyclable plastic products such as textiles and loose packaging (Ignatyev et al. 2014; Liu et al. 2018). Plastic products also contain thousands of different chemicals, and the majority of these are toxic (Groh et al. 2018). These additives increase health risks during the recycling process and may result in reduced quality of the recycled material. A specific recycling process is required for each type of plastic to maintain the quality of the recycled product. Marine plastics have unknown chemical composition because the actual chemical additives may have leached down into the ocean and environmental pollutants may have adsorbed on the surface of plastic (Hirai et al. 2011). Recycling of mixed polymers is extremely problematic because of the different polymers and unknown compositions of the end products. Abovementioned challenges have resulted in a limited focus on plastic recycling programs (Pietrelli et al. 2017).

The problems in recycling plastic waste can be eliminated by an increase in recycling rates, improvement in the condition of recycled products, and the technical progress in the recycling procedure. Due to lesser environmental impacts, the process of recycling is favorable for waste management (Arena et al. 2003; Ross and Evans 2003; Lee and Xu 2005; Chilton et al. 2010; Bernardo et al. 2016). The recycling of polyethylene and polyethylene terephthalate requires only half of the energy utilized for the synthesis of virgin plastics (Arena et al. 2003). Incineration is also favorable because of the higher contamination risk and lower ability of recycled plastics to replace newly synthesized plastic products (Lazarevic et al. 2010). Waste management strategies prioritize recycling as it saves energy, creates job opportunities, reduces environmental pollutants, and reduces the need for landfills.

4.5.2 Coastal Zone Improvements

The gap between usage and disposal of plastics with respect to current cleanup strategies to reduce the adverse effects of plastic pollution is resulting in loading marine environment with plastics (Prata et al. 2019). The strategies to minimize the entry of plastics into the ocean include reducing inputs of plastics and managing them through global multidisciplinary approaches. Coastal zone improvements include the restriction on open ocean aquaculture (Hollaway et al. 2020; Prata et al. 2019). Land-based sustainable aquaculture systems can help decrease plastic litter associated with coastal aquaculture. Campaigns such as Adopt-a-Beach, Blue Flag, and Project Aware can play role in reducing plastic waste on beaches (Hollaway et al. 2020).

4.5.3 Improvement in Production Efficiency

Single-use plastics consumed in the medical and food sector should not be completely banned; rather, improvement in design and biodegradable constituents could be a better alternative. The recyclability of plastic products can be enhanced by using a finite number of polymers and additives (Hopewell et al. 2009; Thompson et al. 2009; Browne et al. 2011; Walker 2018; Liu et al. 2018). The design of plastic bottles can be improved by making bottles with inseparable caps (Brennholt et al. 2018). Plastic usage in cotton buds and microbeads must be replaced with biodegradable materials to minimize the plastic pollution due to cosmetics (Browne et al. 2011; Poeta et al. 2016; Dauvergne 2018; Walker 2018). Companies should reduce the pre-production plastic waste by using meshes in drains (Prata et al. 2019; Schmaltz et al. 2020). Waste recycling can be enhanced by the accountability of corporations for waste generated by their plastic products by introducing deposit schemes and paying for recycling (Milios et al. 2018; Diggle and Walker 2020).

4.5.4 Reduced Consumption of Plastic

Reduction in plastic consumption followed by efficient production can help minimize plastic waste, but it is difficult to achieve because of the diverse applications of plastic in food packaging and medicinal purposes (Beitzen-Heineke et al. 2017). However, the choice of eco-friendly materials, avoiding extra packaging, and increasing public awareness could help minimize the use of plastic (Santos et al. 2005; Chang 2015; Ambrose et al. 2019).

4.5.5 Wastewater Management

Sustainable wastewater infrastructure is a key component of waste management strategies. Wastewater contains toxic chemicals, pathogens, and plastics (Nnorom and Osibanjo 2008; Diggle and Walker 2020). It is hazardous for humans, wildlife, and the environment especially when it enters the natural reservoirs of water (Laganà, Bacaloni et al. 2004; Costanzo et al. 2005; Stewart et al. 2008). Drain filtration and trash collection prevent the entry of microplastics and macroplastics into rivers and ultimately the ocean (Hollaway et al. 2020).

4.5.6 Improvement in Waste Collection and Disposal

Proper waste management is based on 3R's principle which includes reduce, reuse, and recycle. Reduced production and reusing plastic products are a priority; however, waste is also generated between the synthesis and consumption cycle, and it needs to be properly managed by integrated waste management strategies (Thompson et al. 2009; Schneider and Ragossnig 2015). Waste collection is the first step in waste management strategies which includes source collection and post-separation (Bing et al. 2014). Source collection is done by consumers, and it is a cheaper and effective method as it reduces contamination. It includes drop-off centers, door-to-door collection, curbside collection, and buy-back centers (Bing et al. 2014; Liu et al. 2018).

Economic incentives can have a positive effect on the recycling rates of plastics as in buy-back programs, per package money is received by the consumer (Siddique et al. 2010). Container Deposit Legislation (buy-back program) in the USA and Australia has resulted in the reduced deposition of beverage containers on the coast (Schuyler et al. 2018). Door-to-door collection with economic incentives is effective in improving recycling participation, and these programs have been successfully executed in Germany and San Francisco, USA. Storage of waste in homes is a drawback of door-to-door collection, and longer collection routes by fossil fuel vehicles have negative environmental effects (Tanskanen et al. 1998). The use of smart trash containers which store a limited volume of waste could result in curbside waste collection and implementation of the "pay as you throw" rule. Variety of waste containers and their number collected during curbside collection result in improved collection and disposal of waste (Lavee 2010).

4.5.7 Energy and Feedstock Conversion of Plastic Waste

Conversion of waste to energy involves the synthesis of heat, steam, fuel, or electric current from the waste. It allows to recover energy and subjugate the limitations in recycling, i.e., sorting into similar polymer groups. Conversion of plastic to feedstock involves the separation of constituents that can be used for the synthesis of fuels, i.e., gasoline, lubricants, and carbon black (Baytekin et al. 2013; Schneider and Ragossnig 2015; Poulikakos et al. 2017; Liu et al. 2018). Pyrolysis is an effective process for the production of fuel from plastic waste that has properties similar to gasoline (Baytekin et al. 2013). Apart from manufacturing fuel, reclamation from the by-products such as the use of carbon black for improving the quality of asphalt (Poulikakos et al. 2017) as well as the production of carbon nanotubes from plastic litter with lower carbon dioxide emission has been carried out (Zhuo and Levendis 2014).

Incineration of plastic waste is considered to be highly calorific (43.3 MJ kg^{-1} for polyethylene) and produces heat, steam, and electricity causing a reduction in waste volume (Calcott and Walls 2000; Mokrzycki and Uliasz-Bocheńczyk 2003; Chilton et al. 2010; Baytekin et al. 2013; Schneider and Ragossnig 2015). Incineration is usually done for contaminated or mixed plastic polymers that are difficult to be recycled (Calcott and Walls 2000; Ignatyev et al. 2014). Incineration results in the production of ash which can be used for metal recovery and deposited in landfills (Calcott and Walls 2000) or used as aggregates in the construction of roads (Ignatyev et al. 2014; Poulikakos et al. 2017). Plastic litter is used as a source of energy in steel-making furnaces (Baytekin et al. 2013), ovens for melting metals (Ignatyev et al. 2014), and cement furnaces (Mokrzycki and Uliasz-Bocheńczyk 2003).

Advantages of converting waste to energy include higher energy saving (Bernardo et al. 2016), no requirement for pretreatment, processing of contaminated or mixed polymers, and fossil fuel replacement (Eriksson and Finnveden 2009). The drawbacks of incineration include the release of toxic substances into the atmosphere (Schneider and Ragossnig 2015; Liu et al. 2018) which can be reduced by the use of additional combustion chamber (Walker and Xanthos 2018) and supporting linear economy (Ignatyev et al. 2014). Effective air pollution control measures are required for incineration due to the release of toxic substances, massive quantities of CO_2, and waste heterogeneity (Eriksson and Finnveden 2009). Bottom and fly ash are the massively contaminated types of ash produced by incineration (Eriksson and Finnveden 2009; Brunner and Rechberger 2015). After pretreatment techniques, ash can be used in the production of glasses and ceramics and road pavement or metal recovery to allow their reintroduction in a circular economy (Brunner and Rechberger 2015).

4.5.8 Landfilling

Landfills constitute linear management of wastes, and these are also a part of integrated management if the residues are a product of recycling or incineration (Singh and Ruj 2015). Landfills are considered as the last option because of high contamination, resource depletion, and land requirement (Liu et al. 2018). Landfilling is the last resort for the accommodation of waste generated by incineration, recycling, and the conversion of waste to energy. Reclamation of plastic buried in landfills of China may result in the production of 38% of the annual energy being consumed. Conversion of waste to energy could be a part of waste management strategies, and it can be done by the provision of gasification facilities and thermogravimetric analysis to evaluate the calorific value. Limitations of this technology include heterogeneity, the difference in the thermal behavior of waste, impurities, moisture, and ash formation (Burlakovs et al. 2019).

4.5.9 Biobased and Biodegradable Synthetic Polymers

The increased demand for sustainability has resulted in fossil fuels being replaced by renewable energy resources, increased recycling rates, and effective waste management (Iriarte et al. 2009; Lambert and Wagner 2017; Napper and Thompson 2019). Biobased plastics are polymers produced from renewable constituents, i.e., biomass (Kimura 2009; Harding et al. 2017; Prata et al. 2019). Degradable polymers can decompose when triggered by a stimulus like composting of biodegradable polymers and UV exposure for photochemically degradable plastics which produce microplastics (Ren 2003; Tokiwa et al. 2009).

Biodegradable polymers decompose into the water, carbon dioxide, and organic matter. This is processed by microbial activity (fungi, algae, bacteria) and may require specific environmental conditions (Song et al. 2009; Iwata 2015). Biodegradable polymers are the solution to products with a short life span and where composting is possible, i.e., single-use cutlery, packing, and agricultural films (Lambert and Wagner 2017; Rujnić-Sokele and Pilipović 2017).

Various uncertainties regarding biodegradable polymer usage include the requirement of specific facilities, extended time for degradation and release of greenhouse gases, contaminants that reduce quality, release of harmful chemicals to environment, high production cost to avoid degradation to increase product's shelf-life, increased consumption due to the tag of being the biodegradable, and slow rate of degradation in the environment (Ren 2003; Thompson et al. 2009; Vink and Davies 2015; Lambert and Wagner 2017; Rujnić-Sokele and Pilipović 2017; Prata et al. 2019).

4.5.10 Biodegradation of Conventional Plastics

Conventional plastics can be subjected to degradation by using organisms such as insect larvae and various bacterial strains. Biodegradation of plastics manifests an eco-friendly method for the elimination of plastic waste.

4.5.10.1 Insects and Plastic Degradation

Many insect species have been explored for their potential to degrade different types of plastics. Numerous bacterial strains have also been separated from the insect's gut (Liaqat et al. 2020). Details regarding the insect species and their microbial fauna have been discussed in Table 4.1.

Table 4.1 Plastic degradation by insect larvae

Type of plastic	Insect larvae	Microbes in the gut of insect larvae
Polyurethane polystyrene polyvinyl chloride and polylactic acid	Yellow mealworms (*Tenebrio molitor*)	*Exiguobacterium* sp. *YT2* (Yang et al. 2015b), *Citrobacter* and *Kosakonia* (Brandon et al. 2018), *Lactococcus*, *Pantoea*, *Enterobacter*, and *Clostridium* (Wang and Zhang 2015)
Polystyrene	Dark mealworms (*Tenebrio obscurus*)	*Enterobacteriaceae*, *Spiroplasmataceae*, and *Enterococcaceae* (Peng et al. 2019)
Polyethylene, LDPE	Greater waxworms (*galleria mellonella*)	*Acinetobacter* (Cassone et al. 2020)
Polyethylene	Lesser waxworms (*Achroia grisella*)	Not discovered yet
Polyethylene	Indian meal moth larvae (*Plodia interpunctella*)	*Bacillus* sp. *YP1* and *Enterobacter asburiae YT1* (Yang et al. 2015a)
LDPE	Rice moth larvae (*Corcyra cephalonica*)	Not discovered yet
Polyester polyurethane	Giant mealworms (*Zophobas morio*)	*Aspergillus flavus G10* (Schaefer et al. 2020)

Table 4.2 Bacterial species involved in plastic degradation

Type of plastic	Bacteria
Polyhydroxybutyrate	*Streptoverticillium kashmeriense* strain AF1 (Shah et al. 2008)
Polyurethane	*Pseudomonas chlororaphis* (Nakajima-Kambe et al. 1999), *Comamonas acidovorans* TB-35 (Allen et al. 1999)
Polyethylene	*Nocardia asiteroides* (Bonhomme et al. 2003) and *Alcanivorax borkumensis* (Delacuvellerie et al. 2019)
Polylactic acid (Haw, Hawton et al.)	*Amycolatopsis* sp. strain *SCM_MK2–4*, *Amycolatopsis thailandensis*, *A. orientalis*, *Bacillus pumilus*, *B. cereus*, *B. brevis*, *B. licheniformis* (Kawai 2010)
Polyethylene terephthalate (PET)	*Ideonella sakaiensis* (Yoshida et al. 2016), *Bacillus thuringiensis* strain C15, *Pseudomonas* sp. B10, *Pseudomonas* sp. SWI36, *Bacillus albus* strain PFYN01 (Vague et al. 2019; León-Zayas et al. 2019), and *Phaeodactylum tricornutum* (Moog et al. 2019)
Nylon (polyamide)	*Arthrobacter* sp. strain KI72, *Bacillus cereus*, *B. sphaericus*, *Vibrio furnissii*, *Brevundimonas vesicularis*, *Pseudomonas aeruginosa* strain PAO1 (Takehara et al. 2017; Sudhakar et al. 2007; Kanagawa et al. 1993)
Low-density polyethylene (LDPE)	*Alcanivorax borkumensis* (Delacuvellerie et al. 2019)

4.5.10.2 Plastic Degradation by Free-Living Bacterial Species

Numerous free-living bacterial strains have been studied for their potential to degrade different types of plastic. Details regarding plastic degradation by the bacteria have been mentioned in Table 4.2.

4.5.11 Education and Awareness

Education and awareness in the general public can be a successful tool to minimize microplastic waste in oceans, and it has been documented in Brazil where citizens refused the microbead products after an awareness campaign (Santos et al. 2005; Potts et al. 2011). A massive increase of interest in this environmental issue has been observed and it has been supported by Massive Open Online Courses (MOOC) or activities like TechWild (Owens 2018; Prata et al. 2019). Media campaigns like *National Geographic*'s "Planet or Plastic" and BBC's Blue Planet II along with apps like Marine Debris Tracker and Sea Cleaner have a positive influence on the general public (Jambeck and Johnsen 2015; Merlino et al. 2015). The continuously updated knowledge regarding microplastics and microbeads has encouraged the information-seeking behavior of people and has proved Internet and social media as alternative sources for learning and awareness regarding this environmental menace (Selwyn 2007; Dabbagh and Kitsantas 2012).

4.6 Initiatives to Control Marine Plastic Pollution

Sustainable management of marine and coastal sources of plastic pollution needs the integration of different approaches. Over 143 countries have joined 18 regional sea conventions and action plans, and 109 countries committed to a Global Program of Action for the Protection of the Marine Environment from Land-Based Activities. Similarly, 30 countries have joined the *Clean Seas* campaign on marine litter (Unger et al. 2017). Different strategies have been considered to manage plastic pollution which is a global issue. Institutional involvement is a key strategy to treat the current issue, and the contribution of institutes at global, regional, and national scales is necessary to prevent plastic accumulation in aquatic ecosystems across the globe.

4.6.1 Global-Level Initiatives

Law of the Sea is an international initiative to address marine plastic pollution. UN Convention on Law of the Sea (UNCLOS) is a platform to control plastic pollution in the oceans. Articles 207 and 211 include the prevention, control, and reduction of plastic debris in the marine environment. Several declarations have also been delivered by the UN General Assembly to clean the oceans. It includes the resolution for the partnership to spread awareness in the public and business sector regarding the ecological and social impacts of plastic debris and to address these issues by integration with the national strategic framework (Hirai et al. 2011; Cole et al. 2013). This resolution also suggests the involvement of organizations [UN Food and Agriculture Organization (FAO), UN Environment Program (Marine Litter UNEP), International Maritime Organization] working at global, national, and regional levels to find solutions to prevent the deposition of lost fishing gears in the marine environment (Chiappone et al. 2002).

The UN Conference on Sustainable Development (*Rio* de Janeiro, Brazil in 2012) emphasized controlling the accumulation of plastic litter in oceans and also pointed out the necessity to implement the frame of the International Maritime Organization. International Convention for the Prevention of Marine Pollution (MARPOL) focuses on areas that have direct or indirect involvement in the control and prevention of plastic litter in the sea. It prohibits the release of abandoned, lost fishing gears and the dumping of plastic waste in the sea. Article 70 of the Convention on Biological Diversity (CBD) emphasizes controlling the harmful impact of plastic accumulation on marine biodiversity by various strategies such as Strategic Environmental Assessments and Environmental Impact Assessments (EIAs) (Sayied 2007; Jeftic et al. 2009; Warner 2018).

Subsidiary party on Scientific, Technical and Technological Advice (SBSTTA) is the advisory body of CBD which has met 23 times to date and produced a total of 240 recommendations (De Klem et al. 1997). Furthermore, it examines and documents the impact of plastic waste on ecosystem services and biodiversity. Research studies to manage and control plastic, and the regional level programs to eliminate and control plastic-related issues should be conducted (Thevenon et al. 2014).

The Convention of Migratory Species (CMS) works to seek hotspots with high marine debris to monitor the impact of plastic litter on marine life. It also works to identify the methods to control the marine plastic accumulation sources at the regional level and implement an action plan at the national level to diminish plastic pollution (Thevenon et al. 2014).

The Scientific Council recommended the assessment of the plastic effects on migratory species and to identify the management policies for plastic waste generated by maritime ships and vessels (Thevenon et al. 2014). Problems associated with marine plastic pollution can be mitigated by the implementation of an integrated waste management program (UNEP 2011). Basel Convention is an international law against the adverse effects of hazardous waste. Waste that contains solid plastic fragments is considered to be hazardous as it poses severe health risks (UNEP 2005).

The Global Partnership on Waste Management (GPWM) is a gateway for collaborative efforts to manage plastic waste with the involvement of various international and non-government organizations. The action plan of GPWM included the critical analysis of the existing issue, appropriate suggestions to resolve the issue, and involvement of national and international institutes to develop facilities and manage wastes. The Honolulu Strategy is a global collaborative effort to reduce the ecological impacts of marine debris worldwide by devising objectives and strategies having worldwide applicability irrespective of specific conditions or challenges.

Honolulu strategy is a framework to prevent and manage the impact of marine debris in collaboration with the US National Oceanic and Atmospheric Administration (NOAA) and UNEP. European Commission and 64 government authorities agreed with the Manila declaration to accomplish UNEP's global program for management of waste generated by land-based and ocean-based sources.

4.6.2 Regional-Scale Initiatives

UNEP proposed a Regional Seas Program for cleanup activities in 13 regional seas including Baltic Sea, Caspian Sea, Red Sea, East African Sea, Northeast Atlantic Sea, South Asian Sea, Black Sea, Wider Caribbean Sea, Northwest Pacific Sea, Gulf of Aden Sea, East Asian seas, Red Sea, and Southeast Pacific Sea (Bliss-Guest and Keckes 1982; Akiwumi and Melvasalo 1998). As a part of the global project, cleanup activities have been performed in all the abovementioned regions (Oral 2012).

European Union in 2008 established the Marine Strategy Framework Directive (MSFD) which aims to minimize marine litter at the regional level (Borja et al. 2011). This directive utilizes an ecosystem-based approach (Bergmann et al. 2019) for durable resource utilization and ecosystem conservation (Personnic et al. 2014).

A long-term project started by South Korea aimed to monitor, prevent, and eliminate marine litter for a time duration of 10 years (Chen 2015; Ronchi et al. 2019; Madricardo et al. 2020). NOAA/MDP conducted a project for the collection of discarded fishing gears at the coasts of Hawaii and South Africa. The estuarine environment can be improved by tidal restoration and river forcing, and it is the most suitable method for ecosystem restoration (Storm et al. 2005; Slater 2016).

4.6.3 National-Level Initiatives

Solid waste management and a reduction in marine debris are the key issues under the consideration of national-level legislation (McIlgorm et al. 2011). In the USA, the Marine Debris Research, Prevention, and Reduction Act and Marine Plastic Pollution Research and Control Act are the major regulations to reduce plastic pollution (Pettipas, Bernier et al. 2016, Xanthos and Walker 2017).

The Practical Integrated System of Marine Debris in South Korea was established to prohibit the accumulation of plastic waste (Cho 2005; Jung et al. 2010). In 2013, Scottish marine litter was developed by Scotland (Hastings and Potts 2013). In 2017, Sri Lanka introduced regulations for the management of polyethylene and other plastic types (Kaluarachchi et al. 2017). With the 19th amendment in the National Environment Act (Anonymous 1980), this legislation aimed to prohibit the synthesis of polyethylene with or below a diameter of 20 microns, high-density plastic bags, and cutlery manufactured from expanded polystyrene, to prohibit the burning of flammable materials, and to ban the use of polyethylene to manufacture decoration items (Wanyama 2017). In Sri Lanka, MEPA was established under Marine Pollution Prevention Act No. 35 (a national law to prevent, manage, and control marine debris) which is working to find solutions to overcome pollution. From August 2017 to January 2018, MEPA has updated the Action Plan and Policies to tackle the issues of ballast water, invasive species, and maritime traffic with a major focus on plastic pollution in the sea (IUCN 2018).

Among all plastic forms, single-use plastics are of great concern, and various control measures (taxes, charges, and prohibition) have been carried out in different countries including China, South Africa, Congo, Mexico, Hong Kong, Bangladesh, Italy, Japan, Denmark, Romania, New Zealand, Ireland, several states of India, Sao Paolo in Brazil, Kenya, South Korea, Switzerland, and some states of the USA (Commission 2013). Plastic container deposition is another national-level Extended Producer Responsibility (EPR) scheme in Canada, Australia, the USA, Europe, and Asia (Hilton et al. 2019).

4.7 Conclusion and Recommendations

- Data gaps in waste generation, characterization, and management have been identified in various reports of UNEP and other agencies of the world working on the management of plastic pollution.
- Inconsistent monitoring and lack of assessment systems regarding plastics polymer types, volumes entering into the marine environment, and fate in oceans at the global, regional, and national levels are the major reasons for elevated levels of plastic litter in the oceans.
- Understanding the fate and mechanisms of biodegradation and fragmentation of plastics in the marine environment is inadequate.
- Many species are adversely affected by plastics in marine ecosystems. But data on the severity and vulnerability of species to plastics is insufficient.
- Both direct and indirect socioeconomic costs of plastics have not been well quantified regarding commercial fishing, recreation, shipping, and tourism.
- The assessment of the sustainability of plastics and their potential alternatives and their impact on socio-cultural life cycles needs comprehensive evaluation models.

Acknowledgments Authors acknowledge the support provided by Mr. Muhammad Umar (Lecturer, Zoology) and Ms. Maryam Khalid (MPhil Scholar) during write-up and improvements.

References

Abreo NAS, Macusi ED, Blatchley DD, Cuenca GC (2016) Ingestion of marine plastic debris by green turtle (*Chelonia mydas*) in Davao Gulf, Mindanao, Philippines. Philippine J Sci 145:17–23

Abril Ortiz A, Sucozhañay D, Vanegas P, Martínez-Moscoso A (2020) A regional response to a global problem: single use plastics regulation in the countries of the Pacific Alliance. Sustainability 12:8093

Aburto MO, Carvajal M, Barr B, Barbier EB, Boesch DF, Boyd J, Crowder LB, Cudney-Bueno R, Essington T, Ezcurra E (2012) Ecosystem-based management for the oceans. Island Press

ACC (2013) Resin report: the annual statistical report of the North American plastics industry. American Chemistry Council, Washington, DC

Adams SM (2005) Assessing cause and effect of multiple stressors on marine systems. Mar Pollut Bullet 51:649–657

Adelir-Alves J, Rocha GRA, Souza TF, Pinheiro PC, Freire MF (2016) Abandoned, lost or otherwise discarded fishing gears in rocky reefs of Southern Brazil. Braz J Oceanogr 64:427–434

Agboola O, Sadiku R, Mokrani T, Amer I, Imoru O (2017) Polyolefins and the environment. Polyolefin Fibres. Elsevier, pp 89–133

Aisbl P (2018) Plastics–the facts 2018 an analysis of European plastics production. Demand and Waste Data

Aisbl P (2019) Plastics–the facts 2018. Information accessed at http://www.-plasticseurope.org/application

Akiwumi P, Melvasalo T (1998) UNEP's regional seas Programme: approach, experience and future plans. Mar Policy 22:229–234

Alfonso MB, Arias AH, Menéndez MC, Ronda AC, Harte A, Piccolo MC, Marcovecchio JE (2021) Assessing threats, regulations, and strategies to abate plastic pollution in LAC beaches during COVID-19 pandemic. Ocean Coastal Manag 208:105613

Alimba CG, Faggio C (2019) Microplastics in the marine environment: current trends in environmental pollution and mechanisms of toxicological profile. Environ Toxicol Pharmacol 68:61–74

Allen AB, Hilliard NP, Howard GT (1999) Purification and characterization of a soluble polyurethane degrading enzyme from *Comamonas acidovorans*. Int Biodeterior Biodegrad 43:37–41

Allen R, Jarvis D, Sayer S, Mills C (2012) Entanglement of grey seals Halichoerus grypus at a haul out site in Cornwall, UK. Mar Poll Bullet 64:2815–2819

Al-Salem S, Lettieri P, Baeyens J (2009) Recycling and recovery routes of plastic solid waste (PSW): a review. Waste Manag 29:2625–2643

Álvarez-Chávez CR, Edwards S, Moure-Eraso R, Geiser K (2012) Sustainability of bio-based plastics: general comparative analysis and recommendations for improvement. J Clean Prod 23: 47–56

Ambrose KK, Box C, Boxall J, Brooks A, Eriksen M, Fabres J et al (2019) Spatial trends and drivers of marine debris accumulation on shorelines in South Eleuthera, the Bahamas using citizen science. Mar Pollut Bullet 142:145–154

Andrady AL, Neal MA (2009) Applications and societal benefits of plastics. Philos Trans R Soc B Biol Sci 364:1977–1984

Anonymous (1980) Act No. 47 of 1980. Central Environmental Authority. Retrieved from https://www.elaw.org/system/files/National+Environmental+Act+of+1980.pdf. Accessed on 20 October 2020

Arena U, Mastellone ML, Perugini F (2003) Life cycle assessment of a plastic packaging recycling system. Int J Life Cycle Assess 8:92

Arthur C, Sutton-Grier AE, Murphy P, Bamford H (2014) Out of sight but not out of mind: harmful effects of derelict traps in selected US coastal waters. Mar Poll Bullet 86:19–28

Avery-Gomm S, Borrelle SB, Provencher JF (2018) Linking plastic ingestion research with marine wildlife conservation. Sci Total Environ 637:1492–1495

Avery-Gomm S, Provencher JF, Morgan KH, Bertram DF (2013) Plastic ingestion in marine-associated bird species from the eastern North Pacific. Mar Pollut Bull 72:257–259. https://doi.org/10.1016/j.marpolbul.2013.04.021

Awuchi CG, Awuchi CG (2019) Impacts of plastic pollution on the sustainability of seafood value chain and human health. Int J Adv Acad Res 5:46–138

Azzarello MY, Van Vleet ES (1987) Marine birds and plastic pollution. Mar Ecol Prog Ser 37:295–303

Barnes DK, Galgani F, Thompson RC, Barlaz M (2009) Accumulation and fragmentation of plastic debris in global environments. Philos Trans R Soc B Biol Sci 364:1985–1998

Barnes DK, Walters A, Gonçalves L (2010) Macroplastics at sea around Antarctica. Mar Environ Res 70:250–252

Baulch S, Perry C (2014) Evaluating the impacts of marine debris on cetaceans. Mar Poll Bullet 80: 210–221

Baytekin B, Baytekin HT, Grzybowski BA (2013) Retrieving and converting energy from polymers: deployable technologies and emerging concepts. Energy Environ Sci 6:3467–3482

Beaman J, Bergeron C, Benson R, Cook AM, Gallagher K, Ho K, Hoff D, Laessig S (2016) State of the Science White Paper. A Summary of Literature on the Chemical Toxicity of Plastics Pollution to Aquatic Life and Aquatic-Dependent Wildlife. Report EPA-822-R-16-009

Begum SA, Rane AV, Kanny K (2020) Applications of compatibilized polymer blends in automobile industry. Compatibilization of polymer blends. Elsevier, pp 563–593

Beitzen-Heineke EF, Balta-Ozkan N, Reefke H (2017) The prospects of zero-packaging grocery stores to improve the social and environmental impacts of the food supply chain. J Clean Prod 140:1528–1541

Bergmann M, Mützel S, Primpke S, Tekman MB, Trachsel J, Gerdts G (2019) White and wonderful? Microplastics prevail in snow from the Alps to the Arctic. Sci Adv 5:eaax1157

Bernardo C, Simões CL, Pinto LMC (2016) Environmental and economic life cycle analysis of plastic waste management options. A review AIP conference proceedings. AIP Publishing LLC, p 1400011

Bing X, Bloemhof-Ruwaard JM, van der Vorst JG (2014) Sustainable reverse logistics network design for household plastic waste. Flex Serv Manuf J 26:119–142

Bioplastics E (2016) Fact sheet: what are bioplastics? European Bioplastics, Berlin

Bioplastics E (2018) Bioplastics market data 2018. European Bioplastics, Berlin

Bliss-Guest PA, Keckes S (1982) The regional seas programme of UNEP. Environ Conserv 9:43–49

Bond AL, Provencher JF, Elliot RD, Ryan PC, Rowe S, Jones IL et al (2013) Ingestion of plastic marine debris by common and thick-billed Murres in the northwestern Atlantic from 1985 to 2012. Mar Poll Bullet 77:192–195

Bonhomme S, Cuer A, Delort A, Lemaire J, Sancelme M, Scott G (2003) Environmental biodegradation of polyethylene. Polym Degrad Stab 81:441–452

Borja Á, Galparsoro I, Irigoien X, Iriondo A, Menchaca I, Muxika I, Pascual M, Quincoces I, Revilla M, Rodríguez JG (2011) Implementation of the European marine Strategy framework directive: a methodological approach for the assessment of environmental status, from the Basque Country (Bay of Biscay). Mar Poll Bull 62:889–904

Boucher J, Friot D (2017) Primary microplastics in the oceans: a global evaluation of sources. Iucn Gland, Switzerland

Bouwmeester H, Hollman PC, Peters RJ (2015) Potential health impact of environmentally released micro-and nanoplastics in the human food production chain: experiences from nanotoxicology. Environ Sci Technol 49:8932–8947

Brandon AM, Gao SH, Tian R, Ning D, Yang SS, Zhou J et al (2018) Biodegradation of polyethylene and plastic mixtures in mealworms (larvae of *Tenebrio molitor*) and effects on the gut microbiome. Environ Sci Technol 52:6526–6533

Braungart M, McDonough W, Bollinger A (2007) Cradle-to-cradle design: creating healthy emissions–a strategy for eco-effective product and system design. J Clean Prod 15:1337–1348

Brehmer B (2013) Polyamides from biomass derived monomers. Bio-Based Plastics 2013:275–293

Brennholt N, Heß M, Reifferscheid G (2018) Freshwater microplastics: challenges for regulation and management. Freshwater microplastics. Springer, Cham, pp 239–272

Browne MA, Crump P, Niven SJ, Teuten E, Tonkin A, Galloway T et al (2011) Accumulation of microplastic on shorelines worldwide: sources and sinks. Environ Sci Technol 45:9175–9179

Browne MA, Niven SJ, Galloway TS, Rowland SJ, Thompson RC (2013) Microplastic moves pollutants and additives to worms, reducing functions linked to health and biodiversity. Curr Biol 23:2388–2392

Brunner PH, Rechberger H (2015) Waste to energy–key element for sustainable waste management. Waste Manag 37:3–12

Bucci K, Tulio M, Rochman C (2020) What is known and unknown about the effects of plastic pollution: a meta-analysis and systematic review. Ecol Appl 30:e02044

Bugoni L, Krause LG, Petry MV (2001) Marine debris and human impacts on sea turtles in southern Brazil. Mar Poll Bullet 42:1330–1334

Burlakovs J, Kriipsalu M, Porshnov D, Jani Y, Ozols V, Pehme KM et al (2019) Gateway of landfilled plastic waste towards circular economy in Europe. Separations 6:25
Calcott P, Walls M (2000) Can downstream waste disposal policies encourage upstream" design for environment"? Am Econ Rev 90:233–237
Carney Almroth B, Eggert H (2019) Marine plastic pollution: sources, impacts, and policy issues. Rev Environ Econ Policy 13:317–326
Carson HS (2013) The incidence of plastic ingestion by fishes: from the prey's perspective. Mar Poll Bullet 74:170–174
Cassoff RM, Moore KM, McLellan WA, Barco SG, Rotstein DS, Moore MJ (2011) Lethal entanglement in baleen whales. Dis Aqt Org 96:175–185
Cassone BJ, Grove HC, Elebute O, Villanueva SM, LeMoine CM (2020) Role of the intestinal microbiome in low-density polyethylene degradation by caterpillar larvae of the greater wax moth, *Galleria mellonella*. Proc R Soc B 287:20200112
Chae Y, An YJ (2018) Current research trends on plastic pollution and ecological impacts on the soil ecosystem: a review. Environ Pollut 240:387–395
Chandegara V, Cholera S, Nandasana J, Kumpavat M, Pate K (2015) Plastic packaging waste impact on climate Change and its mitigation water management and climate smart agriculture. Adaptation of Climatic Resilient Water Management and Agriculture. Gyan Publishing House
Chang M (2015) Reducing microplastics from facial exfoliating cleansers in wastewater through treatment versus consumer product decisions. Mar Pollut Bullet 101:330–333
Change IP (2007) Climate change 2007: the physical science basis. Agenda 6:333
Chen CL (2015) Regulation and management of marine litter. Marine anthropogenic litter. Springer, Cham, pp 395–428
Chiappone M, White A, Swanson DW, Miller S (2002) Occurrence and biological impacts of fishing gear and other marine debris in the Florida keys. Mar Pollut Bullet 44:597–604
Chilton T, Burnley S, Nesaratnam S (2010) A life cycle assessment of the closed-loop recycling and thermal recovery of post-consumer PET. Resour Conserv Recycl 54:1241–1249
Cho DO (2005) Challenges to marine debris management in Korea. Coast Manag 33:389–409
Cole M, Lindeque P, Halsband C, Galloway TS (2011) Microplastics as contaminants in the marine environment: a review. Mar Pollut Bullet 62:2588–2597
Cole M, Lindeque P, Fileman E, Halsband C, Goodhead R, Moger J et al (2013) Microplastic ingestion by zooplankton. Environ Sci Technol 47:6646–6655
Coleman WF (2008) Molecular models of polymers used in sports equipment. ACS Publications
Collignon A, Hecq JH, Glagani F, Voisin P, Collard F, Goffart A (2012) Neustonic microplastic and zooplankton in the North Western Mediterranean Sea. Mar Pollut Bullet 64:861–864
Commission E (2013) Draft impact assessment for a proposal for a directive of the European Parliament and of the council amending directive 94/62/EC on packaging and packaging waste to reduce the consumption of lightweight plastic carrier bags. European Commission
Costanzo SD, Murby J, Bates J (2005) Ecosystem response to antibiotics entering the aquatic environment. Mar Poll Bullet 51: 218–223
CPMA (2015) Indian petrochemical industry: country paper from India 7–8 may 2015. Seoul, Korea
Dabbagh N, Kitsantas A (2012) Personal learning environments, social media, and self-regulated learning: a natural formula for connecting formal and informal learning. Internet High Edu 15 (1):3–8
Das A, Mahanwar P (2020) A brief discussion on advances in polyurethane applications. Adv Ind Eng Polym Res
Dauvergne P (2018) The power of environmental norms: marine plastic pollution and the politics of microbeads. Environ Politics 27:579–597
Day RH, Wehle D, Coleman FC (1985) Ingestion of plastic pollutants by marine birds. Proceedings of the workshop on the fate and impact of marine debris. US Dept. Commerce, pp 344–386

De Klem C, Glowka L, Suman D, Shivlani M, Favre DS, Van Houtte A, Kuemlangan BS, Yamin F, Smith D, Howard ES (1997) Natural Resource Management and Conservation Yearbook of International Environmental Law 7:208

De Stephanis R, Giménez J, Carpinelli E, Gutierrez-Exposito C, Cañadas A (2013) As main meal for sperm whales: plastics debris. Mar Pollut Bullet 69:206–214

Delacuvellerie A, Cyriaque V, Gobert S, Benali S, Wattiez R (2019) The plastisphere in marine ecosystem hosts potential specific microbial degraders including *Alcanivorax borkumensis* as a key player for the low-density polyethylene degradation. J Hazard Mater 380:120899

Derraik JG (2002) The pollution of the marine environment by plastic debris: a review. Mar Pollut Bullet 44:842–852

Diggle A, Walker TR (2020) Implementation of harmonized extended producer responsibility strategies to incentivize recovery of single-use plastic packaging waste in Canada waste management 110:20–23

Donohue M, Foley DG (2007) Remote sensing reveals links among the endangered Hawaiian monk seal, marine debris, and El Niño. Mar Mamm Sci 23:468–473

Donohue MJ, Boland RC, Sramek CM, Antonelis GA (2001) Derelict fishing gear in the northwestern Hawaiian islands: diving surveys and debris removal in 1999 confirm threat to coral reef ecosystems. Mar Poll Bullet 42:1301–1312

Emblem A (2012) Plastics properties for packaging materials. Packaging Technol:287–309

Eriksen M, Lebreton LC, Carson HS, Thiel M, Moore CJ, Borerro JC et al (2014) Plastic pollution in the world's oceans: more than 5 trillion plastic pieces weighing over 250,000 tons afloat at sea. PLoS One 9(12):e111913

Eriksson O, Finnveden G (2009) Plastic waste as a fuel-CO 2-neutral or not? Energy Environ Sci 2: 907–914

Galgani L, Loiselle SA (2020) Plastic pollution impacts on marine carbon biogeochemistry. Environ Pollut 115598

Galgani L, Loiselle SA (2021) How can plastic on sea surface affect our climate? Frontiers for Young Minds. https://kids.frontiersin.org/articles/10.3389/frym.2020.00120

Gall SC, Thompson RC (2015) The impact of debris on marine life. Mar Pollut Bullet 92:170–179

Galloway TS (2015) Micro-and nano-plastics and human health. Marine anthropogenic litter. Springer, Cham, pp 343–366

Geyer R (2020a) A brief history of plastics. Mare plasticum-the plastic sea. Springer, Cham, pp 31–47

Geyer R (2020b) Production, use, and fate of synthetic polymers. Plastic waste and recycling. Elsevier, pp 13–32

Geyer R, Jambeck JR, Law KL (2017) Production, use, and fate of all plastics ever made. Sci Adv 3: e1700782

Gilbert JM, Reichelt-Brushett AJ, Bowling AC, Christidis L (2016) Plastic ingestion in marine and coastal bird species of southeastern Australia. Mar Ornithol 44:21–26

Grishanov S (2011) Structure and properties of textile materials. Handbook of textile and industrial dyeing. Elsevier, pp 28–63

Groh KJ, Backhaus T, Carney-Almroth B, Geueke B, Inostroza PA, Lennquist A et al (2018) Chemicals associated with plastic packaging: inventory and hazards. Peer J Preprints 6: e27036v1

Gu L, Ozbakkaloglu T (2016) Use of recycled plastics in concrete: a critical review. Waste Manag 51:19–42

Güven O, Gökdağ K, Jovanović B, Kıdeyş AE (2017) Microplastic litter composition of the Turkish territorial waters of the Mediterranean Sea, and its occurrence in the gastrointestinal tract of fish. Environ Pollut 223:286–294

Hagen PE (1989) International community confronts plastics polluting from ships: Marpol annex V and the problem that Won't go away. Am UJ Int'l L Pol'y 5:425

Halmann MM, Steinberg M (1998) Greenhouse gas carbon dioxide mitigation: science and technology, CRC Press

Han M (2019) Depolymerization of PET bottle via methanolysis and hydrolysis. Recycling of Polyethylene Terephthalate Bottles. Elsevier, pp 85–108
Hannah R, Max R (2018) Plastic pollution. Published online at OurWorldInData.org. Retrieved from: https://ourworldindata.org/plastic-pollution
Hardesty BD, Holdsworth D, Revill AT, Wilcox C (2015) A biochemical approach for identifying plastics exposure in live wildlife. Methods Ecol Evol 6(1):92–98
Harding K, Gounden T, Pretorius S (2017) "Biodegradable" plastics: a myth of marketing? Procedia Manufact 7:106–110
Hasan M (2018) Plastic waste management scenario in Bangladesh. Daffodil International University
Hastings E, Potts T (2013) Marine litter: progress in developing an integrated policy approach in Scotland marine policy 42:49–55
He P, Suuronen P (2018) Technologies for the marking of fishing gear to identify gear components entangled on marine animals and to reduce abandoned, lost or otherwise discarded fishing gear. Mar Poll Bullet 129:253–261
Heinze C, Meyer S, Goris N, Anderson L, Steinfeldt R, Chang N, Quéré CL, Bakker DC (2015) The ocean carbon sink–impacts, vulnerabilities and challenges. Earth Syst Dynam 6:327–358
Hilton M, Sherrington C, McCarthy A, Börkey P (2019) Extended producer responsibility (EPR) and the impact of online sales. https://www.oecd.org/environment/waste/policy-highlights-extended-producer-responsibility-and-the-impact-of-online-sales.pdf
Hinojosa IA, Thiel M (2009) Floating marine debris in fjords, gulfs and channels of southern Chile. Mar Pollut Bullet 58:341–350
Hirai H, Takada H, Ogata Y, Yamashita R, Mizukawa K, Saha M et al (2011) Organic micropollutants in marine plastics debris from the open ocean and remote and urban beaches. Mar Pollut Bullet 62:1683–1692
Hoellein TJ, Westhoven M, Lyandres O, Cross J (2015) Abundance and environmental drivers of anthropogenic litter on 5 Lake Michigan beaches: a study facilitated by citizen science data collection. J Great Lakes Res 41:78–86
Hollaway L, Jambeck J, Moss E, Dubey B (2020) Ocean. World Resources Institute. Accessed. https://oceanpanel.org/sites/default/files/2020-09/20_HLP_Report_COVID_Blue_Recovery.pdf
Hopewell J, Dvorak R, Kosior E (2009) Plastics recycling: challenges and opportunities. Philos Trans R Soc B Biol Sci 364:2115–2126
Huang J (2019) Effects of the material of running shoes on biomechanical characteristics during running. Mech Adv Mater Struct 26:2017–2022
Ignatyev IA, Thielemans W, Vander BB (2014) Recycling of polymers: a review. ChemSusChem 7:1579–1593
Inaba M, Murata K, Takahara I, Inoue KI (2011) Production of olefins from ethanol by Fe and/or P-modified H-ZSM-5 zeolite catalysts. J Chem Technol Biotechnol 86:95–104
Iñiguez ME, Conesa JA, Fullana A (2017) Microplastics in Spanish table salt. Sci Rep 7:1–7
Iriarte A, Gabarrell X, Rieradevall J (2009) LCA of selective waste collection systems in dense urban areas. Waste Manag 29:903–914
Isangedighi IA, David GS, Obot OI (2020) Plastic waste in the aquatic environment: impacts and management. Analysis of nanoplastics and microplastics in food, CRC Press, pp 15–43
IUCN (2018) International union for conservation of nature. https://portals.iucn.org/library/sites/library/files/documents/2018-007-En.pdf
Iwata T (2015) Biodegradable and bio-based polymers: future prospects of eco-friendly plastics. Angew Chem Int Ed 54:3210–3215
Jeftic L, Sheavly S, Adler E, Keith N (2009) Marine litter: a global challenge. https://stgwedocs.unep.org/bitstream/handle/20.500.11822/31632/MLAGC
Jepsen EM, de Bruyn PN (2019) Pinniped entanglement in oceanic plastic pollution: a global review. Mar Poll Bullet 145:295–305
Jones MM (1995) Fishing debris in the Australian marine environment. Mar Poll Bullet 30:25–33

Jung RT, Sung HG, Chun TB, Keel SI (2010) Practical engineering approaches and infrastructure to address the problem of marine debris in Korea. Mar Poll Bullet 60:1523–1532

Jyothsna TSS, Chakradhar B (2020) Current scenario of plastic waste management in India: way forward in turning vision to reality. Urban mining and sustainable waste management. Springer, Cham, pp 203–218

Kaluarachchi K, Thathsarani M, Nilmini W, Namal Arachchi N, Silva J (2017) Current plastic wastage and introducing new innovations to minimize plastic wastage in Sri Lanka

Kanagawa K, Oishi M, Negoro S, Urabe I, Okada H (1993) Characterization of the 6-aminohexanoate-dimer hydrolase from *Pseudomonas* sp. *NK87*. Microbiol 139:787–795

Kawai F (2010) Polylactic acid (Haw, Hawton et al.)-degrading microorganisms and PLA depolymerases. Green polymer chemistry: biocatalysis and biomaterials. ACS Publications, pp 405–414

Kimura Y (2009) Molecular, structural, and material design of bio-based polymers. Polym J 41: 797–807

Koerner GR, Koerner RM (2018) Polymeric geomembrane components in landfill liners. Solid waste landfilling: concepts, processes, technology. pp 313–341

Koopmans R (2014) Polyolefin-based plastics from biomass-derived monomers bio-based plastics. p 295

Kühn S, Van Franeker JA (2020) Quantitative overview of marine debris ingested by marine megafauna. Mar Poll Bullet 151:110858

Kühn S, Rebolledo ELB, Van FJA (2015) Deleterious effects of litter on marine life. Marine anthropogenic litter. Springer, Cham, pp 75–116

Laad MS (2020) Polymers in sports. Polymer science and innovative applications. Elsevier, pp 485–523

Laganà A, Bacaloni A, De Leva I, Faberi A, Fago G, MarinoA (2004) Analytical methodologies for determining the occurrence of endocrine disrupting chemicals in sewage treatment plants and natural waters. Anal Chim Acta 501:79–88

Laist DW (1987) Overview of the biological effects of lost and discarded plastic debris in the marine environment. Mar Poll Bullet 18:319–326

Laist DW (1997) Impacts of marine debris: entanglement of marine life in marine debris including a comprehensive list of species with entanglement and ingestion records. Marine debris. Springer, pp 99–139

Lambert S, Wagner M (2017) Environmental performance of bio-based and biodegradable plastics: the road ahead. Chem Soc Rev 46:6855–6871

Lavee D (2010) A cost-benefit analysis of a deposit–refund program for beverage containers in Israel. Waste Manag 30:338–345

Law KL (2017) Plastics in the marine environment. Annu Rev Mar Sci 9:205–229

Lawson T, Wilcox C, Johns K, Dann P, Hardesty BD (2015) Characteristics of marine debris that entangle Australian fur seals (*Arctocephalus pusillus doriferus*) in southern Australia. Mar Poll Bullet 98:354–357

Lazar B, Gračan R (2011) Ingestion of marine debris by loggerhead sea turtles, *Caretta caretta*, in the Adriatic Sea. Mar Poll Bullet 62:43–47

Lazarevic D, Aoustin E, Buclet N, Brandt N (2010) Plastic waste management in the context of a European recycling society: comparing results and uncertainties in a life cycle perspective. Resour Conserv Recyc 55:246–259

Lee SG, Xu X (2005) Design for the environment: life cycle assessment and sustainable packaging issues. Int J Environ Technol Manag 5:14–41

León-Zayas R, Roberts C, Vague M, Mellies JL (2019) Draft genome sequences of five environmental bacterial isolates that degrade polyethylene terephthalate plastic. Microbiol Resour Announcements 8:e00237–e00219

Li WC, Tse H, Fok L (2016) Plastic waste in the marine environment: a review of sources, occurrence and effects. Sci Total Environ 566:333–349

Liaqat S, Hussain M, Malik MF, Aslam A, Mumtaz K (2020) Microbial ecology: a new perspective of plastic degradation. Pure Appl Biol 9:2138–2150

Litchfield SG, Schulz KG, Kelaher BP (2020) The influence of plastic pollution and ocean change on detrital decomposition. Mar Poll Bullet 158:111354

Liu Z, Adams M, Walker TR (2018) Are exports of recyclables from developed to developing countries waste pollution transfer or part of the global circular economy? Resour Conserv Recyc 136:22–23

López-Martínez S, Morales-Caselles C, Kadar J, Rivas ML (2021) Overview of global status of plastic presence in marine vertebrates. Glob Chang Biol 27:728–737

Macfadyen G, Huntington T, Cappell R (2009) Abandoned, lost or otherwise discarded fishing gear. FAO Fisheries and Aquaculture Technical Paper No. 523. http://www.fao.org/3/i0620e/i0620e00.htm

Machovsky-Capuska GE, Amiot C, Denuncio P, Grainger R, Raubenheimer D (2019) A nutritional perspective on plastic ingestion in wildlife. Sci Total Environ 656:789–796

Madricardo F, Ghezzo M, Nesto N, Kiver M, Joseph W, Faussone GC, Fiorin R, Riccato F, Mackelworth PC, Basta J (2020) How to deal with seafloor marine litter: an overview of the state-of-the-art and future perspectives. Front Mar Sci 7:830

Mallakpour S, Behranvand V (2017) Using recycled polymers for the preparation of polymer nanocomposites: properties and applications. Hybrid polymer composite materials. Elsevier, pp 197–226

Malve H (2016) Exploring the ocean for new drug developments: marine pharmacology. J Pharm Bioallied Sci 8:83

Markic A, Gaertner JC, Gaertner-Mazouni N, Koelmans AA (2020) Plastic ingestion by marine fish in the wild. Crit Rev Environ Sci Technol 50:657–697

Mato Y, Isobe T, Takada H, Kanehiro H, Ohtake C, Kaminuma T (2001) Plastic resin pellets as a transport medium for toxic chemicals in the marine environment. Environ Sci Technol 35:318–324

McIlgorm A, Campbell HF, Rule MJ (2011) The economic cost and control of marine debris damage in the Asia-Pacific region. Ocean Coastal Manag 54:643–651

McKeen L (2012) Permeability properties of plastics and elastomers, 3rd edn. https://www.elsevier.com/books/permeability-properties-of-plastics-and-elastomers/mckeen/978-1-4377-3469-0

Merlino S, Locritani M, Stroobant M, Mioni E, Tosi D (2015) SeaCleaner: focusing citizen science and environment education on unraveling the marine litter problem. Mar Technol Soc J 49 (4):99–118

Milios L, Christensen LH, McKinnon D, Christensen C, Rasch MK, Eriksen MK (2018) Plastic recycling in the Nordics: a value chain market analysis. Waste Manag 76:180–189

Mokrzycki E, Uliasz-Bocheńczyk A (2003) Alternative fuels for the cement industry. Appl Energy 74:95–100

Moog D, Schmitt J, Senger J, Zarzycki J, Rexer KH, Linne U et al (2019) Using a marine microalga as a chassis for polyethylene terephthalate (PET) degradation. Microb Cell Factories 18:171

Moore MJ, Van der Hoop JM (2012) The painful side of trap and fixed net fisheries: chronic entanglement of large whales. J Mar Biol 2012

Mullan W (2002) Science and technology of modified atmosphere packaging [On-line]. https://www.dairyscience.info/index.php/packaging/117-modified-atmosphere-packaging.html

Mutha NH, Patel M, Premnath V (2006) Plastics materials flow analysis for India. Resour Conserv Recycl 47:222–244

Mwanza BG, Mbohwa C (2017) Drivers to sustainable plastic solid waste recycling: a review. Proc Manufact 8:649–656

Najafi SK (2013) Use of recycled plastics in wood plastic composites–a review. Waste Manag 33: 1898–1905

Nakajima-Kambe T, Shigeno-Akutsu Y, Nomura N, Onuma F, Nakahara T (1999) Microbial degradation of polyurethane, polyester polyurethanes and polyether polyurethanes. Appl Microbiol Biotechnol 51:134–140

Napper IE, Thompson RC (2019) Environmental deterioration of biodegradable, oxo-biodegradable, compostable, and conventional plastic carrier bags in the sea, soil, and open-air over a 3-year period. Environ Sci Technol 53:4775–4783

Nnorom IC, Osibanjo O (2008) Overview of electronic waste (e-waste) management practices and legislations, and their poor applications in the developing countries. Res Conserv Recyc 52: 843–858

Oliveira M, Ribeiro A, Hylland K, Guilhermino L (2013) Single and combined effects of microplastics and pyrene on juveniles (0+ group) of the common goby *Pomatoschistus microps* (Teleostei, Gobiidae). Ecol Indic 34:641–647

Oral N (2012) The UNEP regional seas Programme: bridging the governance gap. http://digitalcommons.law.umaryland.edu/cgi/viewcontent.cgi?article=1089&context=gelc

Owens KA (2018) Using experiential marine debris education to make an impact: collecting debris, informing policy makers, and influencing students. Mar Pollut Bull 127:804–810

Page B, McKenzie J, McIntosh R, Baylis A, Morrissey A, Calvert N, Haase T, Berris M, Dowie D, Shaughnessy PD (2004) Entanglement of Australian sea lions and New Zealand fur seals in lost fishing gear and other marine debris before and after government and industry attempts to reduce the problem. Mar Poll Bullet 49:33–42

Pakistan M (2016) A handbook on Pakistan's coastal and marine resources. MFF Pakistan, Karachi, Pakistan

Palumbi SR, Sandifer PA, Allan JD, Beck MW, Fautin DG, Fogarty MJ, Halpern BS, Incze LS, Leong JA, Norse E (2009) Managing for ocean biodiversity to sustain marine ecosystem services. Front Ecol Environ 7:204–211

Pascault JP, Höfer R, Fuertes P (2012) Polymer science: a comprehensive reference

Patel MK, Bechu A, Villegas JD, Bergez-Lacoste M, Yeung K, Murphy R et al (2018) Second-generation bio-based plastics are becoming a reality–non-renewable energy and greenhouse gas (GHG) balance of succinic acid-based plastic end products made from lignocellulosic biomass. Biofuels Bioprod Biorefin 12:426–441

Pawar PR, Shirgaonkar SS, Patil RB (2016) Plastic marine debris: sources, distribution and impacts on coastal and ocean biodiversity. PENCIL Publ Biol Sci 3:40–54

Peeters JR, Vanegas P, Devoldere T, Dewulf W, Duflou JR (2012) Closed loop recycling of plastic housing for flat screen TVs' 2012 electronics goes green 2012. IEEE, pp 1–6

Peng BY, Su Y, Chen Z, Chen J, Zhou X, Benbow ME et al (2019) Biodegradation of polystyrene by dark (*Tenebrio obscurus*) and yellow (*Tenebrio molitor*) mealworms (Coleoptera: Tenebrionidae). Environ Sci Technol 53:5256–5265

Personnic S, Boudouresque CF, Astruch P, Ballesteros E, Blouet S, Bellan-Santini D, Bonhomme P, Thibault-Botha D, Feunteun E, Harmelin-Vivien M (2014) An ecosystem-based approach to assess the status of a Mediterranean ecosystem, the *Posidonia oceanica* seagrass meadow. PLoS One 9(6):e98994

Pietrelli L, Poeta G, Battisti C, Sighicelli M (2017) Characterization of plastic beach debris finalized to its removal: a proposal for a recycling scheme. Environ Sci Pollut Res 24:16536–16542

Poeta G, Battisti C, Bazzichetto M, Acosta AT (2016) The cotton buds beach: marine litter assessment along the Tyrrhenian coast of Central Italy following the marine strategy framework directive criteria. Mar Pollut Bullet 113:266–270

Possatto FE, Barletta M, Costa MF, JAI S, Dantas DV (2011) Plastic debris ingestion by marine catfish: an unexpected fisheries impact. Mar Poll Bullet 62:1098–1102

Potts JR, Dreyer DR, Bielawski CW, Ruoff RS (2011) Graphene based polymer nanocomposites. Polymer 52:5–25

Poulikakos L, Papadaskalopoulou C, Hofko B, Gschösser F, Falchetto AC, Bueno M et al (2017) Harvesting the unexplored potential of European waste materials for road construction. Resour Conserv Recycl 116:32–44

Prata JC, Silva ALP, DaCosta JP, Mouneyrac C, Walker TR, Duarte AC et al (2019) Solutions and integrated strategies for the control and mitigation of plastic and microplastic pollution. Int J Environ Res Public Health 16:2411

Rajaram S (2009) Plastic additives: the global market. PLS022B, BCC research

Reisser J, Shaw J, Wilcox C, Hardesty BD, Proietti M, Thums M et al (2013) Marine plastic pollution in waters around Australia: characteristics, concentrations, and pathways. PLoS One 8: e80466

Ren X (2003) Biodegradable plastics: a solution or a challenge? J Clean Prod 11:27–40

Revel M, Châtel A, Mouneyrac C (2018) Micro (nano) plastics: a threat to human health? Curr Opin Environ Sci Health 1:17–23

Rhodes CJ (2018) Plastic pollution and potential solutions. Sci Prog 101:207–260

Rochman CM, Hoh E, Kurobe T, Teh SJ (2013) Ingested plastic transfers hazardous chemicals to fish and induces hepatic stress. Sci Rep 3:3263

Roman L, Gilardi K, Lowenstine L, Hardesty BD, Wilcox C (2020) The need for attention to confirmation bias and confounding in the field of plastic pollution and wildlife impacts: comment on clinical pathology of plastic ingestion in marine birds and relationships with blood chemistry. Environ Sci Technol 53:9224–9231

Ronchi F, Galgani F, Binda F, Mandić M, Peterlin M, Tutman P, Anastasopoulou A, Fortibuoni T (2019) Fishing for litter in the Adriatic-Ionian macroregion (Mediterranean Sea): strengths, weaknesses, opportunities and threats. Mar Policy 100:226–237

Rosevelt C, Los Huertos M, Garza C, Nevins H (2013) Marine debris in Central California: quantifying type and abundance of beach litter in Monterey Bay, CA. Mar Pollut Bullet 71: 299–306

Ross S, Evans D (2003) The environmental effect of reusing and recycling a plastic-based packaging system. J Clean Prod 11:561–571

Rujnić-Sokele M, Pilipović A (2017) Challenges and opportunities of biodegradable plastics: a mini review. Waste Manage Res 35:132–140

Ryan PG (2013) A simple technique for counting marine debris at sea reveals steep litter gradients between the Straits of Malacca and the Bay of Bengal. Mar Pollut Bullet 69:128–136

Ryan PG, Moore CJ, Van Franeker JA, Moloney CL (2009) Monitoring the abundance of plastic debris in the marine environment. Phil Trans R Soc B Biol Sci 364:1999–2012

Samavedi S, Poindexter LK, Van Dyke M, Goldstein AS (2014) Synthetic biomaterials for regenerative medicine applications. Regenerative medicine applications in organ transplantation. Elsevier, pp 81–99

Santos IR, Friedrich AC, Wallner-Kersanach M, Fillmann G (2005) Influence of socio-economic characteristics of beach users on litter generation. Ocean Coastal Manag 48:742–752

Savoca MS, McInturf AG, Hazen EL (2021) Plastic ingestion by marine fish is widespread and increasing. Glob Chang Biol 27:2188–2199

Sayied N (2007) Environmental issues in coastal waters-Pakistan as a case study. https://commons.wmu.se/cgi/viewcontent.cgi

Schaefer DA, Khan S, Nadir S, Dong Y, Mortimer PE, Gui H et al (2020) Biodegradation of polyester polyurethane by *aspergillus flavus G10*. bioRxiv

Scherer C, Emberger-Klein A, Menrad K (2018) Consumer preferences for outdoor sporting equipment made of bio-based plastics: results of a choice-based-conjoint experiment in Germany. J Clean Prod 203:1085–1094

Schmaltz E, Melvin EC, Diana Z, Gunady EF, Rittschof D, Somarelli JA, Virdin J, Dunphy-Daly MM (2020) Plastic pollution solutions: emerging technologies to prevent and collect marine plastic pollution. Environ Int 144:106067

Schneider DR, Ragossnig A (2015) Recycling and incineration, contradiction or coexistence? Waste Manage Res 33:693–695

Schuyler Q, Hardesty BD, Lawson T, Opie K, Wilcox C (2018) Economic incentives reduce plastic inputs to the ocean. Mar Policy 96:250–255

Seltenrich N (2015) New link in the food chain? Marine plastic pollution and seafood safety, NLM-Export

Selwyn N (2007) Web 2.0 applications as alternative environments for informal learning-a critical review' Paper for CERI-KERIS international expert meeting on ICT and educational performance. p 17

Senko JF, Nelms SE, Reavis JL, Witherington B, Godley BJ, Wallace BP (2020) Understanding individual and population-level effects of plastic pollution on marine megafauna. Endanger Species Res 43:234–252

Shah A, Hasan F, Hameed A, Ahmed S (2008) A novel poly (3-hydroxybutyrate)-degrading Streptoverticillium kashmirense AF1 isolated from soil and purification of PHB-depolymerase. Acta Biol Hung 59:489–499

Shahnawaz M, Sangale MK, Ade AB (2019) Policy and legislation/regulations of plastic waste around the globe. Bioremediation technology for plastic waste. Springer, Cham, pp 113–126

Siddique R, Khatib J, Kaur I (2008) Use of recycled plastic in concrete: a review. Waste Manag 28: 1835–1852

Siddique SF, Joshi SV, Lupi F (2010) Factors influencing the rate of recycling: an analysis of Minnesota counties. Resour Conserv Recycl 54:242–249

Sigler M (2014) The effects of plastic pollution on aquatic wildlife: current situations and future solutions. Water Air Soil Pollut 225:1–9

Singh RK, Ruj B (2015) Plastic waste management and disposal techniques-Indian scenario. Int J Plastics Technol 19:211–226

Siskind E (1990) Market development for recycled high density polyethylene (HDPE). Massachusetts Institute of Technology

Slater FM (2016) Ecotones, ecoclines and eco-perturbations: the aquatic flora and fauna of S'Albufera Natural Park, Mallorca: a contribution and review. Mediterranean J Biosci 1:120–127

Song J, Murphy R, Narayan R, Davies G (2009) Biodegradable and compostable alternatives to conventional plastics. Philos Trans R Soc B Biol Sci 364:2127–2139

Stewart JR, Gast RJ, Fujioka RS, Solo-Gabriele HM, Meschke JS, Amaral-Zettler LA, Del Castillo E, Polz MF, Collier TK, Strom MS (2008) The coastal environment and human health: microbial indicators, pathogens, sentinels and reservoirs. Environ Health 7:1–14

Storm C, Vander VJ, Kuijpers J (2005) From nature conservation towards restoration of estuarine dynamics in the heavily modified Rhine-Meuse estuary. Large Rivers, The Netherlands, pp 305–318

Strategy P (2018) A European strategy for plastics in a circular economy. Communication from the Commission to the European Parliament, the Council, the European Economic and Social Committee and the Committee of the Regions. Brussels

Sudhakar M, Priyadarshini C, Doble M, Murthy PS, Venkatesan R (2007) Marine bacteria mediated degradation of nylon 66 and 6. Int Biodeterior Biodegrad 60(3):144–151

Takehara I, Kato DI, Takeo M, Negoro S (2017) Draft genome sequence of the nylon oligomer-degrading bacterium *Arthrobacter* sp. strain *KI72*. Genome announcements 5(17)

Takeuchi K (2012) Polycarbonate. https://www.sciencedirect.com/topics/medicine-and-dentistry/polycarbonate

Talsness CE, Andrade AJ, Kuriyama SN, Taylor JA, Vom Saal FS (2009) Components of plastic: experimental studies in animals and relevance for human health. Philos Trans R Soc B Biol Sci 364:2079–2096

Tanskanen JH, Reinikainen A, Melanen M (1998) Waste streams, costs and emissions in municipal solid waste management: a case study from Finland. Waste Manage Res 16:503–513

Teuten EL, Rowland SJ, Galloway TS, Thompson RC (2007) Potential for plastics to transport hydrophobic contaminants. Environ Sci Technol 41:7759–7764

Thevenon F, Carroll C, Sousa J (2014) Plastic debris in the ocean: the characterization of marine plastics and their environmental impacts, situation analysis report. IUCN, Gland, p 52

Thiel M, Hinojosa I, Vásquez N, Macaya E (2003) Floating marine debris in coastal waters of the SE-Pacific (Chile). Mar Pollut Bullet 46:224–231

Thiel M, Luna-Jorquera G, Álvarez-Varas R, Gallardo C, Hinojosa IA, Luna N, Miranda-Urbina D, Morales N, Ory, Pacheco AS (2018) Impacts of marine plastic pollution from continental coasts to subtropical gyres—fish, seabirds, and other vertebrates in the SE Pacific. Front Mar Sci 5:238

Thompson RC, Moore CJ, Vom SFS, Swan SH (2009) Plastics, the environment and human health: current consensus and future trends. Philos Trans R Soc B Biol Sci 364:2153–2166

Thushari G, Senevirathna J (2020) Plastic pollution in the marine environment. Heliyon 6:e04709

Thushari GGN, Chavanich S, Yakupitiyage A (2017) Coastal debris analysis in beaches of Chonburi Province, eastern of Thailand as implications for coastal conservation. Mar Pollut Bullet 116:121–129

Tokiwa Y, Calabia BP, Ugwu CU, Aiba S (2009) Biodegradability of plastics. Int J Mol Sci 10: 3722–3742

Trasande L, Zoeller RT, Hass U, Kortenkamp A, Grandjean P, Myers JP et al (2015) Estimating burden and disease costs of exposure to endocrine-disrupting chemicals in the European Union. J Clin Endocrinol Metabol 100:1245–1255

Tsuji H (2014) Poly (lactic acid). Bio-based plastics: materials and applications. pp 171–239

Ucankus G, Ercan M, Uzunoglu D, Culha M (2018) Methods for preparation of nanocomposites in environmental remediation. New polymer nanocomposites for environmental remediation. Elsevier, pp 1–28

UNEP (2005) Marine litter: an analytical overview. UNEP

UNEP (2011) Waste–investing in energy and resource efficiency. Towards a green economy, p 44

UNEP ML (2009) A global challenge. UNEP, Nairobi

Unger S, Müller A, Rochette J, Schmidt S, Shackeroff J, Wright G (2017) Achieving the sustainable development goal for the oceans. IASS Policy Brief 2017;1:

Vague M, Chan G, Roberts C, Swartz NA, Mellies JL (2019) Pseudomonas isolates degrade and form biofilms on polyethylene terephthalate (PET) plastic. bioRxiv 647321

van Truong N and Bei Ping C (n.d.) Plastic marine debris in Vietnam: source, impacts and management. Environ Policy Law(Preprint):1–15

Vegter AC, Barletta M, Beck C, Borrero J, Burton H, Campbell ML, Costa MF, Eriksen M, Eriksson C, Estrades A (2014) Global research priorities to mitigate plastic pollution impacts on marine wildlife. Endanger Species Res 25:225–247

Vink ET, Davies S (2015) Life cycle inventory and impact assessment data for 2014 Ingeo™ polylactide production. Ind Biotechnol 11:167–180

Wagner M, Lambert S (2018) Freshwater microplastics: emerging environmental contaminants? Springer Nature, Cham

Walker TR (2018) Drowning in debris: solutions for a global pervasive marine pollution problem. A changing fundy environment: emerging issues, challenges and priorities

Walker TR, Xanthos D (2018) A call for Canada to move toward zero plastic waste by reducing and recycling single-use plastics. Resour Conserv Recycl 133:99–100

Wang Y, Zhang Y (2015) Investigation of gut-associated bacteria in *Tenebrio molitor* (Coleoptera: Tenebrionidae) larvae using culture-dependent and DGGE methods. Ann Entomol Soc Am 108: 941–949

Wanyama P (2017) Experimental study of shear behaviour of high density polyethylene reinforced sand under triaxial compression. University of Cape Town

Warner R (2018) Oceans in transition: incorporating climate-change impacts into environmental impact assessment for marine areas beyond national jurisdiction. Ecology LQ 45:31

Wehle D, Coleman FC (1983) Plastics at sea. Nat Hist 92:20

Wilcox C, Heathcote GJ, Gunn R, Peel D, Hardesty BD (2015) Understanding the sources and effects of abandoned, lost, and discarded fishing gear on marine turtles in northern Australia. Conserv Biol 29:198–206

Wilcox C, Mallos NJ, Leonard GH, Rodriguez A, Hardesty BD (2016) Using expert elicitation to estimate the impacts of plastic pollution on marine wildlife. Mar Policy 65:107–114

Xanthos D, Walker TR (2017) International policies to reduce plastic marine pollution from single-use plastics (plastic bags and microbeads): a review. Mar Poll Bullet 118:17–26

Yang Y, Chen J, Wu WM, Zhao J, Yang J (2015a) Complete genome sequence of *Bacillus* sp. *YP1*, a polyethylene-degrading bacterium from waxworm's gut. J Biotechnol 200:77–78
Yang Y, Yang J, Wu WM, Zhao J, Song Y, Gao L et al (2015b) Biodegradation and mineralization of polystyrene by plasticeating mealworms: part 2. Role of gut microorganisms. Environ Sci Technol 49:12087–12093
Yoshida S, Hiraga K, Takehana T, Taniguchi I, Yamaji H, Maeda Y et al (2016) A bacterium that degrades and assimilates poly (ethylene terephthalate). Science 351:1196–1199
Zhang H, Chen M (2014) Current recycling regulations and technologies for the typical plastic components of end-of-life passenger vehicles: a meaningful lesson for China. J Mat Cyc Waste Manag 16:187–200
Zhuo C, Levendis YA (2014) Upcycling waste plastics into carbon nanomaterials: a review. J Appl Polym Sci 131
Zink T, Geyer R (2018) Recycling and the myth of landfill diversion. J Ind Ecol 23:541

Recent Trends to Address Plastic Waste at the Global Level

5

Lida Vivian Carvajal Rodríguez
and Cesar David Benavides Fernández

Abstract

Plastic waste has become a major problem in the world because there are no effective strategies for its management and it ends up accumulating in the environment. In literature, attempts have been made to introduce human awareness of the use, reduction and reuse of plastic products, accompanied by traditional waste management techniques, but the problem of pollution from synthetic plastics continues to increase. Throughout the world, the problem of plastic residues has been handled by traditional techniques such as incineration, landfill and recycling. Traditional techniques have their limitations, such as the production of greenhouse gases that increase climate change which leads to affecting the living organism in both terrestrial and marine environment. Therefore, researchers are now focusing on the development of biological technologies such as composting, phytoremediation and biodegradation using microorganisms and their enzymes. These techniques are important because they mitigate the environmental pollutants although they require a lot of time to obtain results and highly qualified personnel. Attempts have also been made to replace traditional plastic with renewable and recyclable sources, such as bioplastics. Some bacteria, algae, fungi and actinomycetes have the potential to degrade these polymers because their metabolism allows them to use them as a nutrient and carbon resource. There are also genetically modified microorganisms that acquire the ability to degrade this material, as well as other recent techniques such as nanoremediation that are very promising. In the present chapter, an attempt was made under the following lines: (i) to overview the techniques used to mitigate the plastic waste, (ii) to highlight the advantages and disadvantages of each

L. V. Carvajal Rodríguez (✉) · C. D. Benavides Fernández
Universidad Santiago de Cali, Department of Microbiology, Cali, Colombia
e-mail: lida.carvajal01@usc.edu.co

M. Shahnawaz et al. (eds.), *Impact of Plastic Waste on the Marine Biota*,
https://doi.org/10.1007/978-981-16-5403-9_5

mitigation method and (iii) to discuss the biodegradation of synthetic plastics by microorganisms as an environmentally friendly alternative for the appropriate elimination of plastic waste as well as the implementation of microbial consortia to increase the biodegradation performance at commercial level.

Keywords

Plastics · Recycling · Biodegradation · Landfill · Incineration

5.1 Introduction

In the last 65 years, plastics generate a large amount of waste that has been improperly managed and has been accumulating in terrestrial and aquatic ecosystems (Geyer et al. 2017). Strategies are being adopted around the world to safely and effectively reduce plastic waste and properly manage plastic waste (Bhardwaj et al. 2013; Borrelle et al. 2020). Disposal of non-biodegradable single-use plastic waste is one of the most challenging jobs worldwide, resulting in an unprecedented accumulation of plastic waste in all terrestrial and aquatic ecosystems. This alarming accumulation of plastic waste represents a latent danger to the environment and living beings (Shumba et al. 2020).

Most of the commonly used plastics are not biodegradable because they come from non-renewable sources such as petroleum and its derivatives, so they accumulate instead of decomposing in dumps or in the environmental media (Barnes et al. 2009).

A study conducted in 2020 reported that in 2016 about 11% of the plastic waste generated by 97% of the world's population entered aquatic ecosystems; if this situation continues, in 2030 approximately 90 million tons per year of plastic would enter these ecosystems ending their natural balance (Borrelle et al. 2020).

There are mainly three options for the fate of plastic waste. They can be recycled or reprocessed into another material known as a 'secondary material' (Mutha et al. 2006; Kuczenski and Geyer 2010); they can also be destroyed by incineration or with pyrolysis extracting fuel to recover energy; and finally they can accumulate in landfills or in the environment without any kind of environmental management (Geyer et al. 2017).

Studies report that 30% of all plastics produced to date are currently used primarily in packaging (single-use products) or construction (Jambeck et al. 2015; Geyer et al. 2017; EPA 2018; Lebreton and Andrady 2019). In the period from 1950 to 2015, about 6300 metric tons (Mt) of primary and recycled plastic waste was accumulated, of which 12% (800 Mt) was incinerated, 9% (600 Mt) was recycled and 60% accumulated in landfills or in ecosystems (Geyer et al. 2017; Greenpeace 2019), and of the recycled plastics, 10% have been recycled more than once (PlasticsEurope 2016; Geyer et al. 2017).

Non-governmental organizations (NGOs) and private companies such as the United Nations Program, Clean Seas, the Conferences on our Oceans, the G7

Plastics Charter and the European Union Strategy are leading in several countries the zero production of plastic waste and the consumption of single-use plastic products such as water and juice packaging (Abalansa et al. 2020; Borrelle et al. 2020). Also, together with the government, they are investigating the management of plastic waste to try to enter into a circular economy (Borrelle et al. 2020).

In countries such as India and Colombia, part of the plastic residues generated are retrieved by the informal workers (garbage collectors, recyclers and waste vendors), which means that a certain amount of this waste does not reach ecosystems or landfills. A study conducted in India estimated that this waste recovered by the informal sector represents between 50% and 80% of the plastic waste generated (ACC 2013; Geyer et al. 2017).

The objective of this chapter is to establish how plastic waste is currently being managed worldwide, briefly describing each technique and highlighting its environmental advantages and disadvantages.

5.2 Methods of Plastics Disposal

At present, there are three main methods for managing plastic waste: landfill, incineration and recycling; a fourth method is bioremediation, but it is in the research stage (Fig. 5.1). These methods are belatedly used to manage municipal plastic waste around the world, although each has its limitations and significance (Zhang et al. 2004; Webb et al. 2012).

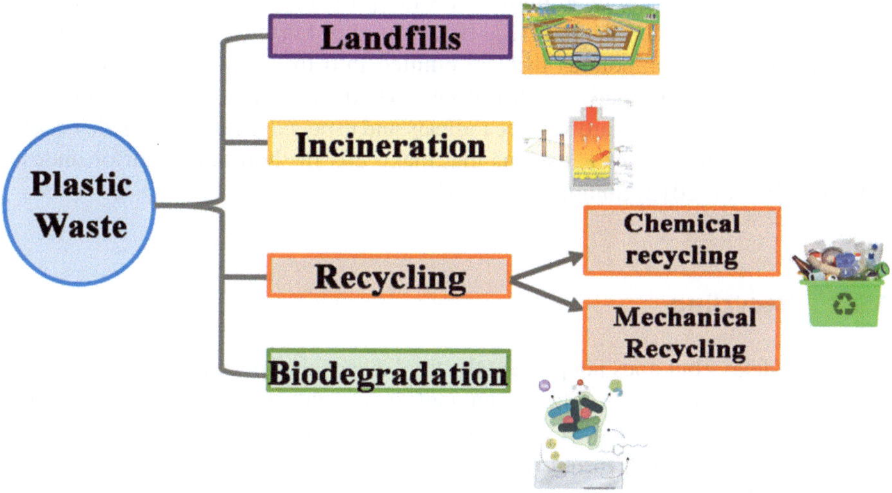

Fig. 5.1 Plastic waste disposal methods. Landfill, incineration and recycling are the traditional methods, and biodegradation is still in its infancy but is a very promising technique (Zhang et al. 2004; Webb et al. 2012; Meereboer et al. 2020) (created in BioRender.com)

5.2.1 Landfills

Landfills are the oldest and most common form of waste management in the world. These are used for temporary storage, consolidation and transfer of waste (Weitz et al. 1999).

The plastic components of landfill waste persist for more than 20 years, due to the anaerobic environment of landfills (Webb et al. 2012; Tollner et al. 2011) that limits the thermoxidative degradation that some plastics present (Andrady 2011), occupying for a long time large tracts of land that are destined for landfills, which takes away the opportunity to use those lands in productive activities such as agriculture and livestock (Zhang et al. 2004). The disposal of plastic waste in landfills produces secondary environmental contaminants that are eliminated as gases or leachates, affecting the atmosphere and terrestrial ecosystems, respectively (Zhang et al. 2004). Examples of these are benzene, toluene and xylene mainly, as well as BPA and hydrogen sulphide, which are potentially lethal (Urase et al. 2008; Xu et al. 2011; Tsuchida et al. 2011). The low level of biodegradability of polymers, the high cost of sustaining landfills for plastic waste (Shumba et al. 2020) and the damage to the environment caused by the emission of greenhouse gases have made legislation for landfill management more stringent, making this method of waste disposal undesirable and unsustainable (Geyer et al. 2017).

5.2.2 Incineration

The incineration of plastic waste has an advantage over landfills as it does not occupy significant space; in addition it has the possibility of recovering energy by generating heat (Webb et al. 2012). However, the burning of plastics produces numerous noxious compounds that are emitted into the atmosphere (Zhang et al. 2004). Polycyclic aromatic hydrocarbons (PAHs), polychlorinated biphenyls (PCBs), heavy metals, hazardous free radicals and especially carbon dioxide become significant greenhouse gas emissions that make this method unpopular or undesirable (Webb et al. 2012; Shumba et al. 2020).

5.2.3 Recycling

Recycling delays but does not prevent the final disposal of plastic waste (Geyer et al. 2015; Geyer et al. 2017). This is an inefficient and costly process, but unlike incineration and landfilling, it is an ecofriendly method (Yamada-Onodera et al. 2001; Zhang et al. 2004); currently, recycling is more economically favourable (Kale et al. 2007). The presence of contaminants or impurities and additives in the waste to be recovered reduces the efficiency of the technique and prevents obtaining a good product (Zhang et al. 2004; Awaja and Pavel 2005). Recycling is currently more economically favourable (Kale et al. 2007).

Recycling emerged as a better method; the recycling industry started as a result of environmental pressure to improve waste management (Edge et al. 1991; Awaja and Pavel 2005). However, only 9–10% of total plastic waste is recycled, leaving the rest for incineration and landfilling (EPA 2016; PlasticsEurope 2016; Geyer et al. 2017; Lebreton and Andrady 2019).

There are several options available for plastic waste management which include primary recycling, secondary or mechanical recycling, chemical or tertiary recycling and quaternary recycling (Shumba et al. 2020).

Chemical rendering and mechanical rendering are the two main techniques used for waste plastic recycling management.

5.2.4 Chemical Rendering

Chemical rendering (chemolysis) consists of complete depolymerization of the plastic to obtain its monomers or incomplete depolymerization to obtain oligomers (Awaja and Pavel 2005; Van Krevelen and Te Nijenhuis 2009). Chemicals used for waste depolymerization include methanol (methanolysis) and ethylene glycol EG (glycolysis). The disadvantages of chemolysis are its high cost, the toxic waste it produces and its low efficiency (Awaja and Pavel 2005).

5.2.5 Mechanical Rendering

Mechanical rendering involves the removal of contamination by classification and cleaning, drilling and treating plastic waste (Awaja and Pavel 2005). Recycling post-consumer plastics is an increasingly important technique because it has advantages over incineration and landfilling such as reducing the amount of plastic waste that should be disposed of, reducing the use of non-renewable sources such as fossil fuels, mitigating the energy impact and reducing the emission of harmful atmospheric gases such as carbon dioxide. In contrast, landfilling and incineration are dangerous for the environment as they produce harmful effects such as greenhouse gases (Basta and Johnson 1990).

In the recycling of plastics, waste undergoes two initial stages, which are melting and shredding or also called granulation. Different applications are also being given to post-consumer plastics by transforming them into fuel through processes such as pyrolysis and gasification (Singh and Sharma 2008); in obtaining oil through liquefaction and pyrolysis; and in cement-based products for construction (Shah et al. 1999).

Recycling and incineration gained strength in most countries of the world as a method for plastic waste management after 1980, increasing their utilization rates slowly and reaching in 2014 a rate of 18% for recycling and 24% incineration (Geyer et al. 2017). In the same year, the highest recycling rates worldwide were obtained in Europe and China with a percentage of 30% and 25%, respectively, contrary to the United States where a recycling rate of 9% has been recorded since 2012

(PlasticsEurope 2016; Geyer et al. 2017). The United States is one of the countries that produces the largest amount of plastic waste, and only 5% of the 1 billion bags produced per year are recycled (Sivan 2011). The picture is better in the case of incineration; in 2014 Europe and China recorded rates of 40% and 30%, respectively (PlasticsEurope 2016); however according to World Bank data, in these two countries, between 70% and 85% of municipal waste is currently poorly managed (Hoornweg and Bhada-Tata 2012; Lebreton and Andrady 2019).

Landfill, incineration and recycling are the traditional methods that have been implemented so far for the management of plastic waste in many countries, and each method has their advantages and disadvantages.

Currently, many researchers are working on the biodegradation of synthetic plastics such as polyethylene (LDPE, MDPE, HDPE and LLDPE), poly (ethylene terephthalate) (PET), polystyrene (PS), polyvinyl chloride (PVC) and polyurethane (PUR) (Geyer et al. 2015; Yang et al. 2020; Janczak et al. 2020; Ji et al. 2020; Kumar et al. 2019; Sarkhel et al. 2020; Urbanek et al. 2021), which are the most widely used plastics because of their high environmental impact. These are the most widely used plastics due to their durability, stability and strength.

Biodegradation is a more recent technique that is currently the subject of many studies; one of its main advantages is that it does not harm the environment or other organisms, which is why it is called an environmentally friendly technique (Webb et al. 2012). There are many works on the subject using microorganisms such as fungi, bacteria and microalgae (Table 5.1), but it has not been implemented in any form at a commercial level.

Although biodegradation is discussed in more detail in another chapter of this book, we will now describe the most relevant current aspects of biodegradation.

5.2.6 Biodegradation

Biodegradation is a natural process that can be carried out, thanks to the action of microorganisms, which are living beings with the metabolic capacity to grow using different substrates, in this case synthetic and natural polymers, to transform them into simpler compounds that can be used by other organisms (Lucas et al. 2008; Bhardwaj et al. 2013). Microorganisms degrade polymers such as polyethylene and polyurethane, using them as a substrate for growth and as a source of carbon and energy (Pramila et al. 2012; Devi et al. 2016).

It is taken into consideration to be the safest technique of breakdown with much less poisonous aspect products (Fesseha and Abebe 2019).

A polymer can be degraded according to its chemical structure; the more similar it is to a substance or compound in nature, the greater its biodegradability (Kijchavengkul and Auras 2008). The intracellular and extracellular enzymes produced by microorganisms recognizes specific sites in the polymer skeleton that will be susceptible to degradation, in the case of polymers such as polyethylene terephthalate (PET) and biodegradable polyesters, such as poly (lactic acid) (PLA) and

Table 5.1 Recent publications on biodegradation of polymers using microorganisms

Sr. no.	Polymer used	Microorganism	Degradation percentage	Techniques	References
1	Poly (e-caprolactone) (PCL), poly (butylene succinate) (PBS), poly (butylene succinate-co-butylene adipate) (PBSA)	**Bacteria:** *Pseudomonas* and *Bacillus* **Fungi:** *Geomyces, Sclerotinia, Fusarium* **Yeast:** *Hansenula anomala*	6–50% in 1 month	Weight change, SEM	Urbanek et al. (2021)
2	Polypropylene (PP)	*Kluyvera* sp., *Citrobacter* sp., *Enterobacter* sp.	32% in 35 days	TGA, FTIR	Yang et al. (2020)
3	Polyvinyl chloride (PVC)	Marine anaerobic microbes	11.7% in 7 months	Gravimetric weight loss and TGA	Giacomucci et al. (2020)
4	Polylactide (PLA) and polyethylene terephthalate (PET)	*Arthrobacter sulfonivorans, Serratia plymuthica, Laccaria laccata*	Not available	EDX, XRD	Janczak et al. (2020)
5	2,6-Dimethylphenol	*Mycobacterium neoaurum* (B5–4)	99–62% in 72 h	UV/VIS	Ji et al. (2020)
6	Polyethylene terephthalate (PET)	*Rhodococcus* sp. (SSM1)	52–30% in 132 h	Weight loss, SEM	Kumar et al. (2019)
7	PET	*Vibrio* sp., *Aspergillus* sp.	Bacteria, 35%; fungi, 22% in 6 weeks	FTIR, SEM and XRD	Sarkhel et al. (2020)
7	Microplastics: Polypropylene (PP), polyethylene (PE) and PET	Periphytic biofilm: *Deinococcus-thermus, Proteobacteria, Cyanobacteria*	PP: 18.02–9.52%; PE: 14.02–5.95%; PET: 19.72–13.24% in 60 days	FTIR, SEM, GPC	Shabbir et al. (2020)
8	PET	*Spirulina* sp.	48.61% in 112 days	FTIR, SEM, EDX	Khoironii et al. (2019)
9	Polyurethane foams	*Achromobacter, Brucella, pseudomonas, rhizobium, Stenotrophomonas, Chryseobacterium, Herbaspirillum*	71–30% in 12 weeks	SEM	Gunawan et al. (2020)

SEM scanning electron microscopy, *TGA* thermogravimetric analysis, *FTIR* Fourier-transform infrared spectroscopy, *EDX* energy dispersive X-ray analysis, *XRD* X-ray diffraction, *GPC* gel permeation chromatography polyhydroxyalkanoates (PHA), present ester bonds that are very susceptible to attack by microorganisms through the hydrolysis reaction (Kale et al. 2007).

In these plastic biodegradation processes, microbial enzymes are an important tool. Enzymes hydrolyse the amorphous part of polymers such as polyethylene terephthalate (PET) and polyurethane (PU) (Tokiwa et al. 2009; Webb et al. 2012; Restrepo-Flórez et al. 2014; Wei and Zimmermann 2017). Several authors claim that fungi have more plastic-degrading enzymes than bacteria (Bhardwaj et al. 2013; Lucas et al. 2008). Among the most important enzymes used in biodegradation of plastic are hydrolases, cutinases, esterases, lipases, laccases and ureases; as well as the bacterial species that produce them the most are *Vibrio* sp., *Streptococcus* sp., *Serratia* sp. *Bacillus* sp., *Thermobifiba* sp., *Pseudomonas* sp. and *Staphylococcus* sp. and fungal species are *Aspergillus* sp., *Curvularia* sp., *Aureobasidium* sp., *Penicillium* sp., *Alternaria* sp., *Fusarium* sp., *Phanerochaete*, *Pestalotiopsis*, etc. (Wei 2011; Bhardwaj et al. 2013; Devi et al. 2016; Khoironii et al. 2019; Gunawan et al. 2020; Janczak et al. 2020; Ji et al. 2020; Kumar et al. 2019; Sarkhel et al. 2020; Urbanek et al. 2021; Soulenthone et al. 2021).

Different classes of hydrolases are secreted by fungi and bacteria capable of degrading synthetic polymers; these enzymes attach to the polymer surface and cleave it into small fragments of hydroxyl acid and other water-soluble monomers (Lucas et al. 2008; Bhardwaj et al. 2013), which are transported through the cell wall to be metabolized to produce carbon dioxide and water under aerobic conditions and methane when anaerobic conditions are present (Fig. 5.2) (Bhardwaj et al. 2013; Devi et al. 2016; Alshehrei 2017; Taniguchi et al. 2019; Lu et al. 2020; Janczak et al. 2020; Kumar et al. 2019; Gu et al. 2000).

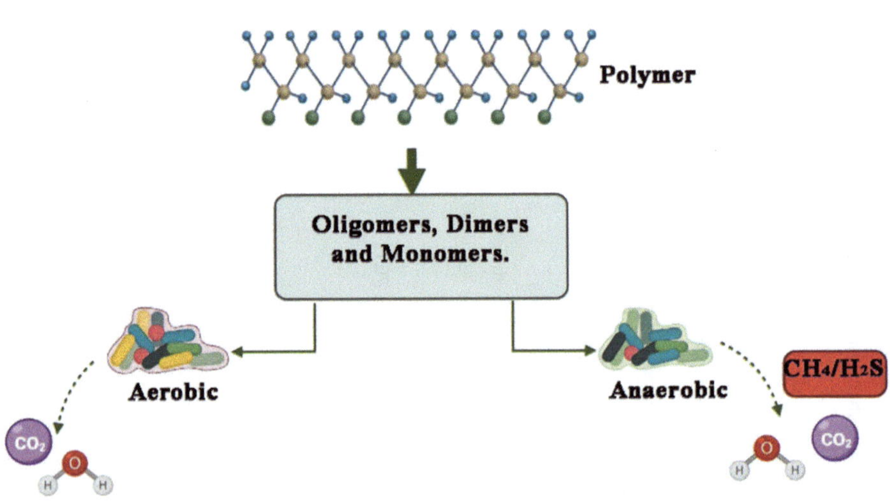

Fig. 5.2 Biodegradation of plastics under aerobic conditions. In an environment with oxygen, the biodegradation of a polymer produces water and carbon dioxide. Under anaerobic conditions, methane or sulphidric acid is also produced (created in BioRender.com)

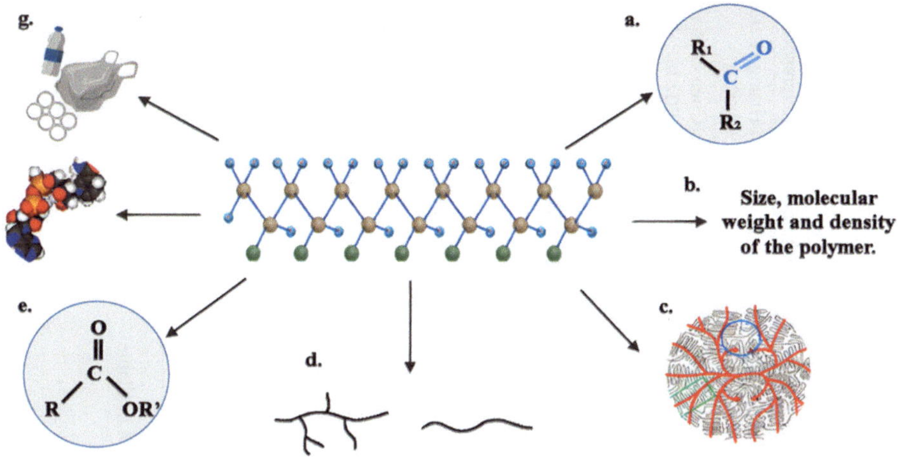

Fig. 5.3 Factors affecting biodegradability of a polymer. (**a**) Availability of functional groups. (**b**) Characteristics of polymer. (**c**) Amorphous regions of polymer. (**d**) Linearity or presence of branching in the polymer. (**e**) Breakable bonds. (**f**) Molecular composition. (**g**) Nature and physical form of polymer such as pellets, films, fibres, etc. (created in BioRender.com)

Biodegradation of synthetic polymers is a very slow and inefficient process (low biodegradation rates) (Arkatkar et al. 2009; Devi et al. 2016), which depends on the characteristics of the polymer, the metabolism of the microorganism and their interactions with the environment (Fig. 5.3) (Gu et al. 2000; Arkatkar et al. 2009; Bhardwaj et al. 2013). To accelerate polymer biodegradation, pretreatments such as the addition of oxidants, surfactants and functional groups or pretreatments with ultraviolet light are used (Arutchelvi et al. 2008; Arkatkar et al. 2009; Restrepo-Flórez et al. 2014).

Many of the microorganisms that are biodegraders of plastics are thermophilic and have been isolated from rhizospheric soil compost (Bhardwaj et al. 2013; Taniguchi et al. 2019). In many recent studies, microbial consortia (bacteria, fungi and protozoa) are being used to increase degradation efficiency and accelerate the process (Taniguchi et al. 2019; Lu et al. 2020; Urbanek et al. 2021).

Increasing the biodegradation rate of polymers is a factor that has gathered great efforts by scientists, who have tried to alter the properties of the polymer and combine two genomes of different microorganisms (Bhardwaj et al. 2013). Using genetic engineering for the modification of microorganisms, genes that express some of the enzymes involved in polymer biodegradation (Table 5.2) are inserted into the genome of another microorganism (Fig. 5.4).

Genes coding for the enzymes such as PETase of *Ideonella sakaiensis* and TfH from *Thermobifida fusca* have been inserted into the genome of *Escherichia coli*, *Pichia pastoris* and *Bacillus* sp. (Müller et al. 2005; Yoshida et al. 2016). These species of microorganisms are not capable of degrading traditional plastic and are therefore genetically modified. PET is the most studied polymer with these

Table 5.2 Publications with genetically modified microorganisms that degrade polymers

Sr. no.	Polymer	Transformed microorganism (OGM)	Gene or enzyme	Donor microorganism	Degradation percentage	Techniques	References
1	Phthalate esters (PAEs)	*Escherichia coli* BL21 (DE3)	*baceso4*	*Bacillus velezensis* SYB	87–32% in 5 h	HPLC	Huang et al. (2020)
2	Polyethylene terephthalate (PET)	*Pichia pastoris*	PETase	*Ideonella sakaiensis* 201-F6	Not available	Not available	Chen et al. (2020)
3	Dibutyl phthalate (DBP)	*Escherichia coli* BL21 (DE3)	*dpeH* y *mpeH*	Bacterial consortium *microbacterium* sp. and *Pandoraea* sp.	83% in 12 h	Clear zone test	Lu et al. (2020)
4	Polyethylene terephthalate (PET)	*Escherichia coli* SoluBL21	PETase	**PETase:** *Ideonella sakaiensis* 201-F6; **RolA:** *Aspergillus oryzae*	35–31% in 5 days	XRD, SEM	Puspitasari et al. (2021)
5	Dimethyl terephthalate (DMT)	*Escherichia coli* BL21 (DE3)	*dmtH*	*Sphingobium* sp.	86% in 7 days	Not available	Cheng et al. (2020)
6	Poly(butylene adipate-co-terephthalate)	*Escherichia coli*	*pbathRf*	*Rhodococcus fascians* NKCM2511	0.34 ± 0.05 mg in 5 days	Clear zone test, SEM	Soulenthone et al. (2021)
7	Aliphatic plastics poly (ε-caprolactone) and poly (1,4-butylene succinate)	*Escherichia coli* BL21 (DE3)	Lipase	*Amycolatopsis mediterranei* (AML)	90–80% in 22 h	HPLC	Tan et al. (2021)
8	Polyethylene terephthalate (PET)	*Bacillus subtilis*	Hidrolase BhrPETasa	Bacteria HR296	Not available	Discovery DSC 2500, TA Instruments	Xi et al. (2021)
9	Polyethylene terephthalate (PET)	*Escherichia coli* BL21 (DE3)	PETasa Mejorada mas peptido señal pelB	*Ideonella sakaiensis* 201-F6	Not available	HPLC, SEM	Shi et al. (2021)

10	Polycaprolactone (PCL)	*Escherichia coli* BL21 (DE3)	*SM14est*	*Streptomyces* sp.	Not available	Not available	Almeida et al. (2019)
11	Polyethylene terephthalate (PET)	Green algae, *Chlamydomonas reinhardtii*	Commercial PETasa	Undetermined	35–17% in 4 weeks	HPLC	Kim et al. (2020)

HPLC high-performance liquid chromatography, *XRD* X-ray diffraction, *SEM* scanning electron microscopy

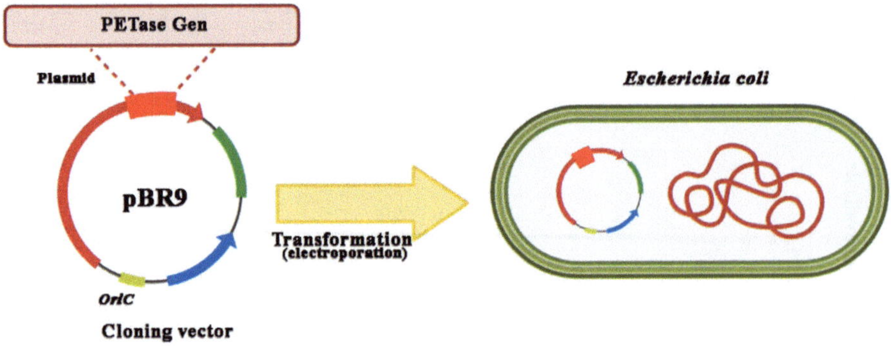

Fig. 5.4 Genetically modified organism. Construction of a cloning vector for transformation of *Escherichia coli* with the PETase gene (created in BioRender.com)

genetically modified organisms because it is a polymer that has in its structure several amorphous regions that facilitate its biodegradation or attack by microorganisms (Tokiwa et al. 2009). PET also has a high level of production and commercialization since it is the material for single-use products such as water and soft drink containers or bottles (Eubeler et al. 2010).

As mentioned above, this biodegradation method is a very promising technique that many researchers are currently working on. The important thing now is to deploy the use of microorganisms on a large scale in the degradation of synthetic polymers to mitigate the impact that plastic waste is generating in the environment.

5.2.7 Phytoremediation

In phytoremediation, plants are used to decontaminate environments through their root absorption capacity and the accumulation and/or degradation of various soil and water components (Kathi and Khan 2011). In this technique, plants are grown in substrates contaminated with some compound, facilitating its sequestration and subsequent degradation (Paz-Alberto and Sigua 2013). With phytoremediation, erosion and leaching are avoided, preventing the spread of toxic substances to other areas of the environment (USEPA 2004).

Through phytoremediation, compounds such as chlorinated solvents, aromatic hydrocarbons, pesticides and surfactants can be remediated at low cost (Kathi and Khan 2011). Although there are no reports so far of phytoremediation on plastics, one could combine this technique with mycoremediation, taking advantage of the potential of fungi to degrade various polymers and the microbial processes associated with plants (Kumar et al. 2019; Paz-Alberto and Sigua 2013).

Phytoremediation is gaining strength as a method of environmental remediation because it is a low-cost technique that does not cause adverse damage or harmful substances to other living beings, taking advantage of the capacity of plants to metabolize environmental pollutants in their tissues and roots. However, this biological method is still under development, and its large-scale application is still limited (Paz-Alberto and Sigua 2013).

This technique has been known for more than 300 years (Lasat 2000). Since then, humans have exploited certain capacities of plants to survive in contaminated areas and help remove pollutants from the soil, and it could be studied to mitigate the impact of plastic waste in the environment.

5.2.8 Nanoremediation

At the environmental level, nanotechnology has prevented the spread of pollutants in aquatic and terrestrial environments and has allowed the treatment and recovery of water bodies and mainly agricultural soils as well as the management of landfills to prevent the spillage of leachates and other hazardous wastes into the surrounding ecosystems (Karn et al. 2009). The advantage it has as a technique over other traditional techniques is its potential for large-scale action at a lower cost, with shorter times and greater efficiency, also allowing in situ work and reducing many pollutants to their lowest concentration (Karn et al. 2009).

Nanotechnology uses reactive nanomaterials to recover and/or transform all kinds of pollutants in different environments; this is called nanoremediation. These materials allow in situ work, thanks to their innovative properties such as their tiny size and special coatings, which allow them to penetrate the subsoil and remain in groundwater as suspended particles (Tratnyek and Johnson 2006). Among the nanoscale materials, most commonly used in remediation are zeolites, noble metals, enzymes, carbon fibres and iron nanoparticles. The latter have been very effective in the detoxification of organochlorine pesticides, polychlorinated biphenyls (PCBs) and organic solvents (Zhang 2003).

There are more sophisticated nanoparticles that can react at the same time with different pollutants in hydrophilic and hydrophobic environments; in the future these same particles can be reprogrammed so that once their remediation activity is completed, they destroy themselves without causing damage to the environment, having also a broader spectrum of remediation of pollutants, including synthetic plastic waste (Karn et al. 2009).

5.3 Alternative to the Plastic

5.3.1 Bioplastics

The future of plastics lies in bioplastics or bio-based polymers. These polymers have attracted increasing attention mainly from consumers and entrepreneurs, who seek to mitigate the environmental damage caused by synthetic plastics (Meereboer et al. 2020). These new polymers are produced from renewable sources, such as cellulose and starch, to be easily recycled or degraded by biological techniques such as composting (Kijchavengkul and Auras 2008).

The application of bioplastics is in short-lived products that do not require much resistance or durability, such as cosmetic containers, food disposables and plastics for agricultural use. These biopolymers also have their disadvantage, and that is in terms of their production, during which greenhouse gases are released that increase global warming. This leads us to think that it is better to design and implement sustainable waste management where the reduction and reuse of these materials are a priority rather than their disposal (Doímal et al. 2007).

This is a developed field that has not had economic support to be implemented on a commercial level (Meereboer et al. 2020); in addition they have higher costs than traditional plastics, and the consumer society is not very aware of their use and thus lacks general acceptance (Kijchavengkul and Auras 2008).

To reduce the cost, bioplastics such as polyhydroxyalkanoates (PHAs) and poly (lactic acid) (PLA) can be used in biocomposite materials, in which bio-based agro-residues are incorporated while maintaining performance in certain applications. In order to obtain new plastic formulations that are intended to be biodegradable in natural environments, work has been carried out on the ability of soil macro-invertebrates such as crickets, cockroaches, termites and bugs to degrade polyethylene (PE) films with different percentages of corn starch (starch/PE) (Tsao et al. 1993), in collaboration with rhizospheric microorganisms such as *Rhizobium meliloti* (Gautam and Kaur 2013) capable in transforming these polymers without causing adverse environmental impact. The use of corn starch and cassava starch additives, in different proportions combined with polyethylene, is a method being used to reduce the longevity of the polymer content of municipal solid waste (MSW) in the environment (Raheem et al. 2017).

5.3.1.1 Compostability of Biopolymers

The attention of industries, consumers and governments of countries worldwide on bio-based polymers has focused on using them as an alternative and efficient way to reduce municipal solid waste, for which these biopolymers must be recovered by techniques such as recycling, composting or incineration to recover their energy (Kijchavengkul and Auras 2008; Hoornweg and Bhada-Tata 2012; Bhardwaj et al. 2013; Geyer et al. 2015; EPA 2016; Alshehrei 2017; Fesseha and Abebe 2019; Borrelle et al. 2020; Gunawan et al. 2020).

In the last 5 years, composting legislation and industry initiatives have fought against the paradigm and scepticism towards biopolymers on the part of end consumers and composting facilities in accepting this type of material among the wastes to be composted (Kijchavengkul and Auras 2008).

The composting's technique consists of transforming organic waste such as plant or animal remains into humus or compost by the activity of microorganisms under predominantly aerobic conditions reaching high temperatures throughout the process (ASTM 2004).

When waste cannot be recycled, these products could be replaced by others made with biodegradable polymers, so that they are degraded by composting and the environment is protected (Kijchavengkul and Auras 2008).

There are several limitations for the implementation of composting as a technique for plastic waste management; on the other hand, there is a lack of clear and relevant legislation, the time for material degradation and obtaining humus is very long, the necessary equipment is expensive and requires trained personnel (Kale et al. 2007), there is no adequate system for material collection and there are not enough adequate facilities to carry out the process (Kale et al. 2007; Kijchavengkul and Auras 2008).

5.4 Conclusion

The management of plastic waste is a wasteful task that requires joint efforts between countries for the implementation of policies that lead to the proper management of plastic waste. As discussed by many authors of publications on this topic, it would be appropriate to combine policies that create awareness in the final consumers of plastic products in general, such as reutilization and reduction of their use and production, with biological technologies that do not generate greater damage to the environment and organisms. There are several very promising technologies that are worth developing further to achieve their maximum efficiency and implementation, such as phytoremediation and nanoremediation, as well as very old techniques such as composting that can be refined to make them more sustainable and usable. Likewise, the fight for the implementation of bioplastics cannot be left aside, providing this industrial sector with greater support to reduce costs and avoid the accelerated increase of pollution with synthetic plastics. Finally, it is also appropriate that all countries enter into the so-called circular economy of plastics so that at the end of the useful life of these products they do not become waste but rather their raw material is recoverable.

Acknowledgement We thank the Universidad Santiago de Cali and our editor Dr. Mohd. Shahnawaz for their support and the opportunity to make this publication possible. We also thank all our family and friends for their collaboration and encouragement.

References

Abalansa S, El Mahrad B, Vondolia GK, Icely J, Newton A (2020) The marine plastic litter issue: a social-economic analysis. Sustainability 12:8677

Almeida EL, Rincón AFC, Jackson SA, Dobson ADW (2019) In silico screening and heterologous expression of a polyethylene terephthalate hydrolase (PETase)-like enzyme (SM14est) with Polycaprolactone (PCL)-degrading activity, from the marine sponge-derived strain *Streptomyces* sp. SM14. Front Microbiol 10:1–14

Alshehrei F (2017) Biodegradation of synthetic and natural plastic by microorganisms. J Appl Environ Microbiol 5(1):8–19

American Chemistry Council (ACC) (2013) Resin review: the annual statistical report of the North American plastics industry

Andrady AL (2011) Microplastics in the marine environment. Mar Pollut Bull 62:1596–1605

Arkatkar A, Arutchelvi J, Muniyasami S, Bhaduri S, Uppara PV, Doble M (2009) Approaches to enhance the biodegradation of Polyolefins. Open Environ Eng J 2:68–80

Arutchelvi J, Sudhakar M, Arkatkar A, Doble M, Bhaduri S, Uppara PV (2008) Biodegradation of polyethylene and polypropylene. Indian J Biotechnol 7:9–22

ASTM (2004) ASTM D 6400–04 Standard specification for compostable plastics. ASTM International, West Conshohocken, PA

Awaja F, Pavel D (2005) Recycling of PET. Eur Polym J 41:1453–1477

Barnes DKA, Galgani F, Thompson RC, Barlaz M (2009) Accumulation and fragmentation of plastic debris in global environments. Philos Trans R Soc B Biol Sci 364(1526):1985–1998

Basta N, Johnson E (1990) Plastics recycling picks up momentum. Chem Eng News 30

Bhardwaj H, Gupta R, Tiwari A (2013) Communities of microbial enzymes associated with biodegradation of plastics. J Polym Environ 21:575–579

Borrelle S, Ringma J, Lavender Law K et al (2020) Predicted growth in plastic waste exceeds efforts to mitigate plastic pollution. Science 369:1515–1518

Chen Z, Wang Y, Cheng Y, Wang X, Tong S, Yang H, Wang Z (2020) Efficient biodegradation of highly crystallized polyethylene terephthalate through cell surface display of bacterial PETase. Sci Total Environ 2020:709

Cheng X, Dong S, Chen D, Rui Q, Guo J, Wang D, Jiang J (2020) Potential of esterase DmtH in transforming plastic additive dimethyl terephthalate to less toxic mono-methyl terephthalate. Ecotoxicol Environ Saf 187:1–10

Devi RS, Kannan VR, Natarajan K, Nivas D, Kannan K, Chandru S, Antony AR (2016) The role of microbes in plastic degradation. In environ waste manage. Taylor and Trancis Group, CRC Press, United States, pp 341–370. Chapter 12

Doímal P, Hoffmann J, Druzbík M (2007) Evaluating the aerobic biodegradability of plastics in soil environments through GC and IR analysis of gaseous phase. Czech Republic, Zlín

Edge M, Hayes M, Mohammadian M, Allen NS, Jewitt TS, Brems K, Jones K (1991) Aspects of poly (ethylene terephthalate) degradation for archival life and environmental degradation. Polym Degrad Stab 32(2):131–153

EPA United States Environmental Protection Agency (2016) Advancing sustainable materials management: 2014 Fact sheet. Assessing trends in material generation, recycling, composting, combustión with energy recovery and landfilling in the United States. United States Environmental Protection Agency. Office of Land and Emergency Management, Washington DC

EPA United States Environmental Protection Agency (2018) Advancing Sustainable Materials Management: Facts and Figures Report. https://www.epa.gov/facts-and-figures-about-materials-waste-and-recycling/advancing-sustainable-materials-management. Accessed 15 June, 2020

Eubeler JP, Bernhard M, Knepper TP (2010) Environmental biodegradation of synthetic polymers II. Biodegradation of different polymer groups. Trends Anal Chem 29(1):84–100

Fesseha H, Abebe F (2019) Degradation of plastic materials using microorganisms: a review. Public Health Open J 4(2):57–63

Gautam N, Kaur I (2013) Soil burial biodegradation studies of starch grafted polyethylene and identification of *Rhizobium meliloti* therefrom. Acad J 5(6):147–158

Geyer R, Kuczenski B, Zink T, Henderson A (2015) Common misconceptions about recycling. J Ind Ecol 20:1010–1017

Geyer R, Jambeck JR, Law KL (2017) Production, use, and fate of all plastics ever made. Sci Adv 3: 1–5

Giacomucci L, Raddadi N, Soccio M, Lotti N, Fava F (2020) Biodegradation of polyvinyl chloride plastic films by enriched anaerobic marine consortia. Mar Environ Res 158:1–10

Greenpeace (2019) How does plastic reach the oceans and what happened then? Greenpeace Spain. https://es.greenpeace.org/es/trabajamos-en/consumismo/plasticos/como-llega-el-plastico-a-los-oceanos-y-que-sucede-entonces/. Accessed 30 May 2020 (In Spanish)

Gu JD, Ford TE, Mitton DB, Mitchell R (2000) Microbial degradation and deterioration of polymeric materials. In: Review (ed) The Uhlig corrosion handbook, 2nd edn. Wiley, New York, pp 439–460

Gunawan NR, Tessman M, Schreiman AC, Simkovsky R, Samoylov AA, Neelakantan NK, Bemis TA, Burkart MD, Pomeroy RS, Mayfield SP (2020) Rapid biodegradation of renewable polyurethane foams with identification of associated microorganisms and decomposition products. Bioresour Technol Rep 11:1–9

Hoornweg D, Bhada-Tata P (2012) What a waste: a global review of solid waste management. World Bank

Huang L, Meng D, Tian Q, Yang S, Deng H, Guan Z et al (2020) Characterization of a novel carboxylesterase from *Bacillus velezensis* SYBC H47 and its application in degradation of phthalate esters. J Biosci Bioeng 129(5):588–594

Jambeck JR, Geyer R, Wilcox C, Siegler TR, Perryman M, Andrady A, Narayan R, Law KL (2015) Plastic waste inputs from land into the ocean. Science 347:768–771

Janczak K, Dąbrowska GB, Raszkowska-Kaczor A, Kaczor D, Hrynkiewicz K, Richert A (2020) Biodegradation of the plastics PLA and PET in cultivated soil with the participation of microorganisms and plants. Int Biodeterior Biodegrad 155:105087

Ji J, Zhang Y, Liu Y, Zhu P, Yan X (2020) Biodegradation of plastic monomer 2,6-dimethylphenol by *Mycobacterium neoaurum* B5-4. Environ Pollut 258:1–10

Kale G, Kijchavengkul T, Auras R, Rubino M, Selke S, Singh SP (2007) Macromol Biosci 7:255

Karn B, Kuiken T, Otto M (2009) Nanotechnology and in situ remediation: a review of the benefits and potential risks. Environ Health Perspect 117(12):1823–1831

Kathi S, Khån AB (2011) Phytoremediation approaches to PAH contaminated soil. Indian J Sci Technol 4(1):56–63

Khoironii A, Anggoro S, Sudarno (2019) Evaluation of the interaction among microalgae *spirulina* sp, plastics polyethylene terephthalate and polypropylene in freshwater environment. J Ecol Eng 20(6):161–173

Kijchavengkul T, Auras R (2008) Perspective compostability of polymers. Polym Int 57:793–804

Kim JW, Park SB, Tran QG, Cho DH, Choi DY, Lee YJ, Kim HS (2020) Functional expression of polyethylene terephthalate-degrading enzyme (PETase) in green microalgae. Microb Cell Factories 19(1)

Kuczenski B, Geyer R (2010) Material flow analysis of polyethylene terephthalate in the US, 1996–2007. Resour Conserv Recycl 54:1161–1169

Kumar A, Chaturvedi AK, Yadav K, Arunkumar KP, Malyan SK et al (2019) Fungal phytoremediation of heavy metal-contaminated resources: current scenario and future prospects. In: Yadav A, Singh S, Mishra S, Gupta A (eds) Recent advancement in white biotechnology through fungi. Fungal biology. Springer, Cham

Lasat MM (2000) Phytoextraction of metals from contaminated soil: a review of plant/soil/metal interaction and assessment of pertinent agronomic issues. J Hazard Substance Res 2(5):1–25

Lebreton L, Andrady A (2019) Future scenarios of global plastic waste generation and disposal. Palgrave Commun 5:6

Lu M, Jiang W, Gao Q, Zhang M, Hong Q (2020) Degradation of dibutyl phthalate (DBP) by a bacterial consortium and characterization of two novel esterases capable of hydrolyzing PAEs sequentially. Ecotoxicol Environ Saf 195:1–9

Lucas N, Bienaime C, Belloy C, Queneudec M, Silvestre F, Nava-Saucedo JE (2008) Polymer biodegradation: mechanisms and estimation techniques – a review. Chemosphere 73(4): 429–442

Meereboer KW, Misra M, Mohant AK (2020) Review of recent advances in the biodegradability of polyhydroxyalkanoate (PHA) bioplastics and their composites. Green Chem 22:5519–5558

Müller RJ, Schrader H, Profe J, Dresler K, Deckwer WD (2005) Enzymatic degradation of poly (ethylene terephthalate): rapid hydrolyse using a hydrolase from *T. fusca*. Macromol Rapid Commun 26:1400–1405

Mutha NH, Patel M, Premnath V (2006) Plastics material flow analysis for India. Resour Conserv Recycl 47:222–244

Paz-Alberto AM, Sigua GC (2013) Phytoremediation: a green technology to remove environmental pollutants. Am J Clim Chang 2:71–86

PlasticsEurope (2016) Plastics—The Facts 2016: an analysis of European plastics production, demand and waste data. PlasticsEurope

Pramila, R., Padmavathy, K., Ramesh, K.V. & Mahalakshmi, K. (2012). *Brevibacillus parabrevis, Acinetobacter baumannii* and *Pseudomonas citronellolis*-potential candidates for biodegradation of low density polyethylene (LDPE). African J Bacteriol Res, 4(1): 9–14

Puspitasari N, Tsai S-L, Lee C-K (2021) Class I hydrophobins pretreatment stimulates PETase for monomers recycling of waste PETs. Int J Biol Macromol 2021:157–164

Raheem A, Aremo B, Adeoye MO (2017) A starch-based biodegradable polyethylene for municipal solid waste mitigation. Mater Sci Appl 8:26–36

Restrepo-Flórez J-M, Bassi A, Thompson MR (2014) Microbial degradation and deterioration of polyethylene a review. Int Biodeterior Biodegradation 88:83–90

Sarkhel R, Sengupta S, Das P, Bhowal A (2020) Comparative biodegradation study of polymer from plastic bottle waste using novel isolated bacteria and fungi from marine source. J Polym Res 27(1)

Shabbir S, Faheem M, Ali N, Kerr PG, Wang LF, Kuppusamy S, Li Y (2020) Periphytic biofilm: an innovative approach for biodegradation of microplastics. Sci Total Environ 717

Shah N, Rockwell J, Huffman GP (1999) Conversion of waste plastic to oil: direct liquefaction versus pyrolysis and hydroprocessing. Energy Fuel 13:832–838

Shi L, Liu H, Gao S, Weng Y, Zhu L (2021) Enhanced extracellular production of is PETase in *Escherichia coli* via engineering of the pelB signal peptide. J Agric Food Chem 69:2245–2252

Shumba T, Moyo M, Mhlanga S (2020) Feasibility and optimization of plastic waste energy recovery process as a plastic waste management tool in Zimbabwe. Proceedings of the 2nd African international conference on industrial engineering and operations management. Harare, Zimbabwe

Singh B, Sharma N (2008) Mechanistic implications of plastic degradation. Polym Degrad Stab 93(3):561–584

Sivan A (2011). New perspectives in plastic biodegradation. In current opinion in biotechnology, vol. 22. Elsevier Current Trends, pp 422–426

Soulenthone P, Tachibana Y, Suzuki M, Mizuno T, Ohta Y, Kasuya K (2021) Characterization of a poly (butylene adipate-co-terephthalate) hydrolase from the mesophilic actinobacteria *Rhodococcus fascians*. Polym Degrad Stab 184:1–12

Tan Y, Henehan GT, Kinsella GK, Ryan BJ (2021) An extracellular lipase from *Amycolatopsis mediterranei* is a cutinase with plastic degrading activity. Comput Struct Biotechnol J 19:869–879

Taniguchi I, Yoshida S, Hiraga K, Miyamoto K, Kimura Y, Oda K (2019) Biodegradation of PET: current status and application aspects. ACS Catal 9(5):4089–4105

Tokiwa Y, Calabia BP, Ugwu CU, Aiba S (2009) Biodegradability of plastics. Int J Mol Sci 10: 3722–3742

Tollner EW, Annis PA, Das KC (2011) Evaluation of strength properties of polypropylene-based polymers in simulated landfill and oven conditions. J Environ Eng 137:291–296

Tratnyek PG, Johnson RL (2006) Nanotechnologies for environmental cleanup. NanoToday 1(2): 44–48

Tsao R, Anderson TA, Coats JR (1993) The influence of soil macroinvertebrates on primary biodegradation of starch-containing polyethylene films. J Environ Polym Degr 1:301–306

Tsuchida D, Kajihara Y, Shimidzu N, Hamamura K, Nagase M (2011) Hydrogen sulfide production by sulfate-reducing bacteria utilizing additives eluted from plastic resins. Waste Manag Res 29, 594:–601

United States Environmental Protection Agency (USEPA) (2004) Hazard summary. Lead compounds. http://www.epa.gov/ttn/atw/hlthef/lead.html

Urase T, Okumura H, Panyosaranya S, Inamura A (2008) Emission of volatile organic compounds from solid waste disposal sites and importance of heat management. Waste Manag Res 26:534–538

Urbanek AK, Strzelecki MC, Mirończuk AM (2021) The potential of cold-adapted microorganisms for biodegradation of bioplastics. Waste Manag 119:72–81

Van Krevelen DW, Te Nijenhuis K (2009) Chemical degradation. In: Properties of polymers, 4th edn. pp 779–786

Webb H, Arnott J, Crawford R, Ivanova E (2012) Plastic degradation and its environmental implications with special reference to poly (ethylene terephthalate). Polymers 5:1

Wei R (2011) Hydrolysis of polyethylene terephthalate by cutinases from *Thermobifida fusca* KW3, PhD thesis. Universitat Leipzig, Leipzig

Wei R, Zimmermann W (2017) Microbial enzymes for the recycling of recalcitrant petroleum-based plastics: how far are we? J Microbial Biotechnol 10:1308–1322. https://doi.org/10.1111/1751-7915.12710

Weitz K, Barlaz M, Ranjithan R, Brill D, Thorneloe S, Ham R (1999) Life cycle management of municipal solid waste. Int J Life Cycle Assess 4:195–201

Xi X, Ni K, Hao H, Shang Y, Zhao B, Qian Z (2021) Secretory expression in *Bacillus subtilis* and biochemical characterization of a highly thermostable polyethylene terephthalate hydrolase from bacterium HR29. Enzym Microb Technol 143:1–10

Xu SY, Zhang H, He PJ, Shao LM (2011) Leaching behaviour of bisphenol a from municipal solid waste under landfill environment. Environ Technol 32:1269–1277

Yamada-Onodera K, Mukumoto H, Katsuyaya Y, Saiganji A, Tani Y (2001) Degradation of polyethylene by a fungus, *Penicillium simplicissimum* YK. Polym Degrad Stab 72:323–327

Yang SS, Ding MQ, He L, Zhang CH, Li QX, Xing DF, Cao GL, Zhao L, Ding J, Ren NQ, Wu WM (2020) Biodegradation of polypropylene by yellow mealworms (*Tenebrio molitor*) and superworms (*Zophobas atratus*) via gut-microbe-dependent depolymerization. Sci Total Environ:1–59

Yoshida S, Hiraga K, Takehana T, Taniguchi I, Yamaji H, Maeda Y, Toyohara K, Miyamoto K, Kimura Y, Oda K (2016) A bacterium that degrades and assimilates poly (ethylene terephthalate). Science 351(6278):1196–1199

Zhang WX (2003) Nanoscale iron particles for environmental remediation: an overview. J Nanopart Res 5:323–332

Zhang J, Wang X, Gong J, Gu Z (2004) A study on the biodegradability of polyethylene terephthalate fiber and diethylene glycol terephthalate. J Appl Polym Sci 93(3):1089–1096

Strategies to Cope with the Plastic Pollution in the Sea

6

Alka Kumari and Doongar R. Chaudhary

Abstract

Seventy percent of the earth's surface is covered by the ocean which is the major contributor to oxygen production in the environment and home to diverse life forms. Plastic discovery has revolutionized many sectors such as industrial, fashion, construction and medical. It is now emerging as an alarming problem and piled up into patches in the ocean affecting coastal communities, biodiversity, climate and human health. The sewage passages, dumping of household, industrial plastic litter, ship docks and fishing gear activities in the oceans are the major source of marine plastic pollution. Plastics are used everywhere globally, but production, consumption and disposal are unequally distributed across the globe. Some common practices, such as restricting microplastic usages and manual clean-up of plastic packages and products across the international ocean boundaries, are being applied to some extent. However, strict legal regulation for the life cycle assessment of plastics which will ensures sustainability during production, manufacturing, consumption, trade and disposal. Implementing target measures based on available knowledge will be a foundation step. Bioremediation technologies are being developed for the in situ application of microbes with enhanced plastic-degrading ability. Nonetheless, overall remediation of plastics from the ocean considering sustainability and the economy with current practices is challenging without technological, social and legal unification. This chapter elaborates on the current situation, adopted strategies and future outlooks of plastic pollution in the sea.

A. Kumari · D. R. Chaudhary (✉)
Plant Omics Division, CSIR-Central Salt and Marine Chemicals Research Institute, Bhavnagar, India
e-mail: drchaudhary@csmcri.res.in

M. Shahnawaz et al. (eds.), *Impact of Plastic Waste on the Marine Biota*,
https://doi.org/10.1007/978-981-16-5403-9_6

Keywords

Bioremediation · Microplastics · Plastic · Ocean · Recycling

6.1 Plastics in Sea

Plastic is everywhere, and the common sources are single-use plastic items such as food packages, pet bottles, carry bags, straws, etc. (Li et al. 2016). Structurally plastics have a stable carbon-hydrogen backbone providing non-degradable property (Danso et al. 2019; Ghatge et al. 2020). The remoulding ability of plastic availing it for various applications and favouring a large production leads to huge accumulation in the environment (Monteiro et al. 2018). Global plastic production has reached more than 385 million tons, in which 40% of wastes were packaging litters wherever; this rate of waste generation predicts the production to be doubled in the next 20 years (Jambeck et al. 2015; Groh et al. 2019). Additionally, around 4.8–12.7 million metric tons of plastic waste are annually entering the ocean, whereas 150 million tons of plastics already entered the ocean (Jang et al. 2015; Jambeck et al. 2015). Around 80% of plastic wastes into the oceans comes from developing countries (Jambeck et al. 2015; Tessnow-von Wysocki and Le Billon 2019). These plastic wastes are a mixture of different types of plastics (such as polyethylene, polystyrene, polyvinylchloride, polyethylene terephthalate, etc.), and many are fragmented into minute-sized particles such as micro (>500 μm) or nano (>100) forms (Barnes et al. 2009; Peng et al. 2020). Approximately 2 million tonnes of microfibres and 0.7 million microfleeces are released from synthetic clothing entering the ocean every year, where 1.5 million trillion microfibres had been estimated to be already present in the ocean (Mishra et al. 2019). These fragmented plastic pieces can be categorized into primary and secondary particles, where primary plastic particles are originally manufactured as micro size and secondary plastics have resulted from environmental fragmentation of larger plastic (Barnes et al. 2009). Both types of plastic pieces persist and travel for long distances. These plastic particles are ingested by marine organisms and also act as carriers of toxic chemicals and pathogens which could travel to several life forms through the food chain (Thompson et al. 2009; Li et al. 2016; Law 2017). There are many unknown risks and associated harms from marine plastic debris (Gorycka 2009; Rochman et al. 2013; Law 2017; Barboza et al. 2020a). The major reason for plastic waste accumulation all around is due to the non-degradable property, and sources include public, industrial and harbour litters and lost or discarded fishing gear (ghost nets) (Litter 2005). Marine plastic pollution has been consequences of natural or anthropogenic factors, including shipping, fishing, industries, landfills and dumps located near the coast or waterways, natural events like rain and winds (Thushari and Senevirathna 2020). The plastics in the ocean could risk food security and human health and result in the entanglement of marine organisms mostly from abandoned plastic debris and fishing nets (Gregory 2009). Microfibres are easily mistaken to be food and ingested by marine animals (Possatto et al. 2011). Overloaded waste management and recycling and management systems cannot cope with rising plastic production and continuously increasing plastic litters in the oceans (Singh et al. 2014; Yee et al. 2021).

6.2 Impacts and Challenges of Plastics in Ocean: Overview

The marine environment covers the major area of the earth as well as supports the food web and environmental sustainability (Albouy et al. 2019). Plastics persist in the marine environment for hundreds to thousands of years, depending on the polymer type (Álvarez-Noriega et al. 2020). Marine plastic debris travel along the ocean boundary is a crucial problem affecting coastal communities and biodiversity as well as risks climate and human health around the world (Mendenhall 2018). Plastic debris has been detected worldwide in significant amounts in marine habitats, in all sizes from nanos, microns to meters (Law 2017). The primary sources of microplastics that directly enter the ocean are not the product of fragmentation but are microbeads used in personal care products (Löhr et al. 2017). The secondary source is plastic particles from the eventual fragmentation of larger plastic materials into fragments (PET, nylon, acrylic fibres) (Malankowska et al. 2021).

Over 700 marine species suffer adversely affected by plastic pollution (Steer and Thompson 2020). A total of 557 species are documented to be entangled or have ingested marine debris (Kühn et al. 2015). Various chemicals used in plastic manufacturing are toxic to animals and humans (Derraik 2002). The presence of anthropogenic debris in fishes and shellfish from the fish market of Indonesia and the USA raises human health concerns (Rochman et al. 2015). Nine fish species from different habitats such as coastal, pelagic and reef-associated from the Arabian Gulf were quantified for the presence of microplastics in their guts, where microplastics are found in an average of 5.71% of total fish collection from Arabian Gulf and polyethylene and polypropylene were observed to be the most abundant polymers (Baalkhuyur et al. 2020). Among 11 commercial marine fishes, *Eleutheronema tridactylum* and *Clarias gariepinus* from the fish market of Seri Kembangan (Malaysia) were found to ingest a high amount of microplastics of size ranging from 200 to 34,900 μm (Karbalaei et al. 2019). In another study, three different commercial fishes from the north-east Atlantic Ocean were analysed for microplastic toxicity, where 49% of fishes had microplastic in different parts such as the digestive tract, gills and dorsal muscle (Barboza et al. 2020b). Along with microplastic presence, the toxicity effects due to the presence of chemical additives or coatings used in plastic manufacturing such as bisphenol A and phthalates had been shown to result in blindness, cancer and endocrine disruption and affect the reproductive system and embryo development; lipid oxidation in gills, muscle and neural system; and acetyl-cholinesterase induction (Koch and Calafat 2009; Hahladakis et al. 2018; Karbalaei et al. 2019; Godswill and Godspel 2019; Barboza et al. 2020b; Kumari et al. 2021). Microplastics are reaching humans through the food chain (Barboza et al. 2020a). In a study, microplastic of sized 50 to 500 μm was found in the human faeces (Schwabl et al. 2019). The plastic-associated chemicals are assayed to have

41% oxidative stress, 32% cytotoxicity, 12% high oestrogen and 27% anti-androgenicity risk (Zimmermann et al. 2019).

6.3 Regulation of Plastic Packaging and Disposal

Around ~85% of total wastes in the ocean are plastic litter in which half of it comprises single-use plastic trash (Shahnawaz et al. 2019). The major part of plastics entering the ocean are from land-based sources (Fig. 6.1); subsequently, the problem of plastics in the sea is emerging as an international issue intended to protect marine habitats and discontinue the loss of exotic habitation (Gregory 2009; Eastman et al. 2020).

Marine plastic pollution needs international and local cooperation for mandated implementation of policies for each sector such as government, public and industrial firms (Wu 2020; Raubenheimer and Urho 2020; Raha et al. 2021). Some manufacturers are encouraged to switch plastics with sustainable options by providing funding to develop alternative sustainable materials for packaging purposes (Guillard et al. 2018).

The government of India had released generalized regulations under the Environment Protection Act of 1986 for goods protection systems and associated problems for producers and consumers in consideration to environmental sustainability; initially, it was not included under specific regulations for plastic waste disposal (Prasad 2006). Later, the 'Recycled Plastics (Manufacture and Usage) rule 1999' was issued as a mandatory guideline for regulating plastic waste recycling (Khanna 2001). The Hazardous Wastes (Management and Handling), 1989, amended in 2003, covers specific plastic waste regulations, laterally with general hazardous waste rules (Prakash and Gowtham 2018). The Plastic Manufacture, Usages, and Waste Management Rules (2009) announced specifically for usage and management

Fig. 6.1 Shows the different source of plastic wastes entering the ocean

of plastic bag for controlled use of colouring agents, recycled or plastics in food packaging or storages (Bhattacharya et al. 2018). Moreover, the use of colours in plastics used in food packaging is restricted following the Bureau of Indian Standards specifications (BIS 2012). The recyclable plastic wastes were legitimate to the registered recycler after collection that follows IS 14534:1998 recycling agreement (Bhaskar and Turaga 2018; Aryan et al. 2019). The US Food and Drug Administration (FDA) has released 'Guidance for Industry: Use of Recycled Plastics in Food Packaging (Chemistry Considerations), recommends that recyclable plastic should consider manufacturer's evaluation for producing material suitability for food-contact applications (Cecon et al. 2021). Similar safety guidelines were also issued by the European Union and European Food Safety Authority for the use of recycled plastic materials in food packaging (Franz 2002; Heckman 2005; Barboza et al. 2020a). Extensive harmonization is required in the existing international law and frameworks for waste management from land-based sources of plastic entering the ocean by targeted monitoring and implementation of policies (Hanisch 2000; Chen 2015). The restricted application of plastic products and manufacturing of reusable and recyclable plastics will encourage the circular economy correspondingly (Deshpande et al. 2020).

The international treaty directed for implementing policies for banning certain plastics item, improved waste management, recycling and regulated manufacturing processes, as extended producer responsibility outlines, but lacking seriousness and adequate action for monitoring the problem (Tessnow-von Wysocki and Le Billon 2019). International cooperation is required to address these challenges to improve plastic production and use, approach a circular economy and minimize the proportion of plastic entering the water body (Table 6.1) (Masi et al. 2018; Raubenheimer and Urho 2020).

The Montreal Protocol provided a model for regulating the international industry for reduced production and minimisation of chemical additives in plastics (Raubenheimer and McIlgorm 2017; Tessnow-von Wysocki and Le Billon 2019). The United Nations Convention on the Law of the Sea (UNCLOS) is the only international agreement regulating marine pollution from land-based sources and establishing suitable laws for monitoring, preventing marine pollution, reporting marine plastic litter and taking strict jural resolution (Marciniak 2017). The UNCLOS lacks implementation guidelines that address the marine plastic debris problem precisely (Marciniak 2017; Wu 2020). The United Nations Environment Programme (UNEP) issued regulations for resolving problems associated with marine microplastics (UNEP 2015; Kershaw et al. 2011; Tiller and Nyman 2018). The third UN Environment Assembly (UNEA-3) assessment suggested a compulsory, voluntary and self-regulatory measure to combat marine plastic litter and microplastics from all sources, especially land-based sources, incorporating international trade and waste generation and defining them for obligatory sustainable production-consumption and disposal of plastics and their chemical additives (Laina 2018).

The General Assembly of the United Nations implemented the end goal for the marine biodiversity affected due to plastic trashes in biodiversity beyond national

Table 6.1 Few international conventions involved in prevention of marine plastic pollution

Statute	Protagonist	Organization
London Convention of 1972 (LC 72)	Prevention of Marine Pollution by Dumping of Wastes and other Matter	United Nations Conference on the Human Environment
MARPOL, 1973	Prevention of Pollution from Ships	International Convention for the Prevention of Pollution from Ships
UN Law of the Sea Convention (UNCLOS), 1982	Straddling Fish Stocks Agreement	United Nations
UN Environment, 2017	Discourage the use of microplastics and single-use plastic	United Nations
International Maritime Organization	Prevention of pollution by ships	United Nations
Biodiversity in areas beyond national jurisdiction (BBNJ), 2017	International agreement on marine biodiversity protection	United Nations and the inter-governmental conference

jurisdiction (BBNJ) (Tiller and Nyman 2018). The International Convention for the Prevention of Pollution from Ships (MARPOL) and International Maritime Organization (IMO) are principal agreements for resolving marine plastic problems with a standard strategy for marine pollution created by ships from operational or accidental causes (Karim 2010; Leary 2016; Jie and Jiaxiang 2018). The International Maritime Organization had coordinated efforts with the United Nations Environment Programme (UNEP) and Food Agriculture Organization (FAO) to develop the protocol and guidelines for states with the relevant international standards and regulation (Wu 2020). The aim of these agreements encourages private sectors to collect the plastic waste followed by sorting and recycling, thereby contribute to the circular economy (Raubenheimer and Urho 2020; Fadeeva and Van Berkel 2021). The fundamentals of regulation for assessing plastic life cycle agreements from every sector focus on plastic litters entering the ocean (Löhr et al. 2017; Bishop et al. 2021). The proposed model is challenging, particularly in terms of cost benefits, but is valuable for the global plastic threats posed to the ocean (Raubenheimer and McIlgorm 2017). The economic unfeasibility in assessing and controlling plastic use and recovery was the major factor for the let-down; these agreements were also perceived due to lacking seriousness by developing countries (Abbott and Sumaila 2019). This has affected the intercontinental agreements to control the spread of plastic in the ocean through trade and tourism (Löhr et al. 2017).

6.4 Life Cycle Assessment of Plastics

The data on production, manufacturing, consumption and disposal of plastics are globally dispersed. The life cycle assessment of plastics can help the authority in designing policy in consideration to the impact of plastics on the environment (Fig. 6.2) (Rebitzer et al. 2004; Hopewell et al. 2009; Turner et al. 2016).

The life cycle assessment would also allow the quantification of greenhouse gas emissions and improvement in the existing waste management policies and approach for a sustainable environment (Turner et al. 2016; Malankowska et al. 2021). The large-scale production and application evidenced greenhouse emission from each stage of the plastic life cycle (Hahladakis et al. 2018; Shen et al. 2020). The problem of marine plastic pollution is not recent, and despite existing efforts, the amount of plastic waste is continuously increasing and finding a way to the ocean through the land and rivers (Raha et al. 2021).

The designing of the polymer production processes and chemicals used, recycling of by-product, appropriate disposal and hazard potential labelling require transparency up to global industry standards including life cycle processes and extended producer responsibility schemes (Tessnow-von Wysocki and Le Billon 2019). The

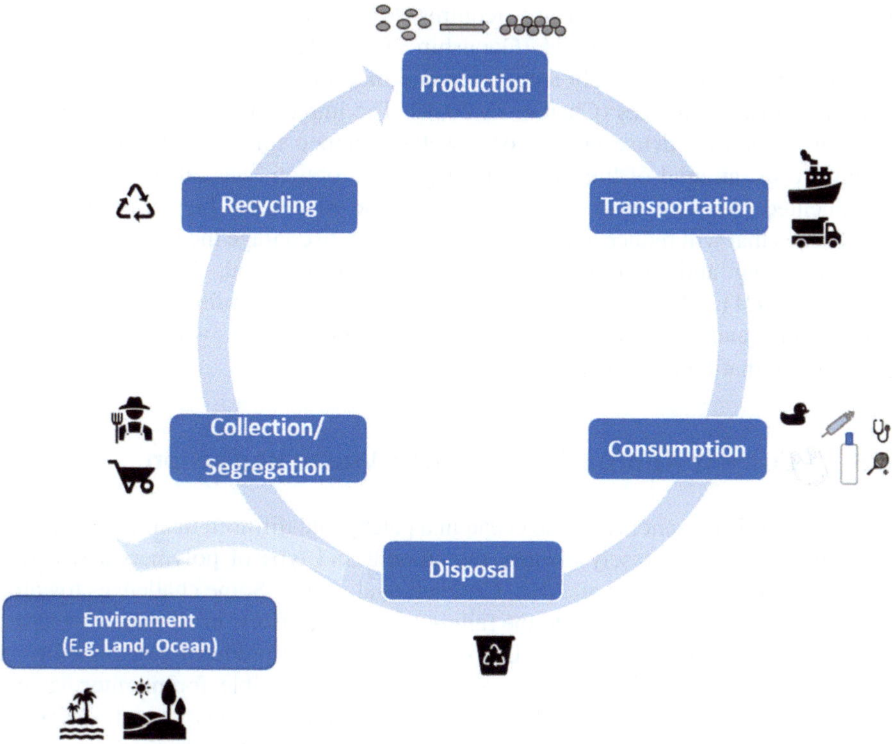

Fig. 6.2 Schematic diagram of life cycle of plastics

properties of polymers were enhanced by using various additives to extend their versatility and protect against physical deterioration such as light, temperature, flame retardancy and colour (Al-Malaika et al. 2017; Groh et al. 2019). The comprehensive information and listing of the risks of chemicals released in the environment during manufacturing, dumping and recycling of plastics were categorised under Classification, Labelling, and Packaging (CLP) regulations (Kumari et al. 2021). The 906 plastic additives are listed in the Chemical associated with Plastic Packaging Database (CPPdb), among them 63 and 68 are hazardous to human health and environment, respectively, where 7 are bioaccumulative and toxic according to the European Chemicals Agency (ECHA) within the CLP regulation (Groh et al. 2018, 2019).

The extended producer responsibility involves management and prevention of plastic packaging use and litter by retailers and industries, raising concern for adequate post-consumer collection and treatment (Fadeeva and Van Berkel 2021).

The selection of the recovery process should follow scientific assessments of the plastics' life cycle (Perugini et al. 2005). The life cycle assessment of plastics does not justify the environmental, economic or health impacts that leak into the environment (Willis et al. 2018; Ahamed et al. 2021). The transparency in exports and imports of plastic waste, as well as standards of recycling method, is lacking (Rosenfeld and Feng 2011; Moharam and Maqtari 2014). However, the manufacturing of virgin plastics requires lower cost, energy and time than re-processing the used material (Goodship 2007). The limitations of life cycle assessment are not adequately recognized and inaccurately emphasized on policies upon industries' practices (Chen et al. 2019). The impact of using various grocery bags in cities with dense populations, well-developed infrastructure and thermal treatment as an end-of-life waste management option was also insufficiently documented (Ahamed et al. 2021). However, the zero-waste hierarchy prioritizes approaches that will reduce and reuse the plastics to discourage the life cycle impacts than recycling and landfilling approaches (Cappucci et al. 2020; Zaman and Newman 2021). This would be successful with better coordination between the governments and industry but also between different ministries within governments through international cooperation.

6.5 Current Approaches in Plastic Waste Reduction

Plastic debris is present across the ocean in a patchy and diffused manner; moreover, it is impossible to precisely identify the amount and type of polymers across this mobile medium (Garaba et al. 2018; Critchell et al. 2019). Some challenges towards marine plastics regulations led to the failure of existing efforts in eliminating marine plastic debris (Raubenheimer and Urho 2020). Extended producer responsibility in which manufacturers or importers of plastic are responsible for monitoring and collecting waste generated in the environment (Andreasi Bassi et al. 2020). It indicates regulations for product design and awareness for the plastic waste origin and upgradation of the waste management systems, circular economy, bans on production and usage of certain plastic products as well as life cycle assessment impacts on the marine plastic pollution (Vince and Hardesty 2017; Fadeeva and Van Berkel 2021). Solid waste segregation and management, reusing, recycling, recovery of plastics and developing alternative biodegradable material are contemporary methods implied to control the dispersal of plastics in different environments. Simultaneously, increase the segregated plastic source, collection rate and the recyclability of all plastic products (Vanapalli et al. 2019). Reusing of plastic products will reduce the plastic wastes and decrease formation of microplastics (Wu et al. 2017). Recycling of used plastics is an effective approach, but reduces quality and increases costs than production of virgin plastics (Vilaplana and Karlsson 2008). Chemical recycling is an infrequent method where polymers are chemically reacted to produce feedstock materials (Rahimi and Garciá 2017). Recycler must approach local bodies for the management of waste generated within the given time frame. Although, recyclers are the weak part of the chain because they are dependent on local and private institutions which deal at various levels (Rahimi and Garciá 2017). There is a risk that plastic waste initially collected for recycling would either be disposed of or exported to developing countries (Andreasi Bassi et al. 2020). However, there is a need for a reduction in the production and use of plastics and developing a sustainable process throughout the life cycle of plastics (Aryan et al. 2019). The implementations of ideas and innovation are hindered due to administrative or political issues. The solution to microplastics is to upgrade the filtering technologies to remove microplastics efficiently and prevent microplastics from entering the water body such as rivers and the ocean (Barcelo and Pico 2020). Filtering technologies could reduce the plastic particles entering the oceans, but removal of already present microplastics requires advanced filtering approach (Tessnow-von Wysocki and Le Billon 2019; Liu et al. 2020).

Pyrolysis involves heating plastics in the absence of oxygen to break down the long polymer chains into small molecules (Almeida and de Fátima Marques 2016). Under mild conditions, polyolefins can yield a petroleum-like oil (Cleetus et al. 2013). The use of plastic waste as an energy source and its recovery for synthesizing crude and valuable products could reduce waste and microplastic formation (Panda et al. 2010). The consumers' perceptions, behaviors and habits should be considered in redesigning of material for packaging or other purposes to reduce the quantity of waste generation by substituting with lighter and stronger materials (Gustavo Jr et al. 2018). Biodegradable plastics such as polylactide and polyhydroxyalkanoates are commercially prevailing polymers that had competed for synthetic plastics to some extent for limited applications such as microbead synthesis in cosmetics, packaging and fashion accessories (Wang et al. 2013). However, the substitution of plastics is not recommended and very difficult because it further requires reconsidering of current practices for its local and global impact of new material (Marsh and Bugusu 2007). However, strategies such as the circular economy approach which emphasises the reduction in use of plastic at the source with clean-up, recovery and recycling of plastic products provide a glimpse of a future in which plastic pollution and its associated risks can be mitigated.

6.6 Implementation and Effective Monitoring

Plastic waste is primarily entering from land sources through rivers and wastewater drainage into the ocean that accumulates in sediments, gyres and deep sea (Chen et al. 2020; Hohn et al. 2020). Estuaries are one of the major coastal ecosystems affected by plastic pollution. Plastic pollution is impacting at the individual level as well as resulting in the various ecological impacts like the release of greenhouse gases during the recycling process. The monitoring of plastic debris source movement and dynamic forces of the ocean is obligatory to trace them (Simon and Schulte 2017). The policymakers and innovators lack comprehensive information about the available technology to target the global urgency of ocean plastic accumulation. Various sectors showed concern over the marine plastic pollution issue but represent different interests and priorities for its implementation (Raha et al. 2021). The implementations of ideas and innovation are hindered due to administrative or political interests. The strategy needs to address plastics entering the ocean from the sea and land-based sources as well as consider all chemical additives used within the life cycle of plastics (Kosior and Crescenzi 2020).

However, the efforts implied for plastic waste collection from the ocean are not enough compared to the plastic debris already present in the environment (Schmaltz et al. 2020). For the successful implication of the existing plastic waste regulation agreements, the policies need assurance to meet the targets for addressing the issue within an adequate scope and not as a short-term incentive. Traditional approaches have been used to define such an agreement in the context of preventing marine litter that does not discuss the possible financial mechanisms for sustainable waste management. The consumer will have to pay for the management of plastic pollution at the global level. The overall goal of a global plastics treaty would impact the reduction, prevention and elimination of marine plastic litter and microplastics. Many developing countries introduced a ban on the use of plastic carry bags and microbeads and are encouraging recycling as well as putting the responsibility on producers through extended producer responsibility schemes (Löhr et al. 2017). The design of a global plastics treaty needs a financial mechanism ensuring that developing countries could afford implementation costs (Vince and Hardesty 2017). The plastic debris movement across the ocean through international boundaries deserves careful management and prevention (Villarrubia-Gómez et al. 2018). International standards of plastic types and quality can reduce the production and facilitate the management of plastics as well as their chemical additives (Tessnow-von Wysocki and Le Billon 2019; Castillo Castillo et al. 2020).

The extended producer responsibility towards manufacturer accountability; eco-friendly strata through an awareness programme and capacity building campaigns for reuse, recycle and reduction of used plastics focusing on the cleaner environment; and scientific studies on emerging environmental issue, sustainability and innovations are effective keys for reducing and controlling the plastics entering the water body (Thushari and Senevirathna 2020). This concept incentivizes producers to change the design of their products according to the collection and recycling systems in place. Recently, various international organizations and non-profit social groups actively work together with the cognizance to save the ocean from plastic pollution in different countries and regions.

6.7 Advancing of Bioremediation Technologies

Presently, there are two frequent approaches that had been followed, i.e. preventing plastic waste from entering the aquatic ecosystem and removal of already entered plastic debris in the ocean or riverine (Schmaltz et al. 2020). These preventive strategies are unable to compete with a high-rate accumulation of plastics entering the ocean, yet clean-up measures are expensive, laborious and inefficient (Irwin 2018). Plastic waste aggregation is the result of modern lifestyle and a direct reflection of state and national governance fragility. This had become a prominent issue due to recent COVID-19 pandemics where the use of disposable plastics and personal protective equipment had surpassed the production and waste collection limit (Sarkodie and Owusu 2020). During the state of emergency, high plastic waste management load was going into landfills and incineration, debarring several plastic waste reduction policies at regional and national levels.

Technological and conceptual progression would uncover the broader opportunities for exploiting microbes that represent an important asset for maintaining a sustainable environment (Bilal and Iqbal 2020). The research on biodegradation studies advances and enables researchers to work towards the unknown and unmapped side of the biodegradation mechanism (Jaiswal and Shukla 2020). The use of microorganisms capable of biodegrading plastics has emerged as a promising low-cost and eco-friendly option (Madsen 1991). Emerging genomic and high-throughput metabolomic technologies of the whole microbial system evidence provide a schematic overview of all possible metabolic pathways in individual microorganisms or consortia (Mishra et al. 2021). The mapping of the whole cell genome and metabolism would provide the way-out solution for plastic bioremediation potential. The exploration of potential biotechnological tools including microbial consortia, molecular genetic engineering and omics techniques might become a crucial part of the eco-remediation of plastic wastes. For instance, plastics are carbon-rich element; perhaps their biodegradation could provide biotechnological industries a source of next-generation carbon feedstock. This will eventually contribute to the success of future negotiations regarding the elimination of plastic pollution in the environment.

6.8 Conclusion

The success of plastic as a material has shaped the development of modern society and challenged previously used polymeric materials in their conventional usage. Now, plastic has become a major component of total wastes in the environment and widely reported within the marine environment. Impacts from plastic debris have been identified as a major global conservation issue with implications for maritime commerce, tourism, marine organisms and human health. Although there are many scientific and even societal findings, hinging on plastics makes it impossible to completely prohibit its use. On the other hand, replacing plastics with an alternative such as biopolymers would need further evaluation for the competing properties, economic feasibility and environmental impacts. Present solutions ease the plastic accumulation in the environment and the challenges to reduce the plastic by volunteering filtration and clean-up programme on a regular basis. A combined agreement of ocean authority and pollution control globally forms a solution for addressing sources of marine plastic pollution by providing transparency in chemical additives used during manufacturing by considering the life cycle of plastics.

However, the challenges of marine plastic pollution policies need to consider all areas concurrently. Developing countries often do not have the financial means to introduce the necessary waste management infrastructure to handle waste sustainably. Efforts have so far failed to adequately address the problem. A legally binding mechanism on the global level could overcome some challenges of marine plastic governance. Future research ought to assess whether microbial genes involved in plastic degradation have begun to spread via horizontal gene transfer in the environment. Research is also needed to develop strategies for in situ biodegradation of microplastics by the microorganisms or by enhanced natural attenuation using native microflora.

Acknowledgement Authors acknowledge the Council of Scientific and Industrial Research (CSIR), New Delhi, for providing financial support (CSC0120-Waste to Wealth). Author AK acknowledges the Department of Biotechnology, Government of India, New Delhi, for providing senior research fellowship.

References

Abbott JK, Sumaila UR (2019) Reducing marine plastic pollution: policy insights from economics. Rev Environ Econ Policy 13:327–336

Ahamed A, Vallam P, Iyer NS, Veksha A, Bobacka J, Lisak G (2021) Life cycle assessment of plastic grocery bags and their alternatives in cities with confined waste management structure: a Singapore case study. J Clean Prod 278:123956. https://doi.org/10.1016/j.jclepro.2020.123956

Al-Malaika S, Axtell F, Rothon R, Gilbert M (2017) Additives for plastics. In: Gilbert M (ed) Brydson's plastics materials. Elsevier, Oxford, pp 27–168. https://doi.org/10.1016/B978-0-323-35824-8.00007-4, https://doi.org/10.1016/B978-0-323-35824-8.00007-4

Albouy C, Archambault P, Appeltans W, Araújo MB, Beauchesne D, Cazelles K, Cirtwill AR, Fortin MJ, Galiana N, Leroux SJ, Pellissier L (2019) The marine fish food web is globally connected. Nat Ecol & Evol 3:1153–1161. https://doi.org/10.1038/s41559-019-0950-y

Almeida D, de Fátima Marques M (2016) Thermal and catalytic pyrolysis of plastic waste. Polimeros 26:44–51

Álvarez-Noriega M, Burgess SC, Byers JE, Pringle JM, Wares JP, Marshall DJ (2020) Global biogeography of marine dispersal potential. Nat Ecol & Evol 4:1196–1203

Andreasi Bassi S, Boldrin A, Faraca G, Astrup TF (2020) Extended producer responsibility: how to unlock the environmental and economic potential of plastic packaging waste? Resour Conserv Recycl 162:105030. https://doi.org/10.1016/j.resconrec.2020.105030

Aryan Y, Yadav P, Samadder SR (2019) Life Cycle Assessment of the existing and proposed plastic waste management options in India: a case study. J Clean Prod 211:1268–1283
Baalkhuyur FM, Qurban MA, Panickan P, Duarte CM (2020) Microplastics in fishes of commercial and ecological importance from the western Arabian Gulf. Mar Pollut Bull 152:110920. https://doi.org/10.1016/j.marpolbul.2020.110920
Barboza LGA, Cunha SC, Monteiro C, Fernandes JO, Guilhermino L (2020a) Bisphenol A and its analogs in muscle and liver of fish from the North East Atlantic Ocean in relation to microplastic contamination. Exposure and risk to human consumers. J Hazard Mater 393:122419. https://doi.org/10.1016/j.jhazmat.2020.122419
Barboza LGA, Lopes C, Oliveira P, Bessa F, Otero V, Henriques B, Raimundo J, Caetano M, Vale C, Guilhermino L (2020b) Microplastics in wild fish from North East Atlantic Ocean and its potential for causing neurotoxic effects, lipid oxidative damage, and human health risks associated with ingestion exposure. Sci Total Environ 717:134625. https://doi.org/10.1016/j.scitotenv.2019.134625
Barcelo D, Pico Y (2020) Case studies of macro-and microplastics pollution in coastal waters and rivers: is there a solution with new removal technologies and policy actions? Case Stud Chem Environ Eng 2:100019
Barnes DKA, Galgani F, Thompson RC, Barlaz M (2009) Accumulation and fragmentation of plastic debris in global environments. Philos Trans R Soc B Biol Sci 364:1985–1998. https://doi.org/10.1098/rstb.2008.0205
Bhaskar K, Turaga RMR (2018) India's e-waste rules and their impact on e-waste management practices: a case study. J Ind Ecol 22:930–942
Bhattacharya RRNS, Chandrasekhar K, Roy P, Khan A (2018) Challenges and opportunities: plastic waste management in India. Energy Resour Inst 24:3–7
Bilal M, Iqbal HMN (2020) Microbial bioremediation as a robust process to mitigate pollutants of environmental concern. Case Stud Chem Environ Eng 2:100011. https://doi.org/10.1016/j.cscee.2020.100011
BIS ISDWS (2012) Bureau of Indian Standards. New Delhi. pp 2–3
Bishop G, Styles D, Lens PNL (2021) Environmental performance comparison of bioplastics and petrochemical plastics: a review of life cycle assessment (LCA) methodological decisions. Resour Conserv Recycl 168:105451
Cappucci GM, Avolio R, Carfagna C, Cocca M, Gentile G, Scarpellini S, Spina F, Tealdo G, Errico ME, Ferrari AM (2020) Environmental life cycle assessment of the recycling processes of waste plastics recovered by landfill mining. Waste Manag 118:68–78. https://doi.org/10.1016/j.wasman.2020.07.048
Castillo Castillo A, Mulakkal M, Bexis P, Brophy K (2020) Enabling a greener plastic future through molecular science
Cecon VS, Da Silva PF, Curtzwiler GW, Vorst KL (2021) The challenges in recycling post-consumer polyolefins for food contact applications: a review. Resour Conserv Recycl 167:105422
Chen CL (2015) Regulation and management of marine litter. In: Bergmann M et al (eds) Marine anthropogenic litter. Springer, Cham, pp 395–428
Chen Y, Awasthi AK, Wei F, Tan Q, Li J (2020) Single-use plastics: production, usage, disposal, and adverse impacts. Sci Total Environ 141772. https://doi.org/10.1016/j.scitotenv.2020.141772
Chen Y, Cui Z, Cui X, Liu W, Wang X, Li X, Li S (2019) Life cycle assessment of end-of-life treatments of waste plastics in China. Resour Conserv Recycl 146:348–357. https://doi.org/10.1016/j.resconrec.2019.03.011
Cleetus C, Thomas S, Varghese S (2013) Synthesis of petroleum-based fuel from waste plastics and performance analysis in a CI engine. J Energy 2013:608797
Critchell K, Bauer-Civiello A, Benham C, Berry K, Eagle L, Hamann M, Hussey K, Ridgway T (2019) Plastic pollution in the coastal environment: current challenges and future solutions. In:

Critchell K (ed) Coasts and estuaries. Elsevier, pp 595–609. https://doi.org/10.1016/B978-0-12-814003-1.00034-4

Danso D, Chow J, Streit WR (2019) Plastics: environmental and biotechnological perspectives on microbial degradation. Appl Environ Microbiol 85:e01095–e01019

Derraik JGB (2002) The pollution of the marine environment by plastic debris: a review. Mar Pollut Bull 44:842–852. https://doi.org/10.1016/S0025-326X(02)00220-5

Deshpande PC, Skaar C, Brattebø H, Fet AM (2020) Multi-criteria decision analysis (MCDA) method for assessing the sustainability of end-of-life alternatives for waste plastics: a case study of Norway. Sci Total Environ 719:137353

Eastman CB, Farrell JA, Whitmore L, Rollinson Ramia DR, Thomas RS, Prine J, Eastman SF, Osborne TZ, Martindale MQ, Duffy DJ (2020) Plastic ingestion in post-hatchling sea turtles: assessing a major threat in Florida near shore waters. Front Mar Sci 7:1–11. https://doi.org/10.3389/fmars.2020.00693

Fadeeva Z, Van Berkel R (2021) Unlocking circular economy for prevention of marine plastic pollution: an exploration of G20 policy and initiatives. J Environ Manage 277. https://doi.org/10.1016/j.jenvman.2020.111457

Franz R (2002) Programme on the recyclability of food-packaging materials with respect to food safety considerations: polyethylene terephthalate (PET), paper and board, and plastics covered by functional barriers. Food Addit Contam 19:93–110

Garaba SP, Aitken J, Slat B, Dierssen HM, Lebreton L, Zielinski O, Reisser J (2018) Sensing ocean plastics with an airborne hyperspectral shortwave infrared imager. Environ Sci Technol 52:11,699–11,707. https://doi.org/10.1021/acs.est.8b02855

Ghatge S, Yang Y, Ahn J-H, Hur H-G (2020) Biodegradation of polyethylene: a brief review. Appl Biol Chem 63:1–14

Godswill AC, Godspel AC (2019) Physiological effects of plastic wastes on the endocrine system (Bisphenol A, Phthalates, Bisphenol S, PBDEs, TBBPA). Int J Bioinforma Comput Biol 4:11–29

Goodship V (2007) Plastic recycling. Sci Prog 90:245–268

Gorycka M (2009) Environmental risks of microplastics. Fac der Aard - en Levenswetenschappen. Vrije Universiteit Amsterdam, Amsterdam, p 171

Gregory MR (2009) Environmental implications of plastic debris in marine settings—entanglement, ingestion, smothering, hangers-on, hitch-hiking and alien invasions. Philos Trans R Soc B Biol Sci 364:2013–2025. https://doi.org/10.1098/rstb.2008.0265

Groh KJ, Backhaus T, Carney-Almroth B, Geueke B, Inostroza PA, Lennquist A, Leslie HA, Maffini M, Slunge D, Trasande L, Warhurst AM (2019) Overview of known plastic packaging-associated chemicals and their hazards. Sci Total Environ 651:3253–3268. https://doi.org/10.1016/j.scitotenv.2018.10.015

Groh KJ, Backhaus T, Carney-Almroth B, Geueke B, Inostroza PA, Lennquist A, Maffini M, Leslie HA, Slunge D, Trasande L, Warhurst M (2018) Chemicals associated with plastic packaging: inventory and hazards. Peer J Prepr 6:e27036v1

Guillard V, Gaucel S, Fornaciari C, Angellier-Coussy H, Buche P, Gontard N (2018) The next generation of sustainable food packaging to preserve our environment in a circular economy context. Front Nutr 5:1–13. https://doi.org/10.3389/fnut.2018.00121

Gustavo JU Jr, Pereira GM, Bond AJ, Viegas CV, Borchardt M (2018) Drivers, opportunities and barriers for a retailer in the pursuit of more sustainable packaging redesign. J Clean Prod 187:18–28

Hahladakis JN, Velis CA, Weber R, Iacovidou E, Purnell P (2018) An overview of chemical additives present in plastics: migration, release, fate and environmental impact during their use, disposal and recycling. J Hazard Mater 344:179–199. https://doi.org/10.1016/j.jhazmat.2017.10.014

Hanisch C (2000) Is extended producer responsibility effective? Environ Sci Technol 34:170A–175A

Heckman JH (2005) Food packaging regulation in the United States and the European Union. Regul Toxicol Pharmacol 42:96–122

Hohn S, Acevedo-Trejos E, Abrams JF, de Moura JF, Spranz R, Merico A (2020) The long-term legacy of plastic mass production. Sci Total Environ 746:141115. https://doi.org/10.1016/j.scitotenv.2020.141115

Hopewell J, Dvorak R, Kosior E (2009) Plastics recycling: challenges and opportunities. Philos Trans R Soc B Biol Sci 364:2115–2126. https://doi.org/10.1098/rstb.2008.0311

Irwin A (2018) How to solve a problem like plastics. New Sci 238:25–31

Jaiswal S, Shukla P (2020) Alternative strategies for microbial remediation of pollutants via synthetic biology. Front Microbiol 11:1–14. https://doi.org/10.3389/fmicb.2020.00808

Jambeck JR, Geyer R, Wilcox C, Siegler TR, Perryman M, Andrady A, Narayan R, Law KL (2015) Plastic waste inputs from land into the ocean. Science 347:768–771. https://doi.org/10.1126/science.1260352

Jang YC, Lee J, Hong S, Choi HW, Shim WJ, Hong SY (2015) Estimating the global inflow and stock of plastic marine debris using material flow analysis: a preliminary approach. J Korean Soc for Mar Environ Energy 18(4):263–273

Jie H, Jiaxiang H (2018) Can free trade agreements enhance MARPOL 73/78 compliance. Tul Mar LJ 43:59

Karbalaei S, Golieskardi A, Hamzah HB, Abdulwahid S, Hanachi P, Walker TR, Karami A (2019) Abundance and characteristics of microplastics in commercial marine fish from Malaysia. Mar Pollut Bull 148:5–15. https://doi.org/10.1016/j.marpolbul.2019.07.072

Karim MS (2010) Implementation of the MARPOL convention in developing countries. Nord J Int Law 79:303–337

Kershaw P, Katsuhiko S, Lee S, Woodring D (2011) Plastic debris in the ocean. United Nations Environment Programme, pp. 20–33

Khanna M (2001) Non-mandatory approaches to environmental protection. J Econ Surv 15:291–324

Koch HM, Calafat AM (2009) Human body burdens of chemicals used in plastic manufacture. Philos Trans R Soc B Biol Sci 364:2063–2078

Kosior E, Crescenzi I (2020) Solutions to the plastic waste problem on land and in the oceans. Elsevier

Kühn S, Bravo Rebolledo EL, van Franeker JA (2015) Deleterious effects of litter on marine life. In: Bergmann M, Gutow L, Klages M (eds) Marine anthropogenic litter. Springer, Cham. https://doi.org/10.1007/978-3-319-16510-3_4

Kumari A, Chaudhary DR, Jha B (2021) Microbial degradation of plastics and its biotechnological advancement. In: Gothandam K et al (eds) Environmental biotechnology, vol 3. Springer, Cham, pp 1–30

Laina E (2018) Expert group identifies barriers and response options on marine litter and microplastics. Environ Policy Law 48:168–171. https://doi.org/10.3233/EPL-180071

Law KL (2017) Plastics in the marine environment. Ann Rev Mar Sci 9:205–229. https://doi.org/10.1146/annurev-marine-010816-060409

Leary D (2016) International Maritime Organization (IMO). Yearb Int Environ Law 27:459–462

Li WC, Tse HF, Fok L (2016) Plastic waste in the marine environment: a review of sources, occurrence and effects. Sci Total Environ 566–567:333–349. https://doi.org/10.1016/j.scitotenv.2016.05.084

Litter BM (2005) UNEP Regional Seas programme, marine litter and abandoned fishing gear

UNEP (2015) Biodegradable plastics and marine litter. Misconceptions, concerns and impacts on marine environments. United Nations Environment Programme (UNEP), Nairobi

Liu F, Nord NB, Bester K, Vollertsen J (2020) Microplastics removal from treated wastewater by a biofilter. Water (Switzerland) 12:1085. https://doi.org/10.3390/W12041085

Löhr A, Savelli H, Beunen R, Kalz M, Ragas A, Van Belleghem F (2017) Solutions for global marine litter pollution. Curr Opin Environ Sustain 28:90–99. https://doi.org/10.1016/j.cosust.2017.08.009

Madsen EL (1991) Determining in situ biodegradation. Environ Sci Technol 25:1662–1673. https://doi.org/10.1021/es00022a001
Malankowska M, Echaide-Gorriz C, Coronas J (2021) Microplastics in marine environment: a review on sources, classification, and potential remediation by membrane technology. Environ Sci Water Res Technol 7:243–258. https://doi.org/10.1039/d0ew00802h
Marciniak KJ (2017) New implementing agreement under UNCLOS: a threat or an opportunity for fisheries governance? Mar Policy 84:320–326. https://doi.org/10.1016/j.marpol.2017.06.035
Marsh K, Bugusu B (2007) Food packaging-roles, materials, and environmental issues. J Food Sci 72:R39–R55
Masi F, Rizzo A, Regelsberger M (2018) The role of constructed wetlands in a new circular economy, resource oriented, and ecosystem services paradigm. J Environ Manage 216:275–284
Mendenhall E (2018) Oceans of plastic: a research agenda to propel policy development. Mar Policy 96:291–298
Mishra S, Charan Rath C, Das AP (2019) Marine microfiber pollution: a review on present status and future challenges. Mar Pollut Bull 140:188–197
Mishra S, Lin Z, Pang S, Zhang W, Bhatt P, Chen S (2021) Recent advanced technologies for the characterization of xenobiotic-degrading microorganisms and microbial communities. Front Bioeng Biotechnol 9:31. https://doi.org/10.3389/fbioe.2021.632059
Moharam R, Maqtari MA (2014) The impact of plastic bags on the environment: a field survey of the City of Sana'a and the surrounding Areas, Yemen. Int J Eng Res Rev 2(4):61–69
Monteiro RCP, do Sul JAI, Costa MF (2018) Plastic pollution in islands of the Atlantic Ocean. Environ Pollut 238:103–110
Panda AK, Singh RK, Mishra DK (2010) Thermolysis of waste plastics to liquid fuel: a suitable method for plastic waste management and manufacture of value added products: a world prospective. Renew Sustain Energy Rev 14:233–248
Peng L, Fu D, Qi H, Lan CQ, Yu H, Ge C (2020) Micro-and nano-plastics in marine environment: source, distribution and threats: a review. Sci Total Environ 698:134254. https://doi.org/10.1016/j.scitotenv.2019.134254
Perugini F, Mastellone ML, Arena U (2005) A life cycle assessment of mechanical and feedstock recycling options for management of plastic packaging wastes. Environ Prog 24:137–154. https://doi.org/10.1002/ep.10078
Possatto FE, Barletta M, Costa MF, do Sul JA, Dantas DV (2011) Plastic debris ingestion by marine catfish: an unexpected fisheries impact. Mar Pollut Bull 62:1098–1102. https://doi.org/10.1016/j.marpolbul.2011.01.036
Prakash R, Gowtham M (2018) Hazardous waste and its treatment process. Elsevier Ltd.
Prasad PM (2006) Environment protection: role of regulatory system in India. Econ Polit Wkly:1278–1288
Raha UK, Kumar BR, Sarkar SK (2021) Policy framework for mitigating land-based marine plastic pollution in the gangetic delta region of Bay of Bengal: a review. J Clean Prod 278:123409. https://doi.org/10.1016/j.jclepro.2020.123409
Rahimi AR, Garciá JM (2017) Chemical recycling of waste plastics for new materials production. Nat Rev Chem 1:1–11. https://doi.org/10.1038/s41570-017-0046
Raubenheimer K, McIlgorm A (2017) Is the Montreal Protocol a model that can help solve the global marine plastic debris problem? Mar Policy 81:322–329. https://doi.org/10.1016/j.marpol.2017.04.014
Raubenheimer K, Urho N (2020) Rethinking global governance of plastics—the role of industry. Mar Policy 113:103802. https://doi.org/10.1016/j.marpol.2019.103802
Rebitzer G, Ekvall T, Frischknecht R, Hunkeler D, Norris G, Rydberg T, Schmidt WP, Suh S, Weidema BP, Pennington DW (2004) Life cycle assessment: part 1: framework, goal and scope definition, inventory analysis, and applications. Environ Int 30:701–720. https://doi.org/10.1016/j.envint.2003.11.005
Rochman CM, Hoh E, Kurobe T, Teh SJ (2013) Ingested plastic transfers hazardous chemicals to fish and induces hepatic stress. Sci Rep 3:1–7. https://doi.org/10.1038/srep03263

Rochman CM, Tahir A, Williams SL, Baxa DV, Lam R, Miller JT, Teh FC, Werorilangi S, Teh SJ (2015) Anthropogenic debris in seafood: plastic debris and fibers from textiles in fish and bivalves sold for human consumption. Sci Rep 5:1–10. https://doi.org/10.1038/srep14340
Rosenfeld PE, Feng LGH (2011) The export of hazardous waste. Risks Hazard Wastes:169–184. https://doi.org/10.1016/b978-1-4377-7842-7.00013-1
Sarkodie SA, Owusu PA (2020) Impact of COVID-19 pandemic on waste management. Environ Dev Sustain:1–10
Schmaltz E, Melvin EC, Diana Z, Gunady EF, Rittschof D, Somarelli JA, Virdin J, Dunphy-Daly MM et al (2020) Plastic pollution solutions: emerging technologies to prevent and collect marine plastic pollution. Environ Int 144:106,067. https://doi.org/10.1016/j.envint.2020.106067
Schwabl P, Köppel S, Königshofer P, Bucsics T, Trauner M, Reiberger T, Liebmann B (2019) Detection of various microplastics in human stool: a prospective case series. Ann Intern Med 171:453–457. https://doi.org/10.7326/M19-0618
Shahnawaz M, Sangale MK, Ade AB (2019) Policy and legislation/regulations of plastic waste around the globe. Bioremediation Technol Plastic Waste:113–126. https://doi.org/10.1007/978-981-13-7492-0_11
Shen M, Huang W, Chen M, Song B, Zeng G, Zhang Y (2020) (Micro) plastic crisis: un-ignorable contribution to global greenhouse gas emissions and climate change. J Clean Prod 254:120138. https://doi.org/10.1016/j.jclepro.2020.120138
Simon N, Schulte ML (2017) Stopping global plastic pollution: the case for an international convention. Heinrich-Böll-Stiftiftung
Singh J, Laurenti R, Sinha R, Frostell B (2014) Progress and challenges to the global waste management system. Waste Manag Res 32:800–812
Tessnow-von Wysocki I, Le Billon P (2019) Plastics at sea: treaty design for a global solution to marine plastic pollution. Environ Sci Policy 100:94–104. https://doi.org/10.1016/j.envsci.2019.06.005
Thompson RC, Moore CJ, Saal FSV, Swan SH (2009) Plastics, the environment and human health: current consensus and future trends. Philos Trans R Soc B Biol Sci 364:2153–2166. https://doi.org/10.1098/rstb.2009.0053
Thushari GGN, Senevirathna JDM (2020) Plastic pollution in the marine environment. Heliyon 6: e04709. https://doi.org/10.1016/j.heliyon.2020.e04709
Tiller R, Nyman E (2018) Ocean plastics and the BBNJ treaty—is plastic frightening enough to insert itself into the BBNJ treaty, or do we need to wait for a treaty of its own? J Environ Stud Sci 8:411–415. https://doi.org/10.1007/s13412-018-0495-4
Turner DA, Williams ID, Kemp S (2016) Combined material flow analysis and life cycle assessment as a support tool for solid waste management decision making. J Clean Prod 129:234–248. https://doi.org/10.1016/j.jclepro.2016.04.077
Vanapalli KR, Samal B, Dubey BK, Bhattacharya J (2019) Emissions and environmental burdens associated with plastic solid waste management. In: Plastics to energy. Elsevier, pp 313–342
Vilaplana F, Karlsson S (2008) Quality concepts for the improved use of recycled polymeric materials: a review. Macromol Mater Eng 293:274–297. https://doi.org/10.1002/mame.200700393
Villarrubia-Gómez P, Cornell SE, Fabres J (2018) Marine plastic pollution as a planetary boundary threat—the drifting piece in the sustainability puzzle. Mar Policy 96:213–220. https://doi.org/10.1016/j.marpol.2017.11.035
Vince J, Hardesty BD (2017) Plastic pollution challenges in marine and coastal environments: from local to global governance. Restor Ecol 25:123–128. https://doi.org/10.1111/rec.12388
Wang B, Sharma-Shivappa RR, Olson JW, Khan SA (2013) Production of polyhydroxybutyrate (PHB) by Alcaligenes latus using sugarbeet juice. Ind Crops Prod 43:802–811. https://doi.org/10.1016/j.indcrop.2012.08.011

Steer M, Thompson RC (2020) Plastics and microplastics: impacts in the marine environment. In: Mare plasticum-the plastic sea. Springer, Cham, pp 49–72. https://doi.org/10.1007/978-3-030-38945-1_3

Willis K, Maureaud C, Wilcox C, Hardesty BD (2018) How successful are waste abatement campaigns and government policies at reducing plastic waste into the marine environment? Mar Policy 96:243–249. https://doi.org/10.1016/j.marpol.2017.11.037

Wu H (2020) A study on transnational regulatory governance for marine plastic debris: trends, challenges, and prospect. Mar Policy:103988. https://doi.org/10.1016/j.marpol.2020.103988

Wu WM, Yang J, Criddle CS (2017) Microplastics pollution and reduction strategies. Front Environ Sci Eng 11. https://doi.org/10.1007/s11783-017-0897-7

Yee MSL, Hii LW, Looi CK, Lim WM, Wong SF, Kok YY, Tan BK, Wong CY, Leong CO (2021) Impact of microplastics and nanoplastics on human health. Nanomaterials 11:496. https://doi.org/10.3390/nano11020496

Zaman A, Newman P (2021) Plastics: are they part of the zero-waste agenda or the toxic-waste agenda? Sustain Earth 4:1–16

Zimmermann L, Dierkes G, Ternes TA, Völker C, Wagner M (2019) Benchmarking the in vitro toxicity and chemical composition of plastic consumer products. Environ Sci Technol 53:11,467. https://doi.org/10.1021/acs.est.9b02293

Biodegradation of the Macroplastic Waste Using Microbial Approach

7

Lakshmi Mohan, Elsa Cherian, Jobil J. Arackal, and T. Jayasree Joshi

Abstract

The quantity of plastic debris entering the ocean per annum is growing at an alarming rate. Synthetic plastic waste, both macro- and microplastics, enter the marine environment from fishing, coastal tourism, seafood and other marine industries and other plastic products. Plastic pollution has a drastic effect on all aquatic life. The conventional plastics which turn up in seas and oceans are recalcitrant to biodegradation and end up being around for decades and centuries. Marine biota is attracted to plastic due to its colour and odour and through the algae that develop films on floating plastics which are a significant source of food for marine animals. The most obvious and disturbing impact of pollution of the marine ecosystem with macroplastics is the ingestion, suffocation and subsequent death of hundreds of marine species. Bioremediation is a useful strategy for the control of plastic pollution in water bodies. The microbes which live in the vicinity of plastic waste adapt and grow on the surface of plastic as biofilms. They produce catalytic enzymes which can degrade the plastic. However, the extent of biodegradation of the plastic will depend upon its structure and chemical properties. This chapter deals with the biodegradation of macroplastic waste utilising various microbes and the challenges associated with the approach.

Keywords

Plastic · Recalcitrant · Macroplastics · Bioremediation · Microbes

L. Mohan (✉) · E. Cherian · J. J. Arackal · T. Jayasree Joshi
Department of Food Technology, Saintgits College of Engineering, Kottukulam Hills, Pathamuttam, Kerala, India
e-mail: lakshmi.mohan@saintgits.org

M. Shahnawaz et al. (eds.), *Impact of Plastic Waste on the Marine Biota*,
https://doi.org/10.1007/978-981-16-5403-9_7

7.1 Introduction

Plastics are produced from cellulose, coal, natural gas, crude oil, etc. by chemical processes. The properties of the plastics, such as durability, malleability, reduced cost and strength, increase the broad applications of plastics (Hopewell et al. 2009). Also, these properties permit it to be extruded in various shapes and size. The primary steps in plastic production include polymerisation and polycondensation. In both reactions, monomeric units are joined together in the presence of specific catalysts. These polymeric molecules can be classified mainly into two categories based on temperature, reaction and strength. Thermoplastics are those which will become soft on applying temperature and hardened on reversing the condition. But thermosets are plastics which will never become soft once they are moulded. Thermoplastics include polyethylene, polycarbonate, etc., and thermosets include epoxide, polyurethane, etc. Also, based on the type of polymeric material selected, the plastics can be divided into seven types such as polyethylene, high-density polyethylene, low-density polyethylene, polyethylene terephthalate, polyvinyl chloride, polystyrene and polypropylene. All these types are widely used in different ways (Jambeck et al. 2015).

The utilisation of plastics in human lives is ever increasing. The production rate of plastics is also in a positive direction due to the increased demand of plastic materials. But these plastics are getting accumulated in landfills and water bodies, affecting the survival of life in the worst manner. In most cases, these plastics are consumed by animals and fishes and stored in their bodies and affect their life cycle. The entry of plastics in micro and nano form into the food chain has very harmful health effects. In most cases, plastic particles from land sources finally end up in the ocean differently. The water movement causes all the plastic materials to dump into specific areas in oceans, causing garbage patches (Lebreton et al. 2018).

Plastic recycling and chemical modification of plastics will not provide a permanent solution to the current situation. And to alleviate the severe issues regarding the accumulation of plastics, different biological solutions have to be applied. Biodegradation of plastics is one of the long-lasting solutions to this problem. Exposure of the microbes to the plastics or plastic-contaminated place forms a thick layer of biofilms on the surface upon adaption of the environment (Juwarkar et al. 2010). The enzymes released by the microbes help in the degradation of the polymeric structures. Even algae also play a significant role in the degradation of plastics. They decompose the polymeric chain by releasing toxins or enzymes produced by them or using the plastic chains as the carbon source.

Many factors affect the biodegradation of plastics. The environmental factors such as temperature, pH and salinity greatly influence the degradation rate of plastics. However, the effect of thermal and light in marine conditions is lower (Ho et al. 1999). Along with the environmental factors, the nature of the polymer chain, such as increased molecular weight, complex structure, hydrophobicity, etc., will affect the degradation rate (Hadad et al. 2005). The chapter mainly revolves around the effect of plastic on the environment and the degradation of the same.

7.2 Plastics in the Environment

Plastics, which are organic polymers of synthetic origin, have become an inevitable component in everyday life due to their workability, lightweight, versatility, stability and economic aspects. Plastic products have become an indispensable component in many applications. Post consumption waste along with the utility grid will further end up in landfills and water bodies (Thompson et al. 2009). According to Eriksen et al. (2016), around five trillion plastic waste, weighing more than 260,000 tonnes, is hovering over the ocean surfaces globally, which has become a serious concern. According to the size range, plastic debris is categorised into microplastics, ≤5 mm; mesoplastics, ranging from 5 mm to 2.5 cm; and macroplastics that are >2.5 cm (Cheshire et al. 2009). The most frequent polymers found in the marine environment include polyolefins, polyesters, polystyrene, polyvinyl chloride and polyamides. The following segments discuss the features of the polymers commonly found in the aquatic environment.

7.2.1 Polyolefins

Polyolefins are the most commonly used synthetic polymers, and they include low-density polyethylene (LDPE), linear low-density polyethylene (LLDPE), high-density polyethylene (HDPE) and polypropylene (PP). HDPE is a linear polymer with few side branches, whereas LDPE has many side-chain units (Piergiovanni and Limbo 2016). LDPE is soft, transparent and flexible compared to HDPE, which has improved properties like high hardness, structural properties and chemical inertness. Commonly used grocery bags, wraps, flexible containers and squeezable bottles are made up of LDPE. Polyethylene films are widely used in packaging materials. Most of the bottles for liquid food products, detergents and other similar containers are made up of HDPE (Marsh and Bugusu 2007). Polypropylene, another essential member of polyolefin family, is also used in large amounts by different industries in its uniaxial, biaxial and non-oriented forms (Himma et al. 2016; Piergiovanni and Limbo 2016). Polypropylene is transparent and chemically inert and has high crystallinity, heat distortion temperature and dimensional stability (Shubhra et al. 2013). Polyolefins are one of the most common materials used in fishing gear applications (Timmers et al. 2005). Erni-Cassola et al. found that polypropylene and polyethene waste products were found more in sea surface samples (25% and 42%, respectively) as their density is very low compared to other polymers. They also found that the abundance of polypropylene (PP) and polyethylene (PE) in the water column is lower (less than 5%) (Erni-Cassola et al. 2019).

7.2.2 Polyesters

Synthetic polyesters commonly include polyethylene terephthalate (PET), polyethylene naphthalate, polybutylene terephthalate and polycarbonates. Polyesters are manufactured by reacting organic alcohol with a carboxylic acid. PET is the most extensively used polyester and is exploited in the food packaging industry to package liquid products due to its good physicomechanical and adequate moisture-gas barrier properties (Bandi et al. 2005). Erni-Cassola et al. found that polymers like polyesters and acrylics are found more in deep-sea locations (77% in marine areas) than surface seawater (5%). A range of biodegradable polyesters like polyhydroxyalkanoates, polycaprolactone, polylactide and polyalkylene dicarboxylic acids have been developed to reduce environmental impacts (Erni-Cassola et al. 2019). The amount of biodegradation is dependent on the material processing conditions, the inherent characteristics of the substrate and the range of microbial and environmental factors (Kim et al. 2003; Block et al. 2017).

7.2.3 Polystyrene (PS)

Polystyrene is a durable thermoplastic formed by the polymerisation reaction of styrene monomer units. General-purpose polystyrene, high impact polystyrene, polystyrene foam and expanded polystyrene are their different variants. Polystyrene finds its application in insulation panel, packaging, storage boxes, cutleries, automobile parts, etc. (Block et al. 2017; Sulong et al. 2019). Polystyrene has good strength, insulating properties, excellent optical clarity and chemical stability. Foamed polystyrene represents an essential constituent of polymer waste. According to European data for 2016–2017, the foamed PS waste generation from the construction and packaging industry alone constitute about 530,000 tonnes (Lassen et al. 2019).

7.2.4 Polyvinyl Chloride (PVC)

PVC is a commonly used polymer formed by the polymerisation of vinyl chloride monomer units. They have an extensive array of packaging, electronics, and healthcare industry applications due to reduced cost, stability and physicomechanical properties (Doble and Kumar 2005; Bueno-Ferrer et al. 2010; Reddy et al. 2010). The high contents of chlorine and other additives added while manufacturing polymers poses a significant threat to the environment (Akovali 2012; Glas et al. 2014). The majority of the pipes and fittings, valves, floats, drums, cans, prawn shelters, fish handling crates, etc. used in the fishing sector are made up of PVC, and a significant part of this also ends up in water bodies.

7.2.5 Polyamide (PA)

Polyamides (PAs) are thermoplastic polymers consisting of recurring amide bonds, which are commonly branded under the name 'Nylon'. Nylon 6 and nylon 66 constitute a significant share of industrial polyamide outlay (Diamond et al. 2014). Polyamides find their application in the textile, automotive, electronics and electrical industries due to their mechanical properties, high crystalline melting point, chemical resistance, flexibility and flow characteristics (Rulkens and Koning 2012). Polyamide is an essential synthetic polymer used for the manufacture of fishing gears. Multifilament twisted yarn and single monofilament yarn made up of polyamides are used for netting purpose. Most of the fishing nets are discarded after the fishing season, leading to environmental hazards (Mondal and Manojkumar 2018).

7.3 Effects of Plastics on Marine Biota

Plastic remains have both a direct and an indirect effect on marine biota and ecosystems. Proper knowledge of the impact of plastic waste is indispensable for formulating remedial measures. However, the literature on the effects of plastic debris on marine biota and ecosystems is comparatively less. Hence, efforts have to be made to strengthen the research in this arena (Maximenko et al. 2019; Fauziah et al. 2021). The impact of marine plastic litter on aquatic biota is depicted in Fig. 7.1.

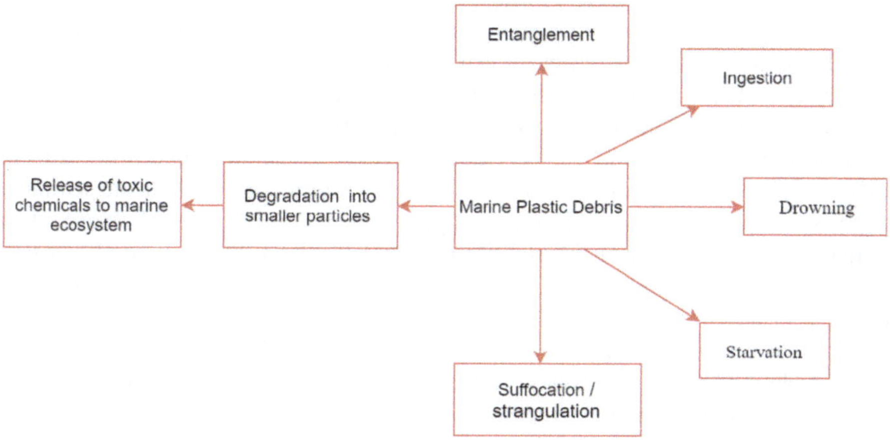

Fig. 7.1 Effect of marine plastic litter on marine biota

7.3.1 Entanglement

Marine plastic debris entanglement is a global issue affecting a significant number of marine animals. Entanglement happens when the openings and loops of debris entrap the marine species. This can cause injuries, damaged organelles, decreased swimming ability and disruption in feeding, reduce their capabilities and even lead to the animal's death by drowning, suffocation or strangulation (Laist 1997; Moore 2008). Many marine species, including birds, turtles, fish, crabs and marine mammals, are entangled worldwide. It has been reported that 92% of entanglement and ingestion cases are due to plastic debris. Derelict fishing gear, also called ghost gear, refers to any discarded or abandoned equipment or related materials used for fishing in the marine environment. They constitute the primary sources of marine entanglement (Cole et al. 2011). Derelict fishing gear includes nets fragments, fishing lines and lures, bait boxes, ropes and bands for strapping (Woodley 2002; Allsopp et al. 2008). Estimates showthat ghost fishing gear alone constitutes about ten percentage of marine waste (Macfadyen et al. 2009). It is possible to reduce the problems caused by the fishing gear to an extent by providing proper guidance to fishers, recycling fishing gears and using biodegradable fishing gear alternatives.

7.3.2 Ingestion

Marine debris ingestion is another one of the main threats to marine biodiversity. Ingestion of plastic waste by diverse organisms, including those listed as endangered species, has been reported (Wright et al. 2013; Deudero and Alomar 2015; Werner et al. 2016). Marine organisms ingest plastic as they find plastic items similar to potential preys. Marine litter also reaches the intestine of animals due to incidental ingestion while feeding or through secondary ingestion (Campani et al. 2013; Fossi et al. 2014; Romeo et al. 2016). Plastic ingestion can result in physical damage such as lacerations and lesions, blockage of internal tracts and pseudo-satiation resulting in reduced food intake (Cole et al. 2011; Wright et al. 2013; Kühn et al. 2015). Chemical toxicants associated with marine plastic debris have shown sublethal effects on aquatic biota and affect the species' reproductive cycle and population dynamics (Thompson et al. 2009). Traces of plastic waste are found in the guts of more than 90% of seabirds (Wilcox et al. 2015). And more than half the sea turtle population across the world has ingested plastic debris (Schuyler et al. 2016). High frequencies of ingestion of macro- and microplastics have been reported in fishes (Eriksen et al. 2016; Ory et al. 2018), sea turtles (Nelms et al. 2016), seabirds (Herling et al. 2005; Luna-Jorquera et al. 2012) and marine mammals (Besseling et al. 2015; Lusher et al. 2015). Inadequate management protocol for the disposal and recycling of plastic debris has drastically augmented the scenario. It has become an urgent need to formulate measures to minimise the problem and to protect biodiversity.

7.3.3 Suffocation and Other Threats

Plastic debris can also result in suffocation, drowning, strangulation and starvation of marine animals (Kühn et al. 2015). Plastic waste can accumulate in the stomach of organisms and obstruct the digestive tract, leading to starvation (Nicolau et al. 2016). It can cause injuries and ulcers within the intestine (Kühn et al. 2015; Acampora et al. 2017). Accumulation of plastic within the intestines affects their buoyancy control, leading to drowning (Nelms et al. 2016; Stelfox et al. 2016). Marine litter modifies the composition of the marine atmosphere and negatively affects the balance of the marine ecosystem (Green et al. 2015). Such changes damage the ecosystem's flora and fauna, leading to desiccation of invertebrates and reproductive disorders (Aloy et al. 2011; Carson et al. 2011; Nelms et al. 2016). Studies show that coral reefs are also primarily affected by this plastic litter (Richards and Maria 2011; Gall and Richard 2015).

7.4 Approaches in Biodegradation of Marine Plastic Waste and its Mechanism of Degradation

The biodegradation of plastics is one of the favourable methods for overcoming the issues of plastic pollution (Moog et al. 2019). Changes in the lifestyle of the human population have caused the deposition of a large number of plastics in different forms and size in the marine environs. The entry of these plastics into the food chain in various forms can harmfully affect the health conditions of all living beings. Biodegradation opens up a way to treat these resistant materials. Many different types of microbial lives support the treatment of both water-soluble and water-insoluble polymers. For the application of microbes for the degradation of various substrates, it is necessary to understand the mechanism behind the reaction.

Many believe that the biodegradation of plastic is a complex and time-consuming process. But a detailed study can help solve many environmental problems. The critical milestones for the degradation include microbial biofilm formation, biodeterioration, bio-fragmentation and mineralisation.

7.4.1 Microbial Biofilm Formation

Microbial film formation refers to the sticking or adherence of microbial cells between and to the surface. The attached cells get entrapped within a slimy layer of ECM (Extracellular matrix) (López et al. 2010). The main components of the matrix, such as proteins, lipids, polysaccharides, etc., are released by the cells creating a three-dimensional assembly (Aggarwal et al. 2015). The microbial slime layer will get attached to the surface of the plastics (plastisphere). The growth and multiplication of the microbial population on the plastisphere initiate the degradation of the plastic polymer (Kirstein et al. 2019). The development of microbial flora on the plastisphere cause changes in the intact structure of the polymer by altering hydrophobicity and buoyancy (Urbanek et al. 2018). Many researchers investigated the microbial biofilm formation on different polymers. The structural changes at the micro level and nature of compactness were analysed using a scanning electron microscope and confocal laser scanning microscopy (Miao et al. 2020).

Research indicated that the microbes such as *Alcanivorax borkumensis* play a significant role in forming denser biofilms in marine ecosystems to degrade LDPE (Delacuvellerie et al. 2019). Studies on the accumulation of biofilm on microplastics along riverine water confirmed that the colonising microbe belonging to the phylum *Proteobacteria* initiated the biofilm formation on microplastics. The pathogens belonging to the strain of *Acinetobacter* also showed a significant presence. But the structure of the biofilm showed a considerable variation based on different environmental factors like season, location, ecological features, alkalinity, total organic carbon, TDS, ionic concentration and pH (Yang et al. 2021). The structure of the biofilms varied according to the type of microbes acting on the plastics. A three-dimensional structure in the form of mushroom was moulded by *Rhodococcus ruber* in the biofilm (Sivan et al. 2006).

7.4.2 Biodeterioration

Following the adherence to the plastics' surface, microbes start releasing certain compounds that aid in the deterioration of the polymeric chain. The action of the colonising microbes will alter the physical, chemical and mechanical properties of the plastics. The different polymeric substances released by the microbial biofilm support the firm attachment of biofilm onto the plastic surface. These compounds also help in forming pores through which microbes can penetrate the inner layers of plastics and colonise, thereby initiating cracks that can reduce the physical bonds' strength (Bonhomme et al. 2003). Even the enzymes secreted by microbes such as endoenzymes and exoenzymes play a significant role in the biodeterioration of plastics. Structural analysis of biofilm of polymer-degrading bacteria such as *Rhodococcus ruber* proved that the number of polysaccharides was almost 2.5 times greater than the protein present (Sivan et al. 2006). Apart from enzymes and EPS, some microbes even release acidic compounds like nitrous acid, sulphuric acid, nitric acid, oxalic acid, gluconic acid and glyoxylic acid, such as *Nitrosomonas* spp., *Thiobacillus* spp., *Nitrobacter* spp. etc. Chemolithotrophic and chemoorganotrophic bacteria release acids that can change the pH in the pores, thereby altering the microstructure of the plastic polymer.

7.4.3 Biofragmentation

Biofragmentation breaks down polymeric compounds into simpler ones, thereby making them ready for assimilation. The exoenzymes and endoenzymes released by the microbes can break down complex carbon backbone so that microbes can assimilate the polymers more easily. The primary mechanism supporting the breakdown is adding oxygen into the carbon skeleton so that the polymer can be made less recalcitrant. Some of the enzymes which play a significant role in the current step include amidases, laccases, peroxidases, etc. (Gómez-Méndez et al. 2018).

7.4.4 Assimilation and Mineralisation

The conversion of the polymeric unit into monomers does not state that microbes assimilate it. For assimilation to happen correctly, it has to cross the plasma membrane and undergo oxidation reaction through catabolic pathways. The pathways can be fermentation and anaerobic and aerobic respiration. Assimilation is a process whereby a combination of atoms occurs inside the microorganisms producing different secondary metabolites that will get exocytosed and will be utilised by some other cells. Mineralisation is the final step in degradation where the polymers undergo complete breakdown for the formation of products like carbon dioxide, nitrogen, methane, water, etc. The level of formation of these compounds determines the biodegradation rate of the polymers.

7.5 Potential Microbes in Plastic Degradation

Plastic is a substance that is difficult to destroy and degrade once it has been produced, which is contrary to the rules of nature and thus a disaster for the whole world. Plastic is a broad term for different natural polymers with high molecular weight and is usually derived from various petrochemical merchandises. Most plastics are not biodegradable, and a few are degradable but very slowly. Therefore, plastics persist in nature for a long time without deformation as they are inert and resistant to microbial degradation. Consequently, there was a need to develop biodegradable polymers that degrade quickly upon disposal due to the action of living microbes. These polymeric materials are capacity resources for carbon and strengthen microorganisms such as bacteria and fungi, which may be heterotrophic. Recently, it has been reported that numerous microorganisms produce degradative enzymes that provide by-products after decomposition that are non-toxic not only to nature but also to resident organisms (Table 7.1). It is considered the safest technique of degradation with much less toxic by-products and has the potential for biogeochemical cycling.

Microorganisms have developed over millennia to convert and mineralise numerous chemicals, including xenobiotics, and hence play a crucial part in maintaining numerous environmental processes. They are at the forefront of reducing bioaccumulation of different chemicals by consuming them and recycling them into natural-recyclable molecules. Microbial communities adapt to changing environmental conditions by modifying their genomes to facilitate the inclusion of novel chemicals into their metabolic pathways and therefore into biogeochemical cycles. As a result, microorganisms' ability to adapt to the metabolism of various

Table 7.1 Microbes responsible for degrading of the different kinds of plastic

Sl. No	Type of plastics	Degrading microbe	Reference
1.	Polylactide (PLA) and polyethylene terephthalate (PET)	*S. plymuthica*	Janczak et al. (2020)
2.	HDPE	*Methylobacter* sp.	Muenmee et al. (2015)
3.	Polyphenylene sulphide plastic beads	*Pseudomonas* sp.	Li et al. (2020)
4.	LDPE	*Streptomyces* sp.	Deepika and Jaya (2015)
5.	Natural and Synthetic Polyethylene	*Pseudomonas* sp.	Nanda et al. (2010)
6.	Polythene and Plastic	*Aspergillus glaucus*	Kathiresan (2003)
7.	High-density polyethylene films	*Aspergillus oryzae*	Konduri et al. (2010)
8.	Disposable plastic films	*Streptomyces strains, M. rouxii NRRL 1835 and Aspergillus flavus*	El-Shafei et al. (1998)
9.	Polythene carry bags	*Serratia marcescens*	Aswale and Ade (2009)
10.	Low-density polythene and polythene	*Pseudomonas stutzeri*	Sharma and Sharma (2004)
11.	Polythene	*Aspergillus terreus* strain MANGF1/WL and *Aspergillus sydowii* strain PNPF15/TS	Sangale et al. (2019)
12.	Polythene	*Lysinibacillus fusiformis* strain VASB14/WL and *Bacillus cereus* strain VASB1/TS	Shahnawaz et al. (2016)
13.	Poly(ethylene terephthalate)	*Ideonella sakaiensis* 201-F6	Yoshida et al. (2016)

anthropogenic substances has been discovered to be dependent on natural selection of mutants with the required degradative enzymes but less specific substrate specificities and possibly unique metabolic pathways.

7.5.1 Bacteria

Bacteria are said to be the 'engine' of the Earth's nutrition supply since they are at the forefront of nutrient transformation and cycling. Their involvement in decomposition, like that of other microbes, guarantees that carbon and nutrients are liberated from a variety of complex polymers, both natural and manmade in origin. They've been investigated for their important function in bioremediation, and they've been demonstrated to break down a variety of pollutants utilised in the Anthropocene era, including antibiotics, metal compounds, petroleum, plastics and other substances. Diverse bacterial species from the genera *Pseudomonas*, *Escherichia* and *Bacillus* have been demonstrated to have great ability to break down plastic polymers using various methodologies such as metagenomics, cloning, pure cultures and even computational methods (Gan and Zhang 2019). These plastic-degrading bacteria have been found in a variety of biological environments, including landfills, recycling sites (Yoshida et al. 2016) and cold marine environments (Urbanek et al. 2018). Because bacteria's capacity to dissolve plastic is based on their inherent propensity to digest long-chain fatty acids, it's no surprise that *Pseudomonas* is the most well-studied and well-known bacterial genus in terms of plastic polymer breakdown (Wilkes and Aristilde 2017). Biofilm formation has been discovered to play a significant role in the bacterial breakdown of plastics by enhancing colony adherence and persistence on the plastic surface (Puglisi et al. 2019). Thermoplastics are known to be more resistant to microbial biodegradation due to their homopolymeric composition. The ability of *I. sakaiensis*, on the other hand, is one of the most significant results on plastic breakdown. *I. sakaiensis*, a new species identified from a group of landfill bacteria, was able to destroy PET because it utilised it as a primary source of energy and carbon (Yoshida et al. 2016). Other common thermoplastics destroyed by *Pseudomonas aeruginosa*, *Bacillus megaterium*, *Rhodococcus ruber*, *Serratia marcescens*, *Staphylococcus aureus*, *Streptococcus pyogenes* and other bacterial species include polystyrene and polycarbonate (Ho et al. 2018; Arefian et al. 2020).

Similarly, bacteria from several groups, such as *Bacillus*, *Pseudomonas* and *Micrococcus*, have been found to destroy thermosetting polymers, primarily polyurethane (Espinosa et al. 2020). Although most studies have focused on the biodegradability of single bacterial strains, bacteria in nature frequently work together in consortia, as has been established in a number of investigations (Lwanga et al. 2018). Some bacteria, such as *Pseudomonas aeruginosa*, *Burkholderia seminalis* and *Stenotrophomonas pavanii*, demonstrated an improvement in the rate of PE breakdown when certain additives, such as food-grade dye-sensitised nanoparticles and starch, were added (Mehmood et al. 2016). Other pretreatment methods have recently been shown to significantly increase the rate of bacterial degradation of various plastic polymers, including the addition of anionic surfactants (Mukherjee et al. 2017), thermal treatment (Savoldelli et al. 2017) and UV pretreatment (Montazer et al. 2018).

7.5.2 Fungi

Fungi, together with bacteria, are the most important organisms for maintaining Earth's biogeochemical cycles and critical nutrients. Because of their ability to use synthetic polymers as a primary/alone carbon or energy source, numerous fungus species have been identified as having the potential to break down diverse plastic polymers. A number of fungal strains from various classes, ecologies and morphologies have been shown to destroy plastics in this manner. Most recent investigations have demonstrated that the genus *Aspergillus* is the most dominant fungal group when it comes to the biodegradation of manmade polymers. *A. clavatus* (Gajendiran et al. 2016), *A. fumigatus* and *A. niger* (Osman et al. 2018) are some prominent *Aspergillus* species isolated from various terrestrial habitats that have demonstrate the ability to degrade PE, PU and PP, respectively. Endophytic fungi isolated from diverse plants have been reported to degrade PU to diverse degrees under both solid-state and submerged fermentation conditions in one research (Russell et al. 2011). *Fusarium solani, Alternaria solani, Spicaria* spp., *Geomyces pannorum, Phoma* sp., *Penicillium* spp., etc. are other fungal species with noteworthy plastic degradability (Muhonja et al. 2018; Zhang et al. 2021). The importance of fungal enzymes, particularly de polymerases, has been underlined in all of these research, as it has been in all biological processes. Furthermore, these enzymes' broad specificity, which allows them to break down a variety of polymers, is critical (da Luz et al. 2019). Fungal hyphae's dispersion and penetrating capacity, as well as their ability to release hydrophobins for enhanced attachment of hyphae to hydrophobic substrates, were revealed to be important factors in their initial colonisation prior to eventual depolymerisation (Sánchez 2020). Pretreatment of diverse substrates with diverse parameters such as phototreatment and temperature (Corti et al. 2010), acid pretreatment (Mahalakshmi and Andrew 2012) and different additives (Jeyakumar et al. 2013) has been shown to improve fungal biodegradation of plastics.

7.5.3 Actinomycetes

Actinomycetes are a complex genus of filamentous bacteria found in soil, plant tissues and the sea that are noted for their metabolic versatility and a wide range of biotechnological applications, including bioremediation, medicine and the food industry. *Streptomycetes, Rhodococcus ruber and Actinomadura* spp. are among the actinomycetes, and the thermophilic *Thermoactinomyces* species have been identified from a variety of ecological zones and proven to have substantial plastic biodegradation capacity (Auta et al. 2018; Jabloune et al. 2020). Their ability to create a wide range of hydrolytic enzymes and other bioactive metabolites has already been mentioned (Gohain et al. 2020). These hydrolytic enzymes are one of the most important components in their capacity to grow on different plastic polymers and reduce high molecular weight molecules into simpler ones. They're also known to create extracellular polymers like dextran, glycogen, levans and N-acetylglucosamine-rich mucilage polysaccharides, which help them cling to plastic surfaces for later microbial functions (Pujic et al. 2015). Biofilm production, similar to bacteria, has been proven to be a key element in actinomycetes colonising polymers (Gilan and Sivan 2013). Using an esterase enzyme with a broad substrate specificity, *Streptomyces scabies*, a potato isolate, has been found to degrade PET as well as other polymers such as p-nitrophenyl ester, cutin and suberin (Jabloune et al. 2020). *Nocardiopsis* sp. is an endophytic actinomycete. PE and diesel degradation were also seen in *Hibiscus* isolates (Singh and Sedhuraman 2015). In a microbial consortium with a considerable fraction of actinomycete species degrading polyurethane and several chemical additives (Gaytán et al. 2020), the efficacy of actinomycete plastic degradation was also highlighted.

7.6 Impact of Biotic and Abiotic Factors in Plastic Degradation

The characteristics of the polymer substrate are critical for bacterial or fungal colonisation of the surface. The surface topography can also play a role in the colonisation process. The molecular weight, shape, size and additives of polymers are all distinct factors that can affect biodegradability. Because microbial colonisation relies on surface features that allow microbes to establish a site from which to expand their growth, the molecular weight of a polymer can be quite restrictive. Microbial adhesion to the polymer surface occurs in amorphous parts of the polymer surface and consumes the polymer material; hence, polymer crystallinity may play a significant impact. Polymer additives are low molecular weight organic compounds that, because of their biodegradability, might serve as a starting point for microbial colonisation. The deterioration of most exposed materials is caused by temperature. Moisture in various forms, non-ionising radiation and ambient temperature are all abiotic elements that contribute to these conditions. When wind influences, pollution and atmospheric gases are added, the whole deterioration process becomes quite bendable. Ionising radiation is produced by the ultraviolet (UV) component of the sun spectrum, which plays an important role in activating weathering effects. The weathering process can also be aided by visible and near-infrared radiation. Other components interact with solar radiation in a synergistic manner to dramatically influence weathering processes. The overall impacts are influenced by the quality and quantity of solar radiation, geographic site changes, time of day and season and climatological factors. Both ozone and atmospheric contaminants interact with atmospheric radiation and can cause mechanical stresses such as stiffening and cracking. Microbial colonisation can be aided by the combination of moisture and temperature influences. Biotic contributors can help colonisation by supplying the nutrients needed for microbial growth. Hydrophilic surfaces might be a better choice for colonisation. The degradation process might be started by readily available exoenzymes from the colonised area. Figure 7.2 outlines some factors affecting plastic biodegradation.

7.6.1 Exposure Conditions

Because the complexity of the environment plays a crucial role in the kinetics of biodegradation, microorganisms rely heavily on the initial individual or synergistic impacts of many environmental factors on polymers. Because each habitat has unique properties, the rate of microbial action will differ in a dry environment, wet air, landfill, compost, marine environment and so on. Bond breakdown is aided by factors such as light, heat, humidity, pH and biological activity. They also have an

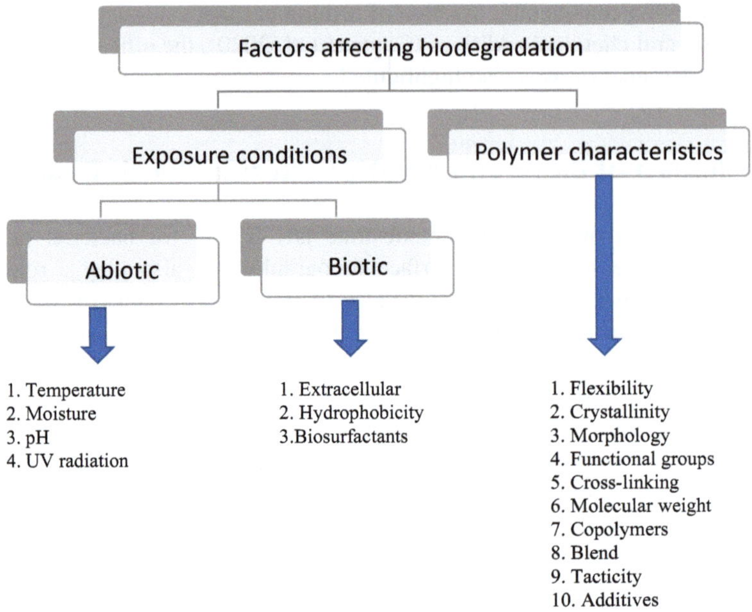

Fig. 7.2 Factors affecting biodegradation of plastics. (Adapted from Kijchavengkul and Auras 2008)

impact on structural homogeneities and the emergence of new functional groups (Siracusa 2019). The presence of moisture in the environment encourages the miniaturisation of plastic polymers by increasing their solubility and hydrolysis rate. As observed by Chamas et al. (2020), more chain scission leads to more microbial targets on polymer chains, resulting in improved biodegradation. Because of their capacity to absorb the greater fraction of tropospheric solar radiation, certain synthetic polymers have been found to be sensitive to electromagnetic radiation. These synthetic polymers absorb high-energy UV radiation, which stimulates their electrons to a higher level of reactivity, resulting in oxidation and cleavage (Brebu 2020). Plastics have also been reported to degrade at high temperatures. Temperatures in landfills have been reported to reach 100 degrees Celsius, a state that accelerates decomposition rates if enough moisture and oxygen are present and available for following thermal oxidative and hydrolytic breakdown pathways (Hao et al. 2017). The kinetic energy of the atoms increases, causing disorder in the polymer structure and molecular scission of the long-chain backbone components (Ray and Cooney 2018). This causes chemical reactions between the various components, resulting in changes in the polymers' physical and optical properties. Thermal degradation, in particular, affects the molecular weights of polymers, reduces ductility and causes embrittlement, initial colour, cracking and other problems. The cleavage of C-C and C H bonds, which occurs as a result of the other environmental factors, causes the polymers to degrade through a series of free radical reactions (Devi et al. 2016).

7.6.2 Characteristics of Polymers

Polymers must be transported across the cell membrane to be metabolised; the rate of microbial degradation has been found to decrease with increasing molecular weight. As a result, smaller polymer units like monomers, dimers and oligomers are more easily degraded and mineralised, as *Rhizopus delemar* lipase has been observed (Tokiwa et al. 2009). The morphology of the polymer, which includes the degree of branching, crystallinity and physical shape, has a significant impact on the rate of degradation of plastic polymers. Microbial breakdown is less absorbed in plastic polymers with a higher proportion of side chains and hence more branching. Non-crystalline polymers are more sensitive to enzymatic degradation because they are loosely packed and accessible, according to studies; hence there is an inverse link between crystallinity and degradation rate (Devi et al. 2016). Under the same conditions, the specific surface degradation rate of polyethylene, a considerably more crystalline polymer, was calculated to be 9.5 m $year^{-1}$ compared to 1105 m $year^{-1}$ of PET in a study comparing the half-lives of different plastic polymers assuming pseudo-Zeroth's order kinetics (Chamas et al. 2020). The melting temperature of a specific polymer, Tm, has a significant impact on its microbial breakdown rate. It was discovered that the T_m and the biodegradation rate had an inverse connection. The change in enthalpy of fusion (H) and the change in melting entropy (S) influence Tm of plastic polymers, as shown in the equation $T_m = H/S$ (Tokiwa et al. 2009). Although there is no evidence on the relationship between the glass transition temperature (T_g) of synthetic plastics and the structural changes occurring at this temperature, it is believed that the structural changes occurring at this temperature are likely to favour microbial infestation (Lucas et al. 2008). The extent of initial microbial colonisation of plastic polymers is increased when the material's hydrophilicity increases, and it's also suggested that higher hydrophobicity inhibits the action of extracellular enzymes. Because of their greater wettability, hydrophilic surfaces have higher surface energy and lower contact angles with water, which encourages microbial adhesion to the polymer surface and speeds up degradation (Chamas et al. 2020). As a result, the presence and development of polar functional groups in plastic polymers due to environmental weathering factors such as UV irradiation have been discovered to result in a decrease in the contact angle with water and hence an increase in hydrophilicity. Molecular dynamics simulations were used to highlight the influence of polymer hydrophilicity/hydrophobicity, with extremely hydrophobic polymeric polypropylene having the lowest biodegradation potential compared to comparably highly hydrophilic nylon (Min et al. 2020).

7.7 Challenges in Macroplastic Biodegradation in Marine Environments

As evident from literature, several microorganisms can at least partially degrade varied kinds of plastic. However, it is rather unfortunate that none of these pathways can be efficiently utilised for the bioremediation of plastics present in the deep seas. Biodegradation of plastic is a process that has immense potential when it is wholly elucidated and, therefore, is an area of research that needs to be prioritised. So far, most attempts at identifying and isolating microbes showed biodegrading activity at temperatures of at least 30 °C. For applying this in the oceans, biodegradability at far lower temperatures is required. Additionally, efficient biofilm formation is needed to colonise the particles before degradation. Genetically engineering an organism that demonstrates good biofilm-forming capability and expresses plastic-degrading enzymes with their maximum activity at low temperatures could be an exciting approach (Schmidt 2017).

Improving the collection and management of solid waste and wastewater appears to be the most robust immediate solution to lowering plastic input, particularly in developing countries. This improvement also leads to social gains in domains such as environmental degradation, human health and financial development. Other matters of concern include improving effluent treatment and decreasing the occurrence of abandoned, lost or otherwise discarded fishing gear (ALDFG). However, the best long-term and sustainable solution will be progressing towards a more circular economy, where there is adequate waste management, and a more balanced consumption pattern. There is enough evidence that marine macro-, micro- and nanoplastics have an unacceptable effect on invoking the precautionary approach. Therefore, we should not wait till there is a clear and quantified evidence of the extent of impact before taking steps to reduce the input of plastic into the oceans. This also needs to be supplemented by an adaptive management approach, which allows for sufficient flexibility to be added to governance frameworks or technical measures to permit modifications as soon as more data becomes accessible.

The focus should be on research areas to better understand the relative consequence of various sources and the fate of marine macro- and microplastics. Filling these knowledge gaps will help guide the efforts directed at reducing plastic debris being disposed in the oceans and lessen the effects of plastic content that is already present.

It is necessary to increase and encourage proper communication of information and expertise, foster a multi-disciplined approach, develop collaborations between the public and private sectors and empower people-led movements. Societies such as the Global Partnerships on Marine Litter (GPML) and Waste Management (GPWM) should be utilised to this end, together with other regional-, national- and global-scale organisations.

Another challenge about plastic biodegradation in marine and terrestrial ecosystems, apart from the limited resources and information about the topic, is knowing how to apply it without producing environmentally adverse by-products.

There exist many analytical and methodical challenges for exploring the interaction between microbes and plastic, their transport and biodegradation pathways in the marine environment. The scientific community accepts the requirement for the development of standard sampling and analytical methods. One particular challenge is resolving and identifying macroplastics scaling from mm to cm and metres in the environment. To apply in field studies, there is a need for standardised and systematic approaches to differentiate macroplastics from the large volume of seawater and sediment in an accurate and effective timeframe. The trouble starts with investigating the plastic and identifying the various polymer types, particle sizes and the distinct phases of degradation. It is also to be noted that all of the methods presently used to detect and isolate plastics from their nearby matrix are very time-consuming (Rogers et al. 2020).

The key challenges in the ecological fate of plastic in marine environments remain the lack of interdisciplinary approaches, understanding the analytical limitations and the proper superimposition of our observations into the ecological context. Future research must improve the awareness of the fate of plastic litter and environmental impact along with necessary information to the public on the environment's macroplastic exposure routes and levels. This data is essential for risk assessment and characterisation and innovative proposals for adequate regulatory measures concerning plastic litter.

7.8 Conclusion and Prospects

Drifting plastic waste has an adverse impact on aquatic species and their ecosystem. We still lack accurate information regarding the sources, quantity, passage, accumulation and the role of plastic debris in the seas. However, the scientific and social awareness of plastic as a universal threat is intensifying. Several actions aim on confronting plastic accumulation by encouraging at consumer, producer, industry and company levels (Urbanek et al. 2018).

Marine plastics are spread from the Arctic to the Antarctic, all over the world's oceans. This prevalence is due to the resilience of plastics, the large-scale nature of potential sources and the action of surface currents which carry drifting plastics. The subtropical gyres in the North and South Atlantic, Indian Ocean and North and South Pacific Ocean have comparatively high concentrations of floating plastics. A higher abundance of macroplastics is seen in coastal waters, especially in regions with high coastal populations but unsatisfactory waste collection and management, rigorous fisheries and high coastal tourism levels. Bigger floating matter is also driven by winds which end up accumulating on mid-ocean islands and shores very distant from the source. Some kinds of plastic are heavier than seawater and will sink once buoyancy is removed. For instance, used up soft drink bottles made of PET plastic are ubiquitous litter objects on shorelines, but they eventually end up on the ocean seafloor. Majority of fishing gear will sink when the floatation buoys are detached.

Engineering the latent capabilities of natural microbial congregations to colonise plastic surfaces is a critical issue in plastic waste management and reduces the extent of plastic debris in the aquatic environment (Syranidou et al. 2019). This chapter presents vital research on the bioremediation of plastic waste by microbes, such as actinomycetes, bacteria and fungi. The mechanism of degradation and the role of the various enzymes involved are also discussed. From literature, it can be concluded that the data on several microbes with potential to degrade plastic is based on pure culture isolates. This indicates that the diversity of microorganisms in different natural habitats such as ocean beds has not been exploited much. We did not find, in particular, any yeast organism which shows potential as a plastic degrader. The use of metagenomics to investigate both culturable and non-culturable microbes will aid in improving the discovery of novel microbes and biocatalysts with plastic-degrading potential. The -omic tools such as genomics, transcriptomics, proteomics and metabolomics can help understand complex biological interactions between genes, transcripts, proteins, metabolites and external ecological factors during biodegradation.

The use of a consortium of microorganisms can bring about better efficiency in plastic degradation due to a synergistic effect between the microbes and enzymes. It will also be advantageous to study the biochemical and structural properties of the enzymes, which will enable modifications by protein engineering tools to facilitate the design of microbial cell factories with enhanced degradation capabilities. Research is also required to study the various pretreatment methods and the effect of additives on the degradation of different kinds of plastics. Taking into consideration the inexhaustible potential of various microorganisms and their continuous evolution with changes in the environment, research in this domain will lead to better and viable biodegradation processes which can be scaled up providing a solution to the world's plastic crisis.

Acknowledgement The authors wish to place on record their gratitude to the editors for giving them an opportunity to contribute to this book.

References

Acampora H et al (2017) Presence of plastic litter in pellets from Great Cormorant (Phalacrocorax carbo) in Ireland. Mar Pollut Bull 117(1-2):512–514

Aggarwal S, Stewart PS, Raymond MJ (2015) Biofilm cohesive strength as a basis for biofilm recalcitrance: are bacterial biofilms overdesigned? Microbiol Insights 8(Suppl 2):29–32

Akovali G (2012) Plastic materials: polyvinyl chloride (PVC). In: Pacheco-Torgal F, Jalali S, Fucic A (eds) Toxicity of building materials. Woodhead, Cambridge, pp 23–53

Allsopp M et al (2008) State of the world's oceans. Springer Science & Business Media, New York

Aloy AB, Vallejo Jr BM, Marie Antonette J (2011) Increased plastic litter cover affects the foraging activity of the sandy intertidal gastropod Nassarius pullus. Mar Pollut Bull 62(8):1772–1779

Arefian M, Tahmourespour A, Mohammadali J (2020) Polycarbonate biodegradation by newly isolated Bacillus strains. Arch Environ Protect:14–20

Aswale PN, Ade AB (2009) Effect of pH on biodegradation of polythene by Serretia marscence. Ecotech 1:152–153

Auta HS et al (2018) Growth kinetics and biodeterioration of polypropylene microplastics by Bacillus sp. and Rhodococcus sp. isolated from mangrove sediment. Mar Pollut Bull 127:15–21

Bandi S et al (2005) The mechanism of color generation in poly (ethylene terephthalate)/polyamide blends. Polym Degrad Stab 88(2):341–348
Besseling E et al (2015) Microplastic in a macro filter feeder: humpback whale Megaptera novaeangliae. Mar Pollut Bull 95(1):248–252
Block C, Brands B, Gude T (2017) Packaging materials 2. Polystyrene for food packaging applications—updated version, ILSI Europe Report Series. ILSI Europe Packaging Materials Task Force, pp 1–36
Bonhomme S et al (2003) Environmental biodegradation of polyethylene. Polym Degrad Stab 81(3):441–452
Brebu M (2020) Environmental degradation of plastic composites with natural fillers—a review. Polymers (Basel) 12(1):166
Bueno-Ferrer C et al (2010) Characterization and thermal stability of poly (vinyl chloride) plasticized with epoxidized soybean oil for food packaging. Polym Degrad Stab 95(11): 2207–2212
Campani T et al (2013) Presence of plastic debris in loggerhead turtle stranded along the Tuscany coasts of the Pelagos Sanctuary for Mediterranean marine mammals (Italy). Mar Pollut Bull 74(1):225–230
Carson HS et al (2011) Small plastic debris changes water movement and heat transfer through beach sediments. Mar Pollut Bull 62(8):1708–1713
Chamas A et al (2020) Degradation rates of plastics in the environment. ACS Sustain Chem Eng 8(9):3494–3511
Cheshire AC et al (2009) UNEP/IOC guidelines on survey and monitoring of marine litter, UNEP regional seas reports and studies. UNEP, Nairobi
Cole M et al (2011) Microplastics as contaminants in the marine environment: a review. Mar Pollut Bull 62(12):2588–2597
Corti A et al (2010) Oxidation and biodegradation of polyethylene films containing pro-oxidant additives: synergistic effects of sunlight exposure, thermal aging and fungal biodegradation. Polym Degrad Stab 95(6):1106–1114
da Luz J, Rodrigues M et al (2019) Plastics polymers degradation by fungi. In: Blumenberg M, Shaaban M, Elgaml A (eds) Microorganisms. IntechOpen. https://doi.org/10.5772/intechopen.88608
Deepika S, Jaya MR (2015) Biodegradation of low density polyethylene by microorganisms from garbage soil. J Exp Biol Agric Sci 3:1–5
Delacuvellerie A et al (2019) The plastisphere in marine ecosystem hosts potential specific microbial degraders including Alcanivorax borkumensis as a key player for the low-density polyethylene degradation. J Hazard Mater 380:120899
Deudero S, Alomar C (2015) Mediterranean marine biodiversity under threat: reviewing influence of marine litter on species. Mar Pollut Bull 98(1-2):58–68
Devi RS et al (2016) The role of microbes in plastic degradation. Environ Waste Manag 341
Diamond GM, Murphy V, Boussie TR (2014) Application of high throughput experimentation to the production of commodity chemicals from renewable feedstocks. In: Hagemeyer A, Volpe AF (eds) Modern applications of high throughput R&D in heterogeneous catalysis, pp 288–309
Doble M, Kumar A (2005) Biotreatment of industrial effluents. Elsevier, Amsterdam
El-Shafei HA et al (1998) Biodegradation of disposable polyethylene by fungi and Streptomyces species. Polym Degrad Stab 62(2):361–365
Eriksen M, Thiel M, Lebreton L (2016) Nature of plastic marine pollution in the subtropical gyres. In: Takada H (ed) Hazardous chemicals associated with plastics in the marine environment. Springer, Cham, pp 135–162
Erni-Cassola G et al (2019) Distribution of plastic polymer types in the marine environment; a meta-analysis. J Hazard Mater 369:691–698
Espinosa MJC et al (2020) Toward biorecycling: isolation of a soil bacterium that grows on a polyurethane oligomer and monomer. Front Microbiol 11:404

Fauziah SH et al (2021) Marine debris in Malaysia: a review on the pollution intensity and mitigating measures. Mar Pollut Bull 167:112,258

Fossi MC et al (2014) Large filter feeding marine organisms as indicators of microplastic in the pelagic environment: the case studies of the Mediterranean basking shark (Cetorhinus maximus) and fin whale (Balaenoptera physalus). Mar Environ Res 100:17–24

Gajendiran A, Krishnamoorthy S, Abraham J (2016) Microbial degradation of low-density polyethylene (LDPE) by Aspergillus clavatus strain Jask1 isolated from landfill soil. 3Biotech 6(1): 52

Gall SC, Richard CJ (2015) The impact of debris on marine life. Mar Pollut Bull 92(1-2):170–179

Gan Z, Zhang H (2019) PMBD: a comprehensive plastics microbial biodegradation database. Database (Oxford) 2019

Gaytán I et al (2020) Degradation of recalcitrant polyurethane and xenobiotic additives by a selected landfill microbial community and its biodegradative potential revealed by proximity ligation-based metagenomic analysis. Front Microbiol 10:2986

Gilan I, Sivan A (2013) Effect of proteases on biofilm formation of the plastic-degrading actinomycete *Rhodococcus ruber* C208. FEMS Microbiol Lett 342(1):18–23

Glas D et al (2014) End-of-life treatment of poly (vinyl chloride) and chlorinated polyethylene by dehydrochlorination in ionic liquids. ChemSusChem 7(2):610–617

Gohain A et al (2020) Actinobacteria: diversity and biotechnological applications. In: De Mandal S, Bhatt P (eds) Recent advancements in microbial diversity. Elsevier, San Diego, pp 217–231

Gómez-Méndez LD et al (2018) Biodeterioration of plasma pretreated LDPE sheets by Pleurotus ostreatus. PLoS One 13(9):e0203786

Green DS et al (2015) Impacts of discarded plastic bags on marine assemblages and ecosystem functioning. PLoS One 49(9):5380–5389

Hadad D, Geresh S, Sivan A (2005) Biodegradation of polyethylene by the thermophilic bacterium *Brevibacillus borstelensis*. J Appl Microbiol 98(5):1093–1100

Hao Z et al (2017) Heat generation and accumulation in municipal solid waste landfills. Environ Sci Technol 51(21):12434–12442

Herling C, Culik BM, Hennicke JC (2005) Diet of the Humboldt penguin (Spheniscus humboldti) in northern and southern Chile. Mar Biol 147(1):13–25

Himma NF et al (2016) Advances in preparation, modification, and application of polypropylene membrane. J Polym Eng 36(4):329–362

Ho BT, Roberts TK, Lucas S (2018) An overview on biodegradation of polystyrene and modified polystyrene: the microbial approach. Crit Rev Biotechnol 38(2):308–320

Ho K-LG et al (1999) Degradation of polylactic acid (PLA) plastic in Costa Rican Soil and Iowa State University compost rows. J Polym Environ 7(4):173–177

Hopewell J, Dvorak R, Kosior E (2009) Plastics recycling: challenges and opportunities. Philos Trans Roy Soc London Ser B Biol Sci 364(1526):2115–2126

Jabloune R et al (2020) Enzymatic degradation of P-nitrophenyl esters, polyethylene terephthalate, cutin, and suberin by Sub1, a suberinase encoded by the plant pathogen *Streptomyces* scabies. Microbes Environ 35(1):ME19086

Jambeck JR et al (2015) Plastic waste inputs from land into the ocean. Science 347(6223):768–771

Janczak K et al (2020) Biodegradation of the plastics PLA and PET in cultivated soil with the participation of microorganisms and plants. Int Biodeter Biodegr 155:105087

Jeyakumar D, Chirsteen J, Doble MJ (2013) Synergistic effects of pretreatment and blending on fungi mediated biodegradation of polypropylenes. Biores Technol 148:78–85

Juwarkar AA, Singh SK, Mudhoo A (2010) A comprehensive overview of elements in bioremediation. Rev Environ Sci Biotechnol 9:215–288

Kathiresan KJ (2003) Polythene and plastics-degrading microbes from the mangrove soil. Rev Biol Trop 51(3-4):629–633

Kijchavengkul T, Auras R (2008) Compostability of polymers. Polym Int 57(6):793–804

Kim DY, Rhee YH (2003) Biodegradation of microbial and synthetic polyesters by fungi. Appl Microbiol Biotechnol 61(4):300–308

Kirstein IV et al (2019) The plastisphere—uncovering tightly attached plastic "specific" microorganisms. PLoS One 14(4):e0215859

Konduri MKR et al (2010) Synergistic effect of chemical and photo treatment on the rate of biodegradation of high density polyethylene by indigenous fungal isolates. Int J Biotechnol Biochem 6(2):157–175

Kühn S, Rebolledo ELB, Van Franeker JA (2015) Deleterious effects of litter on marine life. In: Bergmann M, Gutow L, Klages M (eds) Marine anthropogenic litter. Springer, New York, pp 75–116

Laist DW (1997) Impacts of marine debris: entanglement of marine life in marine debris including a comprehensive list of species with entanglement and ingestion records. In: Coe JM, Rogers D (eds) Marine debris: sources, impacts, and solutions. Springer, New York, pp 99–139

Lassen C et al (2019) Survey of polystyrene foam (Eps and Xps) in the Baltic Sea. COWI A/S, Lyngby

Lebreton L et al (2018) Evidence that the great pacific garbage patch is rapidly accumulating plastic. Sci Rep 8(1):1–15

Li J et al (2020) Rapid biodegradation of polyphenylene sulfide plastic beads by Pseudomonas sp. Sci Total Environ 720:137616

López D, Vlamakis H, Kolter R (2010) Biofilms. Cold Spring Harb Perspect Biol 2(7):a000398

Lucas N et al (2008) Polymer biodegradation: mechanisms and estimation techniques—a review. Chemosphere 73(4):429–442

Luna-Jorquera G et al (2012) Determinants of the diversity of plants, birds and mammals of coastal islands of the Humboldt current systems: implications for conservation. Biodivers Conserv 21(1):13–32

Lusher AL et al (2015) Microplastic and macroplastic ingestion by a deep diving, oceanic Cetacean: the true's beaked whale Mesoplodon mirus. Environ Pollut 199:185–191

Lwanga EH et al (2018) Decay of low-density polyethylene by bacteria extracted from earthworm's guts: a potential for soil restoration. Sci Total Environ 624:753–757

Macfadyen G, Huntington T, Cappell R (2009) Abandoned, lost or otherwise discarded fishing gear. UNEP FAO, Rome

Mahalakshmi V, Andrew SN (2012) Assessment of Physicochemically treated plastic by fungi. Ann Biol Res 3(9):4374–4381

Marsh K, Bugusu B (2007) Food packaging—roles, materials, and environmental issues. J Food Sci 72(3):R39–R55

Maximenko N et al (2019) Toward the integrated marine debris observing system. Front Mar Sci 6: 447

Mehmood CT, et al. (2016) Biodegradation of low density polyethylene (LDPE) modified with dye sensitized titania and starch blend using Stenotrophomonas pavanii. 113:276–286

Miao L et al (2020) Microbial carbon metabolic functions of biofilms on plastic debris influenced by the substrate types and environmental factors. Environ Int 143:106007

Min K, Cuiffi JD, Mathers RT (2020) Ranking environmental degradation trends of plastic marine debris based on physical properties and molecular structure. Nat Commun 11(1):1–11

Mondal S, Manojkumar B (2018) Change in properties of polyamide netting materials exposed to different environments. J Fisheries Soc Taiwan 45(3):145–152

Montazer Z et al (2018) Microbial degradation of UV-pretreated low-density polyethylene films by novel polyethylene-degrading bacteria isolated from plastic-dump soil. *J Polym Environ* 26(9): 3613–3625

Moog D et al (2019) Using a marine microalga as a chassis for polyethylene terephthalate (PET) degradation. Microb Cell Fact 18(1):1–15

Moore CJ (2008) Synthetic polymers in the marine environment: a rapidly increasing, long-term threat. J Environ Res 108(2):131–139

Muenmee S et al (2015) Microbial consortium involving biological methane oxidation in relation to the biodegradation of waste plastics in a solid waste disposal open dump site. Int Biodeter Biodegr 102:172–181

Muhonja CN et al (2018) Biodegradability of polyethylene by bacteria and fungi from Dandora dumpsite Nairobi-Kenya. PLoS One 13(7):e0198446
Mukherjee S et al (2017) Anionic surfactant induced oxidation of low density polyethylene followed by its microbial bio-degradation. Int Biodeter Biodegr 117:255–268
Nanda S et al (2010) Studies on the biodegradation of natural and synthetic polyethylene by Pseudomonas spp. J Appl Sci Environ Manag 14(2)
Nelms SE et al (2016) Plastic and marine turtles: a review and call for research. J Appl Sci Environ Manag 73(2):165–181
Nicolau L et al (2016) Ingestion of marine litter by loggerhead sea turtles, Caretta caretta, in Portuguese continental waters. Mar Pollut Bull 103(1-2):179–185
Ory N et al (2018) Low prevalence of microplastic contamination in planktivorous fish species from the southeast Pacific Ocean. Mar Pollut Bull 127:211–216
Osman M et al (2018) Degradation of polyester polyurethane by Aspergillus sp. strain S45 isolated from soil. Mar Pollut Bull 26(1):301–310
Piergiovanni L, Limbo S (2016) Food packaging materials. Springer, Cham
Puglisi E et al (2019) Selective bacterial colonization processes on polyethylene waste samples in an abandoned landfill site. Sci Rep 9(1):1–13
Pujic P et al (2015) Nocardia and Actinomyces. In: Tang Y-W, Sussman M, Liu D, Poxton I, Schwartzman J (eds) Molecular medical microbiology. Elsevier, Amsterdam, pp 731–752
Ray S, Cooney RP (2018) Thermal degradation of polymer and polymer composites. In: Kutz M (ed) Handbook of environmental degradation of materials. Elsevier, San Diego, pp 185–206
Reddy MS et al (2010) Feasibility study of the separation of chlorinated films from plastic packaging wastes. Waste Manag 30(4):597–601
Richards ZT, Maria J (2011) A quantification of the standing stock of macro-debris in Majuro lagoon and its effect on hard coral communities. Mar Pollut Bull 62(8):1693–1701
Rogers KL et al (2020) Micro-by-micro interactions: how microorganisms influence the fate of marine microplastics. Limnol Oceanogr Lett 5(1):18–36
Romeo T et al (2016) First record of plastic debris in the stomach of Mediterranean lanternfishes. Int J Mar Sci 57:1
Rulkens R, Koning CE (2012) Chemistry and technology of polyamides. In: Matyjaszewski K, Möller M (eds) Polymer science: a comprehensive reference. Elsevier, Amsterdam, pp 431–467
Russell JR et al (2011) Biodegradation of polyester polyurethane by endophytic fungi. Appl Environ Microbiol 77(17):6076–6084
Sánchez C (2020) Fungal potential for the degradation of petroleum-based polymers: an overview of macro-and microplastics biodegradation. Biotechnol Adv 40:107501
Sangale MK, Shahnawaz M, Ade AB (2019) Potential of fungi isolated from the dumping sites mangrove rhizosphere soil to degrade polythene. Sci Rep 9(1):1–11
Savoldelli J, Tomback D, Savoldelli HJ (2017) Breaking down polystyrene through the application of a two-step thermal degradation and bacterial method to produce usable byproducts. Waste Manag 60:123–126
Schmidt M (2017) Microbial degradation of plastics: a tool for cleanup of the oceans? Faculty of Science and Engineering
Schuyler QA et al (2016) Risk analysis reveals global hotspots for marine debris ingestion by sea turtles. Glob Chang Biol 22(2):567–576
Shahnawaz M, Sangale MK, Ade AB (2016) Rhizosphere of Avicennia marina (Forsk.) Vierh. As a landmark for polythene degrading bacteria. Environ Sci Pollut Res 23(14):14,621–14,635
Sharma A, Sharma A (2004) Degradation assessment of low density polythene (LDP) and polythene (PP) by an indigenous isolate of Pseudomonas stutzeri. J Sci Ind Res 63(3):293–296
Shubhra QTH, Alam AKMM, Quaiyyum MA (2013) Mechanical properties of polypropylene composites: a review. J Thermoplast Compos Mater 26(3):362–391
Singh MJ, Sedhuraman P (2015) Biosurfactant, polyethene, plastic, and diesel biodegradation activity of endophytic *Nocardiopsis* sp. mrinalini9 isolated from *Hibiscus rosasinensis* leaves. Bioresour Bioproc 2(1):1–7

Siracusa VJ, Polymers. (2019) Microbial degradation of synthetic biopolymers waste. Polymers 11(6):1066

Sivan A et al (2006) Biofilm development of the polyethylene-degrading bacterium Rhodococcus ruber. Appl Microbiol Biotechnol 72(2):346–352

Stelfox M, Hudgins J, Sweet M, Michael J (2016) A review of ghost gear entanglement amongst marine mammals, reptiles and elasmobranchs. Mar Pollut Bull 111(1-2):6–17

Sulong NHR, Mustapa SAS, Rashid MKA (2019) Application of expanded polystyrene (EPS) in buildings and constructions: a review. J Appl Polym Sci 136(20):47529

Syranidou E et al (2019) Biodegradation of mixture of plastic films by tailored marine consortia. J Hazard Mater 375:33–42

Thompson RC et al (2009) Plastics, the environment and human health: current consensus and future trends. Philos Trans R Soc Lond B Biol Sci 364(1526):2153–2166

Timmers MA, Kistner CA, Donohue MJ (2005) Marine debris of the northwestern Hawaiian islands: ghost net identification. US Department of Commerce, National Oceanic and Atmospheric Administration, Honolulu

Tokiwa Y et al (2009) Biodegradability of Plastics. Int J Mol Sci 10(9):3722–3742

Urbanek AK et al (2018) Degradation of plastics and plastic-degrading bacteria in cold marine habitats. Appl Microbiol Biotechnol 102(18):7669–7678

Werner S et al (2016) Harm caused by marine litter: MSFD GES TG Marine Litter—thematic report. Publications Office of the European Union, Luxembourg

Wilcox C, Van Sebille E, Hardesty BD (2015) Threat of plastic pollution to seabirds is global, pervasive, and increasing. Proc Natl Acad Sci U S A 38:11899–11904

Wilkes RA, Aristilde LJ (2017) Degradation and metabolism of synthetic plastics and associated products by Pseudomonas sp.: capabilities and challenges. J Appl Microbiol 123(3):582–593

Woodley J (2002) Assessing and monitoring floatable debris. DIANE, Collingdale

Wright SL, Thompson RC, Galloway TSJ (2013) The physical impacts of microplastics on marine organisms: a review. Environ Pollut 178:483–492

Yang G et al (2021) Diversity and structure of microbial biofilms on microplastics in riverine waters of the Pearl River Delta, China. Chemosphere 272:129870

Yoshida S et al (2016) A bacterium that degrades and assimilates poly (ethylene terephthalate). Science 351(6278):1196–1199

Zhang K et al (2021) Understanding plastic degradation and microplastic formation in the environment: a review. Environ Pollut:116554

Degradation of Plastic Waste in the Marine Environment

8

Wai Chin Li, Hin Fung Tse, Ho Man Leung, and Ying Kit Yue

Abstract

Annual global plastic production reached 360 million tonnes in 2018, with only 20% of the plastic waste being recycled. It has been predicted that production will double in the next 20 years. This review summarizes the occurrence, effects, and pathways as well as the degradation mechanisms and rates of different types of plastic waste in the marine environment. Approximately 75% of the litter found in the marine environment is mainly plastic, including polyethylene, polypropylene, polyethylene terephthalate, and polymers. Marine plastic waste has been found to accumulate in the open sea and deep sea and on beaches and shorelines of most remote areas, even in Arctic sea ice. Because of its durable and corrosion-resistant properties, plastic waste tends to accumulate and persist in the marine environment. The interaction of plastic waste with environmental conditions results in chemical, physicochemical, and biological degradation of the plastic waste and thus changes in its surface properties. Thus, both physical and chemical impacts from the degradation of plastic waste threaten marine organisms and even human health. Finally, some sustainable practices are suggested for the bioremediation of plastic waste.

Keywords

Plastic waste · Degradation · Marine · Physical effects · Chemical effects

W. C. Li (✉) · H. F. Tse
Department of Science and Environmental Studies, The Education University of Hong Kong, Hong Kong SAR, China
e-mail: waichin@eduhk.hk

H. M. Leung · Y. K. Yue
Department of Biology, Hong Kong Baptist University, Hong Kong SAR, China

M. Shahnawaz et al. (eds.), *Impact of Plastic Waste on the Marine Biota*,
https://doi.org/10.1007/978-981-16-5403-9_8

8.1 1.1 Introduction

Plastic waste pollution is an urgent global issue. According to Plastics Europe (2019), global plastic production reached 359 million tonnes in 2018, with 51% of the plastic being produced in Asia. Table 8.1 indicates the m

ost widely used plastics globally. Plastic types such as polypropylene (PP) and polyethylene (PE) represent approximately 50% of global plastic production. Apart from PP and PE, polyvinyl chloride (PVC) has also been widely used in domestic and industrial products. According to Barnes et al. (2009), in the United States and Europe, disposable products (e.g. packaging, garbage bags, and eating utensils) account for one-third of the use of plastic, which is disposed of within 3 years of their production. The low recycling rate of plastic and mismanaged sewage treatment plants or landfills result in an abundance of plastic accumulating in the marine environment. Plastic waste pollution has attracted increasing attention and has been widely reported in the marine environment, including in water bodies, on beaches and shorelines, and in the deep sea (Cózar et al. 2014; Fok and Cheung 2015; Kanhai et al. 2017). Due to the degradation-resistant properties of plastic, it tends to accumulate and persist in the marine environment. Several studies have noted that a wide range of species are vulnerable to plastic ingestion and plastic entanglement or chemical hazards from plastics (Campbell et al. 2017; Thiel et al. 2018; Gouin 2020). This article will review the sources and occurrence of plastic waste in the marine environment. Moreover, the degradation, fates, and effects of plastic will be summarized. Finally, recommendations are provided for future practices related to the sustainable bioremediation of plastic waste.

8.2 1.2 Sources and Occurrence of Plastic Waste in the Marine Environment

Land-based and sea-based sources contribute to the occurrence of plastics in the marine environment. The majority of plastic waste (~80%) is estimated to be generated from land-based sources. Packaging is an important land-based source of the large plastic observed in seas and on beaches and accounts for 40% of all plastic production (Plastics Europe 2019). Other human activities, such as agricultural activities and industrial activities, can generate plastic waste, including discarded irrigation pipes, planting containers (Gambash et al. 1990), plastic burlap (de Stephanis et al. 2013), and industrial construction materials (Jambeck et al. 2015). The mismanagement of solid waste or landfills might lead to the transport of plastics by winds or during natural disasters such as tsunamis or hurricanes (Jambeck et al. 2015). Sewage treatment plants are known as the primary contributors of microplastics to the marine environment, especially in areas near industrial activities (Conley et al. 2019). Due to their small size, only a small amount or no microplastics can be removed through a sewage system. Microplastics have been widely reported to be used as abrasive cleaners for building surfaces and as blasting media for smoothing, shaping, and roughing (Fendall and Sewell 2009). In

Table 8.1 Types and chemical structures of plastic commonly found in the natural environment

Type	Specific gravity (Li et al. 2016; Andrady 2011)	Structure (Wierckx et al. 2018)	Use/application (Ghosh et al. 2013; Andrady 2011)
Polyester (PES)	1.40	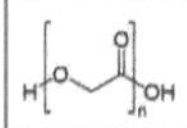	Fibres, textiles
Polyethylene terephthalate (PET)	1.37	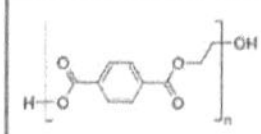	Carbonated drink bottles, peanut butter jars, plastic film, microwavable packaging, tubes, pipes, insulation moulding
Polyethylene (PE)	0.91–0.96	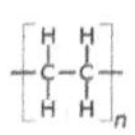	Wide range of inexpensive uses including supermarket bags, plastic bottles
High-density polyethylene (HDPE)	0.94		Detergent bottles, milk jugs, tubes, pipes, insulation moulding
Polyvinyl chloride (PVC)	1.38	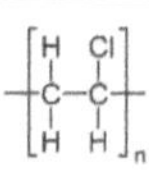	Plumbing pipes and guttering, shower curtains, window frames, flooring, films
Low-density polyethylene (LDPE)	0.91–0.93	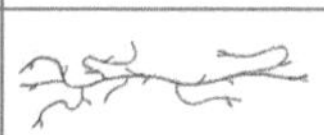	Outdoor furniture, siding, floor tiles, shower curtains, clamshell packaging, films
Polypropylene (PP)	0.85–0.83	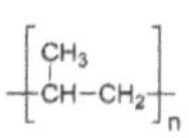	Bottle caps, drinking straws, yogurt containers, appliances, car fenders (bumpers), plastic pressure pipe systems, tanks, and jugs
Polystyrene (PS)	1.05	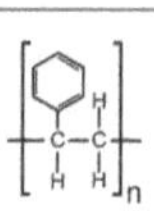	Packaging foam, food containers, plastic tableware, disposable cups, plates, cutlery, CD, cassette boxes, tanks, jugs, building materials (insulation)
High impact polystyrene (HIPS)	1.08	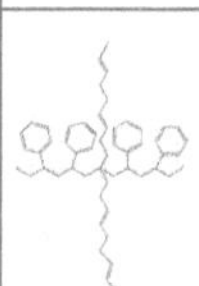	Refrigerator liners, food packaging, vending cups, electronics
Polyamides (PA) (nylons)	1.13–1.35	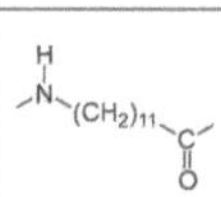	Fibres, toothbrush bristles, fishing line, under-the-hood car engine mouldings, making films for food packaging
Acrylonitrile butadiene styrene (ABS)	1.06–1.08	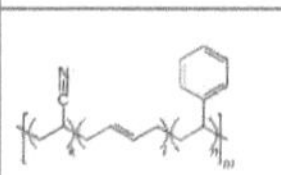	Electronic equipment cases (e.g. computer monitors, printers, keyboards), drainage pipe, automotive bumper bars
Polycarbonate (PC)	1.20–1.22	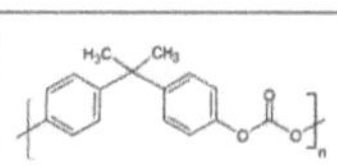	Compact discs, eyeglasses, riot shields, security windows, traffic lights, lenses, construction materials

(continued)

Table 8.1 (continued)

Type	Specific gravity (Li et al. 2016; Andrady 2011)	Structure (Wierckx et al. 2018)	Use/application (Ghosh et al. 2013; Andrady 2011)
Polycarbonate/ acrylonitrile butadiene styrene (PC/ABS)	1.10–1.15	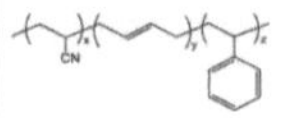	A blend of PC and ABS that creates a stronger plastic. Used in car interior and exterior parts and mobile phone bodies

addition to industrial use, microplastics are used in facial cleansers as physical abrasives (Fendall and Sewell 2009). The remaining approximately 20% are generated from sea-based sources. Fishing activities are the most important sea-based source of microplastics, with approximately 640 thousand metric tonnes of fishing nets abandoned in the sea every year (Greenpeace Germany 2019). Fishing gear such as ropes, floats, and fishing lines can be discarded or lost by accident. Additionally, plastic abrasives commonly used in ship cleaning activities can be another source of plastic pollution (Song et al. 2015).

Plastic waste has been reported in different environments, including in open seas (Zettler et al. 2013; Kanhai et al. 2017; Iannilli et al. 2019) and on beaches, shorelines (Nel et al. 2017; Winton et al. 2020), and the seafloor (Katija et al. 2017; Kane et al. 2020). Ocean gyres and oceanic convergence zones are prone to the accumulation of plastic waste due to the rotational patterns of ocean currents (Fig. 8.1) (Lebreton et al. 2012).

The North Pacific subtropical gyre is referred to as the "Great Pacific Garbage Patch" (Kaiser 2010) and is regarded as one of the largest cluster spots of plastic waste, with nearly 50% being fishing nets (42 thousand metric tonnes) (Lebreton et al. 2018). The high amount of plastic waste reported in the North Pacific Ocean could be the result of intense human activities along the East Asian coast (Cózar et al. 2014). Numerous studies have suggested that the East Asian Sea is the largest hotspot for plastic waste accumulation (Kuo and Huang 2014; Wu et al. 2018). The abundance and surface density of plastic in the East Asian seas around Japan are about 1.7 million pieces km^{-2} and 2400 g km^{-2}, respectively (Isobe et al. 2015). The accumulation of plastic waste in water bodies is not restricted to water columns, including in subsurface waters and the deep sea (Cózar et al. 2014; Courtene et al. 2017). Whether plastic waste sinks or floats depends on its density (Andrady 2011) or the formation of biofilms (Cole 2016). The abundance of microplastics in the sediments of the abysmal sea, ranging from 1,176 to 4,843 m deep in the Atlantic Ocean and the Mediterranean Sea, respectively, was investigated (Van Cauwenberghe et al. 2013). Another study also indicated that there are 2,020 particles per m^2 found in the sediments of the Kuril-Kamchatka Trench area (Fischer et al. 2015).

Apart from water bodies, beaches, beach sediments, and shorelines are subjected to plastic waste pollution worldwide (Fok and Cheung 2015; Klein et al. 2015; Alomar et al. 2016; Iannilli et al. 2020). Studies have reported that the high abundance of plastic waste on beaches is positively correlated with the density of

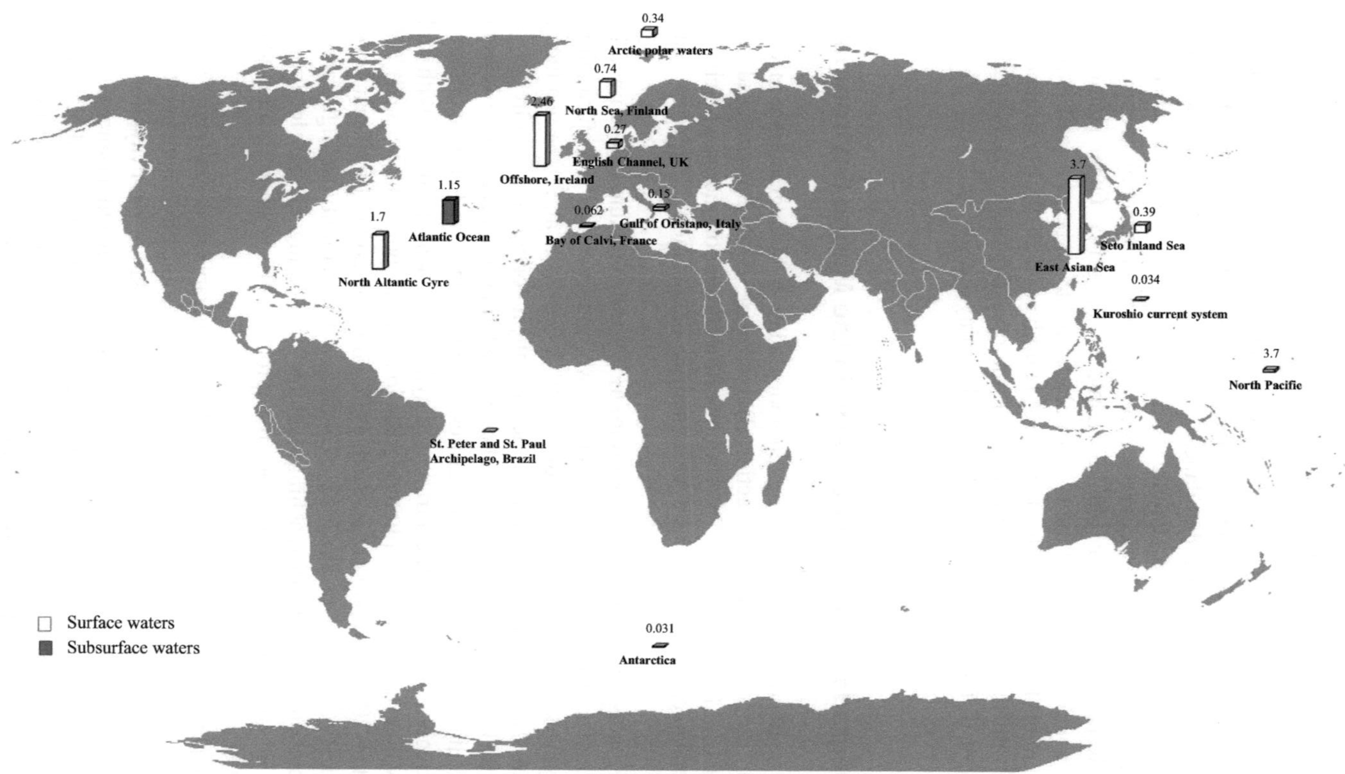

Fig. 8.1 Occurrence (pieces/m^3) of plastics found in water bodies (Yamashita and Tanimura, 2007; Goldstein et al. 2012; Ivar do Sul et al. 2013; Carson et al. 2013; Isobe et al. 2015; Cole et al. 2014; Lusher et al. 2014; de Lucia et al. 2014; Collignon et al.. 2014; Reisser et al. 2014; Magnusson, 2014; Zhao et al. 2014; Isobe et al. 2015; Lusher et al. 2015; Isobe et al. 2017; Kanhai et al. 2017)

human activities (Nor and Obbard 2014; Romeo et al. 2015; Naidoo 2015). For instance, a study reported that a high level of microplastics in regions with intense human activities occurred when compared to levels of microplastics at different control sites (Romeo et al. 2015). However, numerous studies have stated that no remarkable relationship exists between the level of human activities and the number of microplastics in beach sediments or on beaches (Laglbauer et al. 2014; Klein et al. 2015; Alomar et al. 2016). The abundances of plastic waste in locations with a high level of human pollution and remote regions in Australia are not substantially different, possibly because plastic waste travels to the open ocean by ocean currents and wind (Reisser et al. 2013; Shahul Hamid et al. 2018). Other studies have reported seasonal differences in the level of plastic waste on beaches and sediments (Veerasingam et al. 2016; Cheung et al. 2016; Iannilli et al. 2020). For instance, Lee et al. (2013) illustrated that the abundance of microplastics on beaches differs with seasons, with differences occurring before the rainy season and after the rainy season.

8.3 Degradation of Plastic Waste in the Marine Environment

Plastic can be degraded by weathering, photodegradation, biodegradation, mechanical abrasion turbulence, and wave action (Barnes et al. 2009; Browne et al. 2011; Andrady 2017; Min et al. 2020). The structure of a plastic polymer can be changed, and the mechanical integrity can be weakened via degradation (Singh and Sharma 2008).

Photodegradation is induced by exposure to UV radiation and oxygen and is an important mechanism leading to the breakdown and chemical transformation of plastic polymers (Cole et al. 2011; Andrady 2017; Gewert et al. 2018). An increase in temperature and humidity facilitates the rate of photodegradation (Gregory and Andrady 2003). Therefore, the rate of photodegradation in a beach zone is usually higher than that in seawater because plastics are subjected to higher temperatures. It has been suggested that the concentration of hydroxyl radicals increases with higher humidity which enhances the photodegradation rate of poly(l-lactic acid) (PLA) (Copinet et al. 2004) and polyolefins such as PE (Jin et al. 2006), PP (Fernando et al. 2009), and PVC (James et al. 2013). Under seawater conditions, light and UV radiation are prevented from reaching foulants on floating plastic, leading to a reduction in oxidation reactions and the process of weathering degradation (Weinstein et al. 2016).

Plastic types (e.g. PE, PP, PS, and PVC) are vulnerable to degradation by light, which plays an essential role in degradation in open aerobic environments (Gijsman et al. 1999). Figures 8.2 and 8.3 show the photodegradation pathways for PP and PE. When the degradation process starts, it can continue with temperature-dependent thermo-oxidative reactions without further exposure to ultraviolet radiation if oxygen is available (Singh and Sharma 2008). Generally, photodegradation can be separated into three major steps: "initiation", "propagation", and "termination". For initiation, light can break down the chemical bonds in the backbone chain and

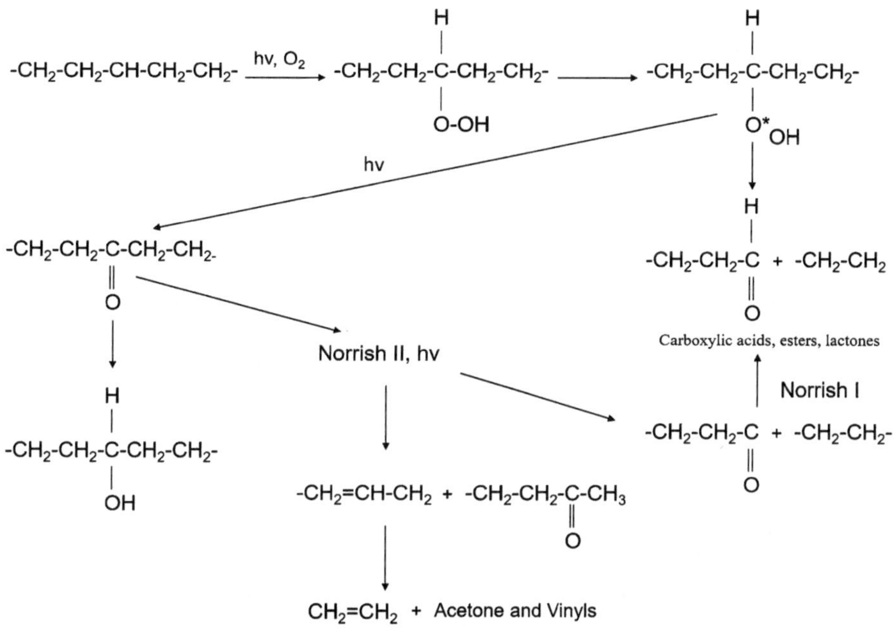

Fig. 8.2 The photodegradation pathway for PE (Gardette et al. 2013)

Fig. 8.3 The photodegradation pathway for PP (Rånby 1989)

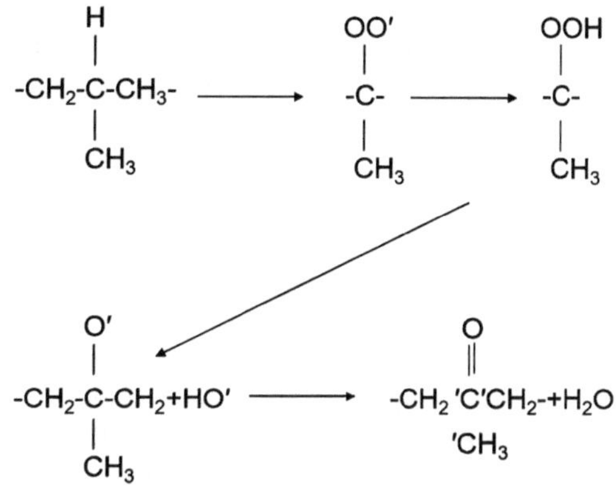

thus produce a free radical. Usually, the unsaturated chromophoric group in the plastic absorbs light energy to induce the photoinitiation process (Gijsman et al. 1999). Although some plastics, such as PE and PP, lack unsaturated hydrocarbons in their backbone chain, flaws, including structural abnormalities or metallic impurities incorporated into the macromolecular structure formed in the manufacturing process, can also lead to the photoinitiation degradation process (Gijsman et al. 1999; Vasile and Pascu 2005; Montazer et al. 2020). Finally, free radicals are formed when the breakdown of carbon-hydrogen bonds on the polymer backbone is induced by light. Second, during the process of propagation, peroxy radicals will be formed when the free radicals react with oxygen (Rabek 1995; Singh and Sharma 2008). Therefore, hydroperoxides will be formed by the reaction between free radicals and oxygen and will eventually result in autoxidation via further complex reactions (Singh and Sharma 2008). In general, cross-links or chain scission will eventually result in the process of propagation (Tolinski 2009). Finally, in the process of termination, the combination of peroxy radicals and free radicals will lead to the formation of inert products, such as olefins, aldehydes, and ketones, which are more prone to the degradation induced by light since they contain unsaturated double bonds (Gewert et al. 2015). The plastic will be brittle since the molecular weight is reduced, thus making it more vulnerable to degradation and fragmentation (Summers and Rabinovitch 1999). Generally, the photodegradation rate of plastics depends on additives such as antioxidants or ultraviolet stabilizers, which can hinder the process of photodegradation (Rånby 1989; Hahladakis et al. 2018). According to Min et al. (2020), plastic without tertiary hydrogens (e.g. polymethyl methacrylate and polytetrafluoroethylene) is often highly stable. On the other hand, plastics that are relatively stable (e.g. PET and PC) or not very stable (e.g. PS, PVC, PI, PBD, PA, PU, PE, and PP) are more susceptible to photodegradation.

Second, hydrolytic degradation is another degradation mechanism that depends on the properties and functional groups of plastics (Pickett and Coyle 2013; Szycher 2013; Gewert et al. 2015). Plastics contain functional groups (e.g. esters, carbonates, and amides) that are subjected to abiotic hydrolysis (Gewert et al. 2015). Ambient conditions, such as the alkalinity of seawater with a pH range from 8 to 8.3 and the presence of hydroxide ions, favour the process of hydrolysis (Marion et al. 2011). Pickett and Coyle (2013) studied the hydrolytic stability of PC, PET, and resorcinol polyacrylate (RPA) films. The study suggested that the degradation rate of hydrolysis is determined by the functional groups and structures of plastics (Pickett and Coyle 2013). For example, PET and PU are likely more prone to hydrolytic degradation than plastics lacking functional groups (e.g. PE and PS) because of their carbon and heteroatoms in the main chain (Müller et al. 2001). The hydrolysis process is the most essential degradation process of PET under low-temperature conditions, although the degradation rate is extremely slow (Allen et al. 1991; Venkatachalam et al. 2012). During the hydrolysis degradation process, alcohol functional groups and carboxylic acids are formed, which reduces the molecular weight of PET (Culbert and Christel 2004). In addition, hydrolysis of the ester bond of PU is regarded as the most common hydrolytic degradation reaction. Apart from the ester bond, urea and urethane bonds are subjected to hydrolytic degradation at slower rates (Lamba 1998; Szycher 2013). In addition, a decrease in humidity levels can reduce the rate of degradation by eliminating hydrolysis. For instance, according to Edge et al. (1991), the chain scission of PET was estimated to be five times lower at 45% relative humidity than at 100% relative humidity. On the other hand, it has been reported that when PET is exposed at temperatures of 80 °C or higher, with increased humidity, the hydrolysis rate of plastic is not significant because temperature-dependent thermo-oxidative reactions become more important (Edge et al. 1991).

Biodegradation is regarded as a process in which microbes, such as bacteria, algae, fungi, actinomycetes, etc., transform or change the chemical structure introduced into the environment through metabolic or enzymatic action (Muthu 2014). The biodegradation rate of plastic in marine environments can be determined by environmental conditions, the characteristics of plastics, and the abilities of microorganisms (Donlan 2002; Wilkes and Aristilde 2017; Urbanek et al. 2018). Plastic characteristics, including hydrophobicity, molecular mobility, flexibility, molecular weight, crystallinity, and the use of additives or plasticizers, control the rate of biodegradation (Gu 2003; Artham and Doble 2008).

Most plastic molecules are non-water soluble, and microorganisms cannot move into plastics through their outer cellular membranes where most biochemical processes take place (Fesseha and Abebe 2019). Therefore, microbes develop a mechanism to excrete exoenzymes to convert polymers into monomers outside the cells to apply such substances as carbon and supply strength (Fesseha and Abebe 2019). The biodegradation mechanisms of microbes under aerobic conditions and anaerobic conditions are different. Under aerobic conditions, oxygen is used by microbes as an electron acceptor, and organic chemicals are broken down into smaller compounds with CO_2 and water by-products, while under anaerobic conditions, elements (e.g. manganese, sulphate, nitrate, iron, and carbon dioxide) are used as electron acceptors to break down organic chemicals into smaller compounds (Muthu 2014). Generally, the steps of biodegradation of plastics involve (1) the attachment of the microorganism to the surface of the plastic; (2) the growth of the microorganism, using the plastic as a carbon source; and (3) ultimately, the degradation of the polymer (Gu 2003; Fesseha and Abebe 2019; Singh and Rawat 2019).

The formation of biofilms and microbial attachments on plastic surfaces is the first step of biodegradation and can be attributed to the different abilities of microorganisms and the properties and surface structures of plastics (Donlan 2002), such as topography, electrostatic interactions of plastic surfaces, free energy, hydrophobicity, and roughness (Rummel et al. 2017). Marine plastic is subjected to the weathering process, resulting in the loss of physical integrity and leading to biofilm formation or colonization by microbes (Rummel et al. 2017). According to the study by Oberbeckmann et al. (2015), the microbial cells found in holes on the surfaces of plastics potentially lead to biodegradation of the plastic surfaces, which would be an important initial process for degradation. However, some studies have indicated that no evidence of potential degradation was observed during early attachment even though plastics can be easily colonized by bacteria (Lobelle and Cunlife 2011). Once microorganisms are affiliated with the surfaces of plastics, they can use plastic as a carbon supply for growth (Fesseha and Abebe 2019). Therefore, exoenzymes are secreted by microbes, leading to the breakdown of plastics and the formation of low molecular weight fragments (e.g. oligomers, dimers, or monomers). Additionally, low molecular weight fragments are further utilized by microbes for energy or electron inputs, or they are assimilated into their microbial cells (Premraj and Doble 2005; Fesseha and Abebe 2019). O'Brine and Thompson (2010) studied the degradation of two oxo-biodegradable plastic bags, compostable plastic bags, and standard polyethylene bags in the marine environment. Biofilms were formed on the surface of the four types of plastics after 4 weeks of exposure, and the thickness of the biofilms continuously increased over time with macrofouling organisms after 8 weeks of exposure. The degradation rate of the biodegradable bags was more significant than that of polyethylene bags, with full degradation of the decomposable material after 16 weeks (Lobelle and Cunlife 2011). Eich et al. (2015) also reported that the formation of biofilms occurred on plastic bags after approximately 4 weeks under marine environmental conditions, and the biofilm increased remarkably within 5 weeks on biodegradable and polyethylene plastic bags under benthic and pelagic conditions. The difference in biofilm formation on plastic could be attributed to different environmental conditions. It has been suggested that types of plastic and environmental conditions could influence biofilm formation in the early formation stage and biofilm composition (Eich et al. 2015).

8.4 Pathways for the Degradation of Plastic Waste in the Marine Environment

Once plastic waste moves into the environment, it persists and accumulates in water bodies or on beaches for a long period. Most commercial plastic is highly durable and corrosion-resistant because of the use of additives (Gewert et al. 2015). Consequently, plastic waste will float in the marine environment and will eventually occur in different oceans or gyres or on beaches or shorelines or be ingested by marine organisms and induce entanglement (Derraik 2002).

Floating plastic debris is usually submerged, but it continues to be buoyant within the top of the water column (~1 m). The density of plastic determines its buoyancy. It has been reported that plastics with higher densities (~1.03 g cm^{-3}) than seawater tend to sink or are submerged, while those with lower densities than seawater will float on the surface of waters (Andrady 2011). However, the process of biofouling will cause low-density plastics to sink (Lobelle and Cunlife 2011). It has been reported that the density of plastic increases with biofouling by an organism, leading to the sinking of plastic waste (Andrady 2011). As long as the density of plastic is greater than that of seawater, it tends to sink to the seafloor (Jorissen 2014). Cole (2016) also noted that zooplankton faecal pellets can attach to plastic waste, altering the sinking rates of the pellets, which is a possible pathway to the benthos. The formation of surface biofouling could limit the rate of plastic degradation because biofilms or colonizing organisms eliminate the amount of ultraviolet radiation reaching the plastic surfaces (Kershaw et al. 2011). According to Weinstein et al. (2016), approximately 4 weeks of biofilm development on the surfaces of plastic debris would reduce the amount of UV radiation reaching the surface by approximately 95%.

8.5 Effects of Plastic Waste Degradation

Numerous organisms (e.g. invertebrates, seabirds, sea turtles, and fish) have been found to have ingested microplastics (Table 8.2) (Gouin 2020). A recent review collected information on the ingestion of plastic by over 800 species representing approximately 87,000 individual organisms (Gouin 2020). It was reported that 20% of these observed individual organisms, approximately 17,500, ingested plastic. Another review by Provencher et al. (2017) reported similar findings and reported that over 690 species of marine organisms had ingested plastics. The effects of plastic ingestion have been well summarized (Plotkin and Amos 1990; Wright et al. 2013; Thompson 2015; Acampora et al. 2016; Brate et al. 2016; Rummel et al. 2016; Rizzi et al. 2019). The presence of ingested plastics can lead to internal or external fray and canker and blockage of the gastrointestinal tract, which can adversely affect the growth of organisms and eventually result in death (Rizzi et al. 2019). It has been reported that microplastic ingestion by larger organisms, such as seabirds, seldom causes death immediately, and the effects are not as serious as those when larger plastic is ingested (Lusher et al. 2015). However, both microplastic and microplastic accumulation causes a reduction in feeding behaviour, resulting in a negative influence on reproduction because microplastic ingestion causes the affected organisms to feel satiated (Thompson 2015).

Feeding habitat or behaviour influences the incidence of plastic ingestion by organisms (Lusher et al. 2013; Carlos de Sa et al. 2015). In fish, according to Lusher et al. (2013), in comparison to other fishes, mackerels are more vulnerable to ingestion of floating or neutrally buoyant plastic because of their unselective filter-feeding behaviour. A study observed that young *Pomatoschistus microps* confuse plastic debris with their natural prey, leading to plastic ingestion, which leads to a remarkable decrease in their hunting capability and prey selection performance because of ontogenic developmental condition changes (de Sá et al. 2015). Of different types of seabirds, species with surface-foraging behaviour (e.g. gulls, shearwaters, and fulmars) are more likely to ingest plastic. Species with mixed diets, such as those in the family Laridae, could be less prone to plastic ingestion since they feed in other areas, such as landfills (Lindborg et al. 2012). Barrett et al. (2007) and Lindborg et al. (2012) reported that most gull species can regurgitate indigestible items including plastic from their stomach, resulting in a reduction in the amount of ingested plastic. However, such regurgitation behaviour in gulls would transfer plastic to juvenile birds (Carey 2011; van Franeker et al. 2011; Kuhn and van Franeker 2012). Several studies have reported that in comparison to mature seabirds, juvenile seabirds, such as northern fulmars, contain more plastic in their gastrointestinal tracts (Kuhn and van Franeker 2012; Acampora et al. 2014; Provencher et al. 2014). This result could be attributed to the regurgitation behaviour of seabirds. In addition, zooplanktivorous seabirds, such as *Alle*, are susceptible to direct ingestion of plastic since it is difficult to distinguish plastic particles from zooplankton. Apart from direct ingestion, abundant studies have also illustrated that microplastics are contained in zooplankton, which could lead to indirect ingestion of plastic by *Alle* via their prey. The increase in the amount of ingested plastic results in

Table 8.2 Occurrence of plastic ingestion found in marine organisms

Species		*N*	Locations	Mean number of particles per individual (±s.d.)	Reference
Seabirds	Alle alle	65	White Bay, Newfoundland	1.11	Fife et al. (2015)
	Northern fulmar	40	Isfjord, Svalbard	15.32	Trevail et al. (2015)
	Short-tailed shearwater	129	North Stradbroke Island, Australia	7.1	Acampora et al. (2014)
	Scopoli's shearwater	49	Catalan coast, Mediterranean	14.6	Codina et al. (2013)
	Sooty shearwater	25	Oregon and Washington beaches	19.5	Terepocki et al. (2017)
	Northern fulmar	143	Oregon and Washington beaches	13.3	Terepocki et al. (2017)
Marine mammal	Harbour seal	100	The Netherlands	8 items	Bravo et al. (2013)
	Humpback whale	-	The Netherlands	17	Besseling et al. (2015)
	Beaked whale	3	North and west coast of Ireland	29	Lusher et al. (2015)
	Fin whale	1	Jeju Island	45	Im et al. (2020)
Turtle	Green turtle	19	Eastern Mediterranean	61.8 ± 15.8	Duncan et al. (2019)
	Loggerhead turtle	102	Central Mediterranean	12 ± 2	Matiddi et al. (2017)
Marine invertebrates	Brown shrimp	110	Belgium	11.5 fibres per 10 g shrimp	Devriese et al. (2014)
	Amphipods	65	New Hebrides Trench	0.9 ± 0.4	Jamieson et al. (2019)
	Copepods	-	Northeast Pacific Ocean	1 particle/every 34 copepods	Desforges et al. (2015)
	Euphausiids	-	Northeast Pacific Ocean	1/every 17 euphausiids	Desforges et al. (2015)

an increase in the body mass of birds, which could lead to an increase in the duration of feeding activities (Adams et al. 2009; Heggøy et al. 2015).

In addition, the colour of plastic plays an essential role in plastic ingestion by marine organisms, such as fish (Egbeocha et al. 2018), sea turtles (Eastman et al. 2020), and seabirds (Cousin et al. 2015; Provencher et al. 2019). Studies have reported that plastics with light colours are ingested by seabirds more than those with dark colours (Verlis et al. 2013; Santos et al. 2016). For instance, in Australia, the most common plastic colour found in the wedge-tailed shearwater *Ardenna pacifica* was off-white/white (37.5%) and green (31.3%). The amount of dark coloured plastic ingested by the short-tailed shearwater was six times lower than that of the lighted colour. This result could be ascribed to the physical characteristics and sensory systems of animals as well as to general physical processes such as background matching (Santos et al. 2016). Apart from seabirds, sea turtles could misidentify plastics as prey and consume them since most sea turtles are visual foragers (Hoarau et al. 2014; Van Houtan et al. 2016). Plastics with a white or transparent colour, e.g. plastic bags, are easily ingested by turtles because they appear similar to jellyfish (Nelms et al. 2015; Clukey et al. 2017). In Australia, Schuyler et al. (2012) also reported that green sea turtles (*Chelonia mydas*) chose to ingest transparent or white plastics with the shape of sheets or ropes rather than harder, coloured pieces of plastic when the turtles were foraging in coastal benthic habitats. In a study by Duncan et al. (2019), green sea turtles ingested black, clear, and green plastic as these colours more closely resemble seagrass in the water. Selective ingestion of plastic by green sea turtles was also suggested by Bjorndal (1980) because sea turtles can choose specific seagrass species over others based on their natural selective feeding behaviour. In addition, various plastic colours affect the incidence of plastic ingestion by different species of fish. Plastics with food-like colours were ingested by fish remarkably more often than other plastics (Choy and Drazen 2013; Alomar and Deudero 2017; Roch et al. 2020). For example, dark coloured plastics such as blue and black were the favoured colour ingested by *Galeus melastomus* (Alomar and Deudero 2017), a fish in the English Channel (Lusher et al. 2013), and fish from the Spanish Atlantic and Mediterranean coasts (Bellas et al. 2016). In contrast, light-coloured plastics were ingested more by mesopelagic fish such as *Alepisaurus ferox* (Choy and Drazen 2013).

Plastic entanglement of pinnipeds, seabirds, and cetaceans in large plastic litter, such as nets, ropes, and fishing gear, was first reported in the early 1970s (Derraik 2002). Derelict fishing gear is the most common type of plastic leading to the entanglement of organisms, although it accounts for less than only 10% of marine litter (Macfayden et al. 2009). The high durability and strong and lightweight properties of fishing materials make it more difficult for entangled animals to escape entanglement in lost fishing gear (Gilardi et al. 2010). A recent review by Thiel et al. (2018) suggested that subtropical gyres are potential hotspots for plastic entanglement of marine organisms because there is a high level of fishing activity in these areas of the open sea, resulting in abundant discarded fishing materials. Entanglement with plastic causes tissue damage or body damage and affects feeding animal behaviour, social interactions, breeding patterns, migration, and hunting, ultimately leading to mortality (Alfaro-Shigueto et al. 2010; Aguilera et al. 2016; Butterworth 2016; Thiel et al. 2018). In addition, Derraik (2002) and Boland and Donohue (2003) reported that drag effects on organisms could be the result of plastic entanglement because it can lead to an increase in problems related to movement restriction, exhaustion, and drowning of the entangled animal. It was estimated that entanglement by approximately 200 g of material will increase food consumption by four times to compensate for the drag caused by interference with water flow.

Marine mammals (e.g. pinnipeds, large baleen whales, dolphins, porpoises, and toothed whales) are the organisms most affected by plastic entanglement (Table 8.3) (Wilcox et al. 2015; Aguilera et al. 2016; Butterworth 2016; Campbell et al. 2017; Thiel et al. 2018). Many reports have noted that large marine mammals are entangled in fishing gear and packing straps during swimming and diving (Alfaro et al. 2010; Thiel et al. 2018). Small marine mammals (e.g. pinnipeds) may die in a short period after entanglement because they cannot escape from the entangled materials, and these air-breathers will consequently drown (Thiel et al. 2018). The most common plastics reported to entangle pinnipeds are packing bags or fishing materials (Townsend and Barker 2014; Lawson et al. 2015; Butterworth 2016; Thiel et al. 2018). In comparison to other pinnipeds, young seals and sea lions are more prone to entanglement due to their curiosity (Butterworth 2016). The entanglement rate of juvenile Antarctic fur seals was five times that of mature seals since the young seals likely interacted and played with plastic materials due to their curiosity (Waluda and Staniland 2013). Seabirds are also affected by entanglement in discarded fishing gear and plastic bags on beaches during foraging activities (Thiel et al. 2011). Lines and rope fragments discarded by fishing activities lead to the entanglement of coastal seabird species such as gulls and cormorants. Reports on plastic entanglement of fish are rare when compared to those on plastic entanglement of marine mammals, seabirds, and sea turtles. The absence of reports on fishes could be ascribed to their mortality at sea and immediate sinking, whereas other marine animals float on the sea surface (Thiel et al. 2018).

Plastic ingestion not only causes severe physical effects on marine organisms but also induces chemical hazards by exposing these organisms to persistent organic pollutants and endocrine disruptor chemicals (e.g. phthalates, bisphenol A, polychlorinated biphenyls (PCBs), dioxins, hexachlorobenzene, antioxidants, UV stabilizers, and pigments) (Gallo et al. 2018). These chemical contaminants are regarded as "forever chemicals", which can bioaccumulate in the tissue of organisms and can travel long distances to remote locations by ocean currents and buoyant plastics, harming marine organisms (Rochman et al. 2013; Zacharia 2019). Plastics, especially microplastics and nanoplastics, are prone to absorb or carry chemical pollutants due to their hydrophobic surface and high surface area-to-volume ratios (Campanale et al. 2020). Teuten et al. (2007) reported that PCB concentrations in polypropylene pellets were 10^6 times higher than those in the surrounding water. In addition, the absorption and desorption rate of pollutants depends on the type of plastic. For example, Rochman et al. (2013) reported that the concentrations of parent PAHs on PS are estimated to be 8–200 times greater than those on PP, HDPE, LDPE, PET, and PVC. It has been suggested that low-molecule plastics

Table 8.3 Occurrence of plastic entanglement by organisms

Species		Location	Entangled material	Entanglement rate	Reference
Seals and sea lions	Northern fur seal	North East Pacific	Fishing line (9%), plastic (37%), plastic net (39%)	0.08–0.35%	Allen and Angliss (2014)
	Southern elephant seal	South West Atlantic	Fishing line (64%), plastic (36%)	0.001–0.002%	Campagna et al. (2007)
	Antarctic fur seals	Bird Island, South Georgia	Packaging bands (43%), synthetic line (25%) or fishing net (17%)	–	Waluda and Staniland (2013)
	Australian fur seals	Southern Australia	Plastic twine or rope (50%), other plastics (20%), monofilament line (17%)	–	Lawson et al. (2015)
	Steller sea lions	Southeast Alaska and northern British Columbia	Packing bands (54%), rubber bands (30%), net (7%), rope (7%)	0.26%	Raum-Suryana et al. (2009)
	Australian fur seals	Seal Rocks, south-eastern Australia	Juvenile seals: Trawl net (52%) pup seals: fishing line (21%), trawl net (19%), and other materials (18%)	0.02–0.19%	McIntosh et al. (2015)
Seabirds	Northern gannets	Spanish Iberia and Mauritania	Fishing ropes (73.5%), nets (11.8%), nylon fishing lines (14.7%)	0.93%	Rodríguez et al. (2013)
	American crow	Davis, California	Synthetic twine, string and rope (77%), plastic strips (10%)	5.6%	Townsend and Barker (2014)
	26 seabird species	California	Monofilament line (78%), commercial fishing (4%), plastic (ring, bands, string, swim goggles) (4.8%)	1.7%	Donnelly et al. (2019)

usually contain higher concentrations of petrogenic PAHs, while high-molecule plastics always contain higher concentrations of pyrogenic PAHs (Rochman et al. 2013). In addition, some plastic types, for example, PP, LDPE, and HDPE, have a higher ability to sorb more organic pollutants than PET and PVC since they reach equilibrium more slowly.

Once organisms ingest polluted plastics, organic pollutants can be assimilated into their tissues (Chua et al. 2014; Wardrop et al. 2016). According to Wardrop et al. (2016), the desorption rate of organic pollutants under gut conditions can be

30 times greater than that under seawater conditions. The potential harmful effects on molluscs, crustaceans, fish, and amphibians have been summarized and reviewed (Talsness et al. 2009; Munn and Goumenou 2013; Avio et al. 2017). Numerous experimental studies have shown that exposure to a low concentration of endocrine disruptor chemicals can cause both temporary and permanent changes to endocrine systems because of the disturbance to endogenous hormones (Oehlmann et al. 2009; Talsness et al. 2009; Munn and Goumenou 2013; Mathieu-Denoncourt et al. 2015; Avio et al. 2017). The disruption of organismal hormones leads to impairments to reproduction, reductions in birth rates, impaired metabolisms and thyroid functions, potential losses of biodiversity, and increased occurrences of hormone-sensitive cancers by the induction of immunotoxicity and oxidative stress in organisms (Gore et al. 2015; Mathieu-Denoncourt et al. 2015). For instance, the PBDE in the tissues of short-tailed shearwaters could lead to endocrine disruption (Tanaka et al. 2013). Additionally, in vivo studies illustrated that BPA could potentially induce disruptions in androgen and oestrogen synthesis and the metabolism of fish such as carp (Mandich et al. 2007), fathead minnow (Brian et al. 2005), and medaka (Ishibashi et al. 2005).

8.6 Suggestions for Future Practices to Sustainably Bioremediate Plastic Waste

Recently, it was reported that organisms have developed abilities to degrade plastic (Sheth et al. 2019). Table 8.4 summarizes the potential enzymes used for plastic degradation. Enzymes with broad substrates can degrade different synthetic and natural plastics because of the similarities in their molecular structures.

First, several researchers have reported that the enzymes cutinase and lipase could be used for the degradation of PET (Austin et al. 2018; Liu et al. 2018; Shirke et al. 2018; Kawai et al. 2019). Cutinases can catalyse the hydrolytic process of ester bonds found in cutin, which is similar to lipases displaying a catalytic triad consisting of Ser-His-Asp (Kawai et al. 2014; Cen et al. 2019). Since enzymes can lack lipase lid structures, the active sites of cutinases are prone to be solvents (Wei et al. 2014a). Generally, cutinases for plastic degradation can be divided into fungi and bacteria cutinases (Baker et al. 2012; Wei et al. 2014a). According to Shirke et al. (2018), cutinases isolated from bacteria located around plant compost show PET degradation activity. Additionally, the mutation of the *Thermobifida fusca* cutinase TfCut2 was investigated (Wei et al. 2016). The degradation activity of the mutated cutinases was 2.7-fold higher than that of the wild-type cutinases, resulting in a 42% weight loss for the PET films after only 50 h of treatment (Wei et al. 2016). Other bacterial cutinases with PET-hydrolysing activity have been reported to have been isolated from different bacterial species, such as the actinomycetes *Saccharomonospora viridis* (Kawai et al. 2014) and *Thermobifida* species (Wei et al. 2014b). Apart from bacterial cutinases, fungal cutinases can be used in the hydrolysis and surface modifications of PET films and fibres (Ronkvist et al. 2009). A study by Ronkvist et al. (2009) showed that in comparison to other fungal

Table 8.4 Putative degrading enzymes and microorganisms of plastics

Polymer name (abbreviation)		Putative enzymes	Microorganisms	Reference
Vinyl polymers (polyolefins, styrenics =acrylates. . .)	Poly(ethylene) (PE)	Laccase (EC 1.10.3.2)	*Rhodococcus ruber*	Santo et al. (2013)
		Manganese peroxidase (MnP, EC 1.11.1.13)	*Amycolatopsis species* *T. versicolor* *P. chrysosporium*	Iiyoshi et al. (1998); Muhamad et al. (2015)
	Poly(propylene) (PP)	Laccase (EC 1.10.3.2)	*Phanerochaete chrysosporium* *Engyodontium album*	Jeyakumar et al. (2013)
	Poly(vinyl chloride) (PVC)	Unknown	*Streptomyces (genus)*	Tokiwa et al. (2009)
	Poly(styrene) (PS)	Hydroquinone peroxidase (EC 1.11.1.7)	*Azotobacter beijerinckii*	Nakamiya et al. (1997)
Polyesters	Poly(ethylene terephthalate) (PET)	Cutinase (EC 3.1.1.74)	*Thermobifida fusca*	Ronkvist et al. (2009), Wei et al. (2014a), b), Yoshida et al. (2016), Wei et al. (2016)
		Lipase (EC 3.1.13)	*Thermomyces lanuginosus*	Eberl et al. (2009)
		Carboxylesterase (EC 3.1.1.1)	*Thermobifida fusca*	Billig et al. (2010)
		Unidentified	*Pseudomonas fluorescens* *Pseudomonas putida* *Ochrobactrum (genus)*	Danko et al. (2004), Orr et al. (2006)
	Poly(caprolactone) (PCL)	Cutinase (EC 3.1.1.74)	*Fusarium (genus)*	Murphy et al. (1996), Wei et al. (2014b); Muhamad et al. (2015)
		Lipase	*Clostridium botulinum*	Abou-Zeid et al. (2001), Tokiwa et al. (2009)

(continued)

Table 8.4 (continued)

Polymer name (abbreviation)		Putative enzymes	Microorganisms	Reference
			Rhizopus arrhizus *Rhizopus delemar*	
		Catalase, protease	*Aspergillus niger*	Fesseha and Abebe (2019)
		Glycosidase	*Aspergillus flavus*	Tokiwa et al. (2009)
	Poly(lactic acid) (PLA)	Lipase, cutinase, carboxylesterase, alkaline protease	*Alcanivorax borkumensis* *Rhodopseudomonas palustris*	Hajighasemi et al. (2016)
		Manganese peroxidase	*Amycolatopsis species*	Muhamad et al. (2015)
	Poly(hydroxyl-alkanoate) (PHA)	PHA depolymerase	*Pseudomonas* spp.	Kim et al. (2007)
		Unidentified	*Brevibacillus borstelensis*	Calabia and Tokiwa (2006)
	Poly(butylene succinate co-adipate) (PBSA)	Cutinase (EC 3.1.1.74)	*Fusarium solani* *Pseudozyma antarctica*	Hu et al. (2016), Shinozaki et al. (2013)
		Lipase (EC 3.1.3)	*Rhizopus arrhizus* *Rhizopus delemar*	Thirunavukarasu et al. (2016), Fesseha and Abebe (2019)
Polyethers	Poly(ethylene glycol) (PEG)	Dehydrogenases	*Stenotrophomonas maltophilia*	White et al. (1996), Kawai (2002)
	Poly(tetramethylene glycol) (PTMEG)	Dehydrogenases	*Alcaligenes denitrificans* subsp. *denitrificans* *Pseudomonas maltophilia*	White et al. (1996), Kawai (2002)
Polyamides	Polyamide 6.6 (PA6.6)	Manganese peroxidase (EC 1.11.1.13)	*Bjerkandera adusta*	Deguchi et al. (1998), Friedrich et al. (2007)
		Laccase (EC 1.10.3.2)	–	Fujisawa et al. (2001)

		Nylon hydrolase (EC 3.5.1.117)	*Agromyces* sp.	Negoro et al. (2012)
	Polyamide 11 (PA11)	–	–	–
Polyurethanes	Polyester-polyol polyurethane	Cutinase (EC 3.1.1.74)		Schmidt et al. (2017)
		Esterase (EC 3.1.1.1)	*Comamonas acidovorans*	Akutsu et al. (1998)
		Aryl acylamidase (EC 3.5.1.13)	*Bacillus sphaericus*	Akutsu et al. (2006)
		Elastase (EC 3.4.21.36)	–	Labo et al. (1996)
		Urethanase (EC 3.5.1.75)	*Pseudomonas chlororaphis* *Alternaria* sp.	Ruiz et al. (1999), Matsumiya et al. (2010)
	Polyether-polyol polyurethane	Aryl acylamidase (EC 3.5.1.13)		Akutsu et al. (2006)
		Elastase (EC 3.4.21.36)		Labo et al. (1996)
		Urethanase (EC 3.5.1.75)	*Pseudomonas chlororaphis* *Alternaria* sp.	Ruiz et al. (1999), Matsumiya et al. (2010)
Polycarbonates	Poly(bisphenol A carbonate) (PBPA)	–	–	–
	Poly(trimethylene carbonate) (PTMC)	Lipase	*Candida antarctica*	Suyama and Tokiwa (1997), Matsumura et al. (2001)

cutinases, *Thermomyces insolens* have the highest activity against PET films with low crystallinity (Ronkvist et al. 2009). Apart from cutinases, lipases can be used to degrade PET (Eberl et al. 2009; Ronkvist et al. 2009). It has been reported that lipases are not very active in PET hydrolysis, which could be attributed to the buried hydrophobic catalytic centre covered by a lid structure. This attribute hinders aromatic polymeric substrates from entering the active site of the enzyme (Zimmermann and Billig 2011).

In addition, laccases, manganese peroxidases, and lignin peroxidases can degrade lignin (Friedrich et al. 2007) and a complex cross-linked aromatic polymer composed of phenylpropanoid units (Ruiz-Dueñas and Martínez 2009). Lignin is a highly hydrophobic, aromatic compound found in the cell walls of plants that is resistant to degradation, similar to many plastics (Sheth et al. 2019). It has been suggested that PE can be biodegraded by these oxidoreductases (Restrepo-Flórez et al. 2014). For example, a thermostable laccase isolated from *Rhodococcus ruber* C208 can biodegrade UV-irradiated PE films by increasing the amount of carbonyl groups and reducing the molecular weight within the amorphous part of the PE films (Santo et al. 2013). Another study reported that adding a "mediator" reagent can increase laccase activity (Fujisawa et al. 2001). For example, nylon with nonphenolic structures could be easily oxidized by *Trametes versicolor* fungi, and its molecular weight was reduced over fivefold in 3 days (Fujisawa et al. 2001).

In addition, petroleum-based plastic is extremely resistant to biodegradation because it contains carbon fractions (Wierckx et al. 2018). Therefore, recently developed thermochemical conversion techniques, such as gasification and pyrolysis, can be applied as suitable alternatives instead of traditional incineration (Tanigaki et al. 2013). Gasification and pyrolysis use high temperatures to decompose waste without using a high amount of oxygen compared to the process of traditional incineration. It has been suggested that the residues of the thermochemical treatments of plastic solid waste could be upgraded to value-added chemicals and biopolymers by using aerobic microbial approaches. For example, biodegradable polymers such as PHA can be produced by metabolically versatile bacteria *Pseudomonas*, which utilize the secondary product as substrates derived from thermochemical process. Another study reported that PHA can be produced via the thermal decomposition process (e.g. pyrolysis) by soil bacteria supplied with a solid fraction containing terephthalic acid (Kenny et al. 2008). In addition, PE can be used as a substrate for pyrolysis-mediated remediation (Guzik et al. 2014). It has been demonstrated that postconsumer PE waste can be converted into medium-chain-length polyhydroxyalkanoate (Guzik et al. 2014).

8.7 Conclusion

Plastic waste is ubiquitous and accumulates on shorelines and throughout the different water columns. Land-based sources include leakage from landfills and industrial activities, direct dumping, and sewage discharge. Abandoned fishing nets are important sea-based sources of plastic that contribute to the high abundance of plastic waste in marine waters. The common physical effects on marine organisms induced by plastic waste are plastic ingestion and entanglement. Due to the high surface area-to-volume ratios of plastic waste, hydrophobic organic contaminants can be adsorbed to plastics and potentially cause endocrine system disruption or hormone alterations once ingested by organisms. The natural degradation mechanisms of plastic waste, such as photodegradation and hydrolytic degradation, are limited because most commercial plastics involve different additives, which are extremely resistant to degradation. However, recent studies have suggested a new sustainable way to utilize different enzymes, such as cutinases, lipases, and laccases, to degrade plastics with high persistence, such as PET and PE.

Acknowledgments The financial support provided by grants from the Dean's Research Fund (Ref: IRS-10-2020) and Seed Funding Grant (Ref: RG 53/2019-2020R) of the Education University of Hong Kong is gratefully acknowledged.

References

Abou-Zeid D-M, Muller R-J, Deckwer W-D (2001) Degradation of natural and synthetic polyesters under anaerobic conditions. J Biotechnol 86:113–126

Acampora H, Lyashevska O, Van Franeker J-A, O'Connor I (2016) The use of beached bird surveys for marine plastic litter monitoring in Ireland. Mar Environ Res 120:122–129

Acampora H, Schuyler Q-A, Townsend K-A, Hardesty B-D (2014) Comparing plastic ingestion in juvenile and adult stranded short-tailed shearwaters (Puffinus tenuirostris) in eastern Australia. Mar Pollut Bull 78(1-2):63–68

Adams J, Scott D, McKechnie S, Blackwell G, Shaffer S, Moller H (2009) Effects of geolocation archival tags on reproduction and adult body mass of sooty shearwaters (Puffinus griseus). N Zeal J Zool 36(3):355–366

Aguilera M-A, Broitman B-R, Thiel M (2016) Artificial breakwaters as garbage bins: structural complexity enhances anthropogenic litter accumulation in marine intertidal habitats. Environ Pollut 214:737–747

Akutsu Y, Nakajima K-T, Nomura N, Nakahara T (1998) Purification and properties of a polyester polyurethane-degrading enzyme from Comamonas acidovorans TB-35. Appl Environ Microbiol 64:62–67

Akutsu S-Y, Adachi Y, Yamada C, Toyoshima K, Nomura N, Uchiyama H, Nakajima K-T (2006) Isolation of a bacterium that degrades urethane compounds and characterization of its urethane hydrolase. Appl Environ Microbiol 70:422–429

Alfaro S-J, Mangel J-C, Pajuelo M, Dutton P-H, Seminoff J-A, Godley B-J (2010) Where small can have a large impact: structure and characterization of small-scale fisheries in Peru. Fish Res 106: 8–17

Allen B-M, Angliss R-P (2014) Northern fur seal (Callorhinus ursinus): Eastern Pacific Stock; annual human-caused mortality and serious injury national oceanic and atmospheric administration. https://media.fisheries.noaa.gov/dam-migration/ak2014_northernfurseal-ep_508.pdf

Allen N-S, Edge M, Mohammadian M, Jones K (1991) Hydrolytic degradation of poly(ethylene terephthalate): importance of chain scission versus crystallinity. Eur Polym J 27:1373–1378

Alomar C, Deudero S (2017) Evidence of microplastic ingestion in the shark Galeus melastomus Rafinesque, 1810 in the continental shelf off the western Mediterranean Sea. Environ Pollut 223:223–229

Alomar C, Estarellas F, Deudero S (2016) Microplastics in the Mediterranean Sea: deposition in coastal shallow sediments, spatial variation and preferential grain size. Mar Environ Res 115:1–10
Andrady A-L (2011) Microplastics in the marine environment. Mar Pollut Bull 62:1596–1605
Andrady A-L (2017) The plastic in microplastics: A review. Mar Pollut Bull 119(1):12–22
Artham T, Doble M (2008) Biodegradation of aliphatic and aromatic polycarbonates. Macromol Biosci 8(1):14–24
Austin H-P, Allen M-D, Donohoe B-S, Rorrer N-A, Kearns F-L, Silveira R-L (2018) Characterization and engineering of a plastic-degrading aromatic polyesterase. Proc Natl Acad Sci 115: 4350–4357
Avio C-G, Gorbi S, Regoli F (2017) Plastics and microplastics in the oceans: from emerging pollutants to emerged threat. Mar Environ Res 128:2–113
Baker P-J, Poultney C, Liu Z, Gross R-A, Montclare J-K (2012) Identification and comparison of cutinases for synthetic polyester degradation. Appl Microbiol Biotechnol 93:229–240
Barnes D-K, Galgani F, Thompson R-C, Barlaz M (2009) Accumulation and fragmentation of plastic debris in global environments. Philos Trans R Soc B Biol Sci 364:1985–1998
Barrett R-T, Camphuysen K, Anker N-T, Chardine J-W, Furness R-W, Garthe S, Veit R-R (2007) Diet studies of seabirds: a review and recommendations. ICES J Mar Sci 64(9):1675–1691
Bellas J, Martnez A-J, Martnez C-A, Besada V, Martnez G-C (2016) Ingestion of microplastics by demersal fish from the Spanish Atlantic and Mediterranean coasts. Mar Pollut Bull 109:55–60
Besseling E, Foekema E-M, Van Franeker J-A, Leopold M-F, Kühn S, Bravo Rebolledo E-L, Heße M-L, Ijzer J, Kamminga P, Koelmans A-A (2015) Microplastic in a macro filter feeder: humpback whale Megaptera novaeangliae. Mar Pollut Bull 95:248–252
Billig S, Oeser T, Birkemeyer C, Zimmermann W (2010) Hydrolysis of cyclic poly(ethylene terephthalate) trimers by a carboxylesterase from Thermobifida fusca KW3. Appl Microbiol Biotechnol 87:1753–1764
Bjorndal K-A (1980) Nutrition and grazing behavior of the green turtle Chelonia mydas. Mar Biol 56:147–154
Boland R-C, Donohue M-J (2003) Marine debris accumulation in the nearshore marine habitat of the endangered Hawaiian monk seal, Monachus schauinslandi 1999-2001. Mar Pollut Bull 46(11):1385–1394
Brate L-N, Eidsvoll D-P, Steindal C-C, Thomas K-V (2016) Plastic ingestion by Atlantic cod (Gadus morhua) from the Norwegian coast. Mar Pollut Bull 112:105–110
Bravo RE-L, van Franeker J-A, Jansen O-E, Brasseur S-M (2013) Plastic ingestion by harbour seals (Phoca vitulina) in The Netherlands. Mar Pollut Bull 67(1):200–202
Brian J-V, Harris C-A, Scholze M, Backhaus T, Booy P, Lamoree M, Pojana G, Jonkers N, Runnalls T, Bonfà A, Marcomini A, Sumpter J-P (2005) Accurate prediction of the response of freshwater fish to a mixture of estrogenic chemicals. Environ Health Perspect 113(6): 721–728
Browne M-A, Crump P, Nivens S-J, Teuten E, Tonkin A, Galloway T, Thompson R (2011) Accumulation of microplastics on shorelines worldwide: sources and sinks. Environ Sci Technol 45(21):9175–9179
Butterworth A (2016) A review of the welfare impact on pinnipeds of plastic marine debris. Front Mar Sci 3:149
Calabia B-P, Tokiwa Y (2006) A novel PHB depolymerase from a thermophilic Streptomyces sp. Biotechnol Lett 28(6):383–388
Campagna C, Falabella V, Lewis M (2007) Entanglement of southern elephant seals in squid fishing gear. Mar Mamm Sci 23:414–418
Campanale C, Massarelli C, Savino I, Locaputo V, Uricchio V-F (2020) A detailed review study on potential effects of microplastics and additives of concern on human health. Int J Environ Res Public Health 17(4):1212
Campbell E, Alfaro S-J, Mangel J-C (2017) Whale entanglements in Peru: frequency and cost, in 22nd Biennial Conference on the Biology of Marine Mammals (Halifax)

Carey M-J (2011) Intergenerational transfer of plastic debris by Short-tailed Shearwaters (Ardenna tenuirostris). Emu 111:229–234

Carlos de Sa L, Luís L, Guilhermino L (2015) Effects of microplastics on juveniles of the common goby (Pomatoschistus microps): confusion with prey, reduction of the predatory performance and efficiency, and possible influence of developmental conditions. Environ Pollut 196:359–362

Carson H-S, Lamson M-R, Nakashima D, Toloumu D, Hafner J, Maximenko N, McDermid K-J (2013) Tracking the sources and sinks of local marine debris in Hawai'i. Mar Environ Res 84: 76–83

Cen Y, Singh W, Arkin M, Moody T-S, Huang M, Zhou J, Wu Q, Reetz M-T (2019) Artificial cysteine-lipases with high activity and altered catalytic mechanism created by laboratory evolution. Nat Commun 10(1):3198

Cheung P-K, Cheung T-O, Fok L (2016) Seasonal variation in the abundance of marine plastic debris in the estuary of a subtropical macro-scale drainage basin in South China. Sci Total Environ 562:658–665

Choy C-A, Drazen J-C (2013) Plastic for dinner? Observations of frequent debris ingestion by pelagic predatory fishes from the central North Pacific. Mar Ecol Prog Ser 485:155–163

Chua C-W, Shibata M, Lei M, Toivanen R, Barlow L-J, Bergren S-K, Badani K-K, McKiernan J-M, Benson M-C, Hibshoosh H, Shen M-M (2014) Single luminal epithelial progenitors can generate prostate organoids in culture. Nat Cell Biol 16(10):951–954

Clukey K-E, Lepczyk C, Balazs G, Work T, Lynch J-M (2017) Investigation of plastic debris ingestion by four species of sea turtles collected as bycatch in pelagic Pacific longline fisheries. Mar Pollut Bull *120(1-2)*:117–125

Codina G-M, Militão T, Moreno J, Gonzalez S-J (2013) Plastic debris in Mediterranean seabirds. Mar Pollut Bull 77:220–226

Cole M (2016) A novel method for preparing microplastic fibers. Sci Rep 6:34519

Cole M, Lindeque P, Halsband C, Galloway T-S (2011) Microplastics as contaminants in the marine environment: a review. Mar Pollut Bull 62(12):2588–2597

Cole M, Webb H, Lindeque P, Fileman E-S, Halsband C, Galloway T-S (2014) Isolation of microplastics in biota-rich seawater samples and marine organisms. Sci Rep 4:4528

Collignon A, Hecq J-H, Galgani F, Collard F, Goffart A (2014) Annual variation in neustonic micro- and meso-plastic particles and zooplankton in the Bay of Calvi (Mediterranean–Corsica). Mar Pollut Bull 79:293–298

Conley K, Clum A, Deepe J, Lane H, Beckingham B (2019) Wastewater treatment plants as a source of microplastics to an urban estuary: removal efficiencies and loading per capita over one year. Water Res X 3:100030. https://doi.org/10.1016/j.wroa.2019.100030

Copinet A, Bertrand C, Govindin S, Coma V, Couturier Y (2004) Effects of ultraviolet light (315 nm), temperature and relative humidity on the degradation of polylactic acid plastic films. Chemosphere 55(5):763–773

Courtene J-W, Quinn B, Gary S-F, Mogg O-M, Narayanaswamy B-E (2017) Microplastic pollution identified in deep-sea water and ingested by benthic invertebrates in the Rockall Trough, North Atlantic Ocean. Environ Pollut 231:271–280

Cousin H-R, Auman H-J, Alderman R, Virtue P (2015) The frequency of ingested plastic debris and its effects on body condition of Short-tailed Shearwater (Puffinus tenuirostris) pre-fledging chicks in Tasmania, Australia. Emu—Aust Ornithol 115:6–11

Cózar A, Echevarria F, Gonzalez-Gordillo J-I, Irigoien X, Ubeda B, Hernandez L-S (2014) Plastic debris in the open ocean. Proc Natl Acad Sci U S A 111:10,239–10,244

Culbert B, Christel A (2004) In: Scheirs J, Long TE (eds) Wiley series in polymer science. John Wiley and Sons, Ltd, Chichester, UK, pp 143–194

Danko A-S, Luo M, Bagwell C-E, Brigmon R-L, Freedman D-L (2004) Involvement of linear plasmids in aerobic biodegradation of vinyl chloride. Appl Environ Microbiol 70:6092–6097

de Sá L-C, Luís L-G, Guilhermino L (2015) Effects of microplastics on juveniles of the common goby (Pomatoschistus microps): confusion with prey, reduction of the predatory performance

and efficiency, and possible influence of developmental conditions. Environ Pollut 196:359–362
de Stephanis GJ, Carpinelli E, Canadas A (2013) As main meal for sperm whales: plastics debris. Mar Pollut Bull 69(1-2):206–214
Deguchi T, Kitaoka Y, Kakezawa M, Nishida T (1998) Purification and characterization of a nylon-degrading enzyme. Appl Environ Microbiol 64:1366–1371
Derraik J-GB (2002) The pollution of the marine environment by plastic debris: a review. Mar Pollut Bull 44:842–852
Desforges J-PW, Galbraith M, Ross PS (2015) Ingestion of microplastics by zooplankton in the northeast Pacific Ocean. Arch Environ Contam Toxicol 69:320–330
Devriese L, Vandendriessche S, Theetaert H, Vandermeersch G, Hostens K, Robbens J (2014) Occurrence of synthetic fibres in brown shrimp on the Belgian part of the North Sea. Platform presentation, International workshop on fate and impact of microplastics in marine ecosystems (MICRO2014). Plouzane (France), 13–15 January 2014
de Lucia G-A, Caliani I, Marra S, Camedda A, Coppa S, Alcaro L, Campani T, Giannetti M, Coppola D, Cicero A-M, Panti C, Baini M, Guerranti C, Marsili L, Massaro G, Fossi M-C, Matiddi M (2014) Amount and distribution of neustonic micro-plastic off the western Sardinian coast (Central-Western Mediterranean Sea). Mar Environ Res 100:10–16
Donlan R-M (2002) Biofilms: microbial life on surfaces. Emerg Infect Dis Jl 8(9):881–890
Donnelly G-E-L, Nevins H-M, Harvey J-M (2019) Entangled seabird and marine mammal reports from citizen science surveys from coastal California (1997–2017). Mar Pollut Bull 149:110,557
Duncan E-M, Arrowsmith J-A, Bain C-E (2019) Diet-related selectivity of macroplastic ingestion in green turtles (Chelonia mydas) in the eastern Mediterranean. Scientific Reports
Eastman C-B, Farrell J-A, Whitmore L, Rollinson Ramia D-R, Thomas R-S, Prine J, Eastman S-F, Osborne T-Z, Martindale M-Q, Duffy D-J (2020) Plastic ingestion in post-hatchling sea turtles: assessing a major threat in Florida near shore waters. Front Mar Sci
Eberl A, Heumann S, Brückner T, Araujo R, Cavaco-Paulo A, Kaufmann F, Kroutil W, Guebitz G-M (2009) Enzymatic surface hydrolysis of poly(ethylene terephthalate) and bis (benzoyloxyethyl) terephthalate by lipase and cutinase in the presence of surface active molecules. J Biotechnol 143:207–212
Edge M, Hayes M, Mohammadian M, Allen N-S, Jewitt T-S, Brems K, Jones K (1991) Aspects of poly(ethylene terephthalate) degradation for archival life and environmental degradation. Polym Degrad Stab 32:131–153
Egbeocha C-O, Malek S, Emenike C-U, Milow P (2018) Feasting on microplastics: ingestion by and effects on marine organisms. Aquat Biol 27:93–106
Eich A, Mildenberger T, Laforsch C, Weber M (2015) Biofilm and diatom succession on polyethylene (PE) and biodegradable plastic bags in two marine habitats: early signs of degradation in the pelagic and benthic zone? PLoS One
Fendall L-S, Sewell M-A (2009) Contributing to marine pollution by washing your face: microplastics in facial cleansers. Mar Pollut Bull 58:1225–1228
Fernando S-S, Christensen P-A, Egerton T-A, White J-R (2009) Humidity dependence of carbon dioxide generation during photodegradation of biaxially oriented polypropylene in oxygen. Polym Degrad Stab 94:83–89
Fesseha H, Abebe F (2019) Degradation of plastic materials using microorganisms: a review. Public Health 4(2):57–63
Fife D-T, Robertson G-J, Shutler D, Braune B-M, Mallory M-L (2015) Trace elements and ingested plastic debris in wintering dovekies (Alle alle). Mar Pollut Bull 91:368–371
Fischer V, Elsner N-O, Brenke N, Schwabe E, Brandt A (2015) Plastic pollution of the Kuril–Kamchatka Trench area (NW pacific). Deep-Sea Res II Top Stud Oceanogr 111:399–405
Fok L, Cheung P-K (2015) Hong Kong at the Pearl River estuary: a hotspot of microplastic pollution. Mar Pollut Bull 99(1-2):112–118
Friedrich J, Zalar P, Mohorčič M, Klun U, Kržan A (2007) Ability of fungi to degrade synthetic polymer nylon-6. Chemosphere 67:2089–2095

Fujisawa M, Hirai H, Nishida T (2001) Degradation of polyethylene and nylon-66 by the laccase-mediator system. J Polym Environ 9:103–108
Gallo F, Fossi C, Weber R, Santillo D, Sousa J, Ingram I, Nadal A, Romano D (2018) Marine litter plastics and microplastics and their toxic chemicals components: the need for urgent preventive measures. Environ Sci Eur 30:13
Gambash S, Kochba M, Avnimelech Y (1990) Studies on slow-release fertilizers .2. A method for evaluation of nutrient release rate from slow-releasing fertilizers. Soil Sci 150(1):446–450
Gardette M, Perthue A, Gardette J-L, Janecska T, Foldes E, Pukanszky B, Therias S (2013) Photo and thermal oxidation of polyethylene. Comparison of mechanisms and influence of unsaturated content. Polym Degrad Stab 98:2383–2390
Gewert B, Plassmann M-M, MacLeod M (2015) Pathways for degradation of plastic polymers floating in the marine environment. Environ Sci: Processes Impacts 17:1513–1521
Gewert B, Plassmann M, Sandblom O, MacLeod M (2018) Identification of chain scission products released to water by plastic exposed to ultraviolet light. Environ Sci Technol Lett 5:272–276
Ghosh S-K, Pal S, Ray S (2013) Study of microbes having potentiality for biodegradation of plastics. Environ Sci Pollut Res 20:4339–4355
Gijsman P, Meijersn G, Vitarelli G (1999) Comparison of the UV-degradation chemistry of polypropylene, polyethylene, polyamide 6 and polybutylene terephthalate. Polym Degrad Stab 65:433–441
Gilardi K-V-K, Carlson-Bremer D, June J-A, Antonelis K, Broadhurst G, Cowan T (2010) Marine species mortality in derelict fishing nets in Puget Sound, WA and the cost/benefits of derelict net removal. Mar Pollut Bull 60:376–382
Goldstein MC, Rosenberg M, Cheng L (2012) Increased oceanic microplastic debris enhances oviposition in an endemic pelagic insect. Biol Lett 8:817–820
Gore A-C, Chappell V-A, Fenton S-E, Flaws J-A, Nadal A, Prins G-S, Toppari J, Zoeller R-T (2015) EDC-2: the endocrine society's second scientific statement on endocrine-disrupting chemicals. Endocr Rev 36(6):E1–E150
Gouin T (2020) Toward an improved understanding of the ingestion and trophic transfer of microplastic particles: critical review and implications for future research. Environ Toxicol Chem 39(6):1119–1137
Greenpeace Germany (2019) Ghost gear: the abandoned fishing nets haunting our oceans. Hamburg
Gregory M-R, Andrady A-L (2003) Plastic in the marine environment. In: Andrady A-L (ed) Plastics and the environment. John Wiley, New York, pp 379–401
Gu J-D (2003) Microbiological deterioration and degradation of synthetic polymeric materials: recent research advances. Int Biodeter Biodegr 52(2):69–91
Guzik M-W, Kenny S-T, Duane G-F, Casey E, Woods T, Babu R-P, Nikodinovic-Runic J, Murray M, O'Connor K-E (2014) Conversion of post consumer polyethylene to the biodegradable polymer polyhydroxyalkanoate. Appl Microbiol Biotechnol 98:4223–4232
Hahladakis J-N, Velis C-A, Weber R, Iacovidou E, Purnell P (2018) An overview of chemical additives present in plastics: migration, release, fate and environmental impact during their use, disposal and recycling. J Hazard Mater 344:179–199
Hajighasemi M, Nocek B-P, Tchigvintsev A, Brown G, Flick R, Xu X, Cui H, Hai T, Joachimiak A, Golyshin P-N, Savchenko A, Edwards E-A, Yakunin A-F (2016) Biochemical and structural insights into enzymatic depolymerization of polylactic acid and other polyesters by microbial carboxylesterases. Biomacromolecules 17:2027–2039
Heggøy O, Christensen-Dalsgaard S, Ranke P-S, Chastel O, Bech C (2015) GPSloggers influence behaviour and physiology in the black-legged kittiwake Rissa tridactyla. Mar Ecol Prog Ser 521: 237–248
Hoarau L, Ainley L, Jean C, Ciccion S (2014) Ingestion and defecation of marine debris by loggerhead sea turtles, Caretta caretta, from by-catches in the South-West Indian Ocean. Mar Pollut Bull 84:90–96
Hu X, Gao Z, Wang Z, Su T, Yang L, Li P (2016) Enzymatic degradation of poly(butylene succinate) by cutinase cloned from Fusarium solani. Polym Degrad Stab 134:211–219

Iannilli V, Corami F, Grasso P, Lecce F, Buttinelli M, Setini A (2020) Plastic abundance and seasonal variation on the shorelines of three volcanic lakes in Central Italy: can amphipods help detect contamination? Environ Sci Pollut Res 27:14,711–14,722

Iannilli V, Pasquali V, Setini A, Corami F (2019) First evidence of microplastics ingestion in benthic amphipods from Svalbard. Environ Res 179:108,811

Iiyoshi Y, Tsutsumi Y, Nishida T (1998) Polyethylene degradation by lignin-degrading fungi and manganese peroxidase. J Wood Sci 44:222

Im J, Joo S, Lee Y, Kim B-Y, Kim T (2020) First record of plastic debris ingestion by a fin whale (Balaenoptera physalus) in the sea off East Asia. Mar Pollut Bull 159:111,514

Ishibashi H, Watanabe N, Matsumura N, Hirano M, Nagao Y, Shiratsuchi H, Kohra S, Yoshihara S, Arizono K (2005) Toxicity to early life stages and an estro- genic effect of a bisphenol A metabolite, 4-methyl-2,4- bis(4-hydroxyphenyl)pent-l-ene on the medaka (Oryzias latipes). Life Sci 11:2643–2655

Isobe A, Uchida K, Tokai T, Iwasaki S (2015) East Asian seas: a hot spot of pelagic microplastics. Mar Pollut Bull 101:618–623

Isobe A, Uchiyama-Matsumoto K, Uchida K, Tokai T (2017) Microplastics in the Southern Ocean. Mar Pollut Bull 114(1):623–626

Ivar do Sul J-A, Costa M-F, Barletta M, Cysneiros F-J-A (2013) Presence of pelagic microplastics around an archipelago of the Equatorial Atlantic. Mar Pollut Bull 75:305–309

Jambeck J-R, Geyer R, Wilcox C, Siegler T-R, Perryman M, Andrady A, Narayan R, Law K-L (2015) Plastic waste inputs from land into the ocean. Science 347(6223):768–771

James S-L, Robinson A-J, Arnold J-C, Worsley D-A (2013) The effects of humidity on photodegradation of poly(Vinyl Chloride) and polyethylene as measured by the CO2 evolution rate. Polym Degrad Stab 98(2):508–513

Jamieson A-J, Brooks L-S-R, Reid W-D-K, Piertney S-B, Narayanaswamy B-E, Linley T-D (2019) Microplastics and synthetic particles ingested by deep-sea amphipods in six of the deepest marine ecosystems on Earth. R Soc Open Sci:6180667

Jeyakumar D, Chirsteen J, Doble M (2013) Synergistic effects of pretreatment and blending on fungi mediated biodegradation of polypropylenes. Bioresour Technol 148:78–85

Jin C, Christensen P-A, Egerton T-A, Lawson E-J, White J-R (2006) Rapid measurement of polymer photo-degradation by FTIR spectrometry of evolved carbon dioxide. Polym Degrad Stab 91(5):1086–1096

Jorissen F-J (2014) Colonization by the benthic foraminifer Rosalina (Tretomphalus) concinna of Mediterranean drifting plastics in CIESM 2014. CIESM Publisher, Monaco, p 180

Kaiser J (2010) The dirt on ocean garbage patches. Science 328(5985):1506

Kane I-A, Clare M-A, Miramontes E, Wogelius R, Rothwell J-J, Garreau P, Pohl F (2020) Seafloor microplastic hotspots controlled by deep-sea circulation. Science:368

Kanhai L-S-L, Officer R, Lyashevska O, Thompson R, O'Connor I (2017) Microplastic abundance, distribution and composition along a latitudinal gradient in the Atlantic Ocean. Mar Pollut Bull:307–314

Katija K, Choy C-A, Sherlock R-E, Sherman A-D, Robison R-H (2017) From the surface to the seafloor: how giant larvaceans transport microplastics into the deep sea. Sci Adv 3(8):e1700715

Kawai F (2002) Microbial degradation of polyethers. Appl Microbiol Biotechnol 58:30–38

Kawai F, Kawabata T, Oda M (2019) Current knowledge on enzymatic PET degradation and its possible application to waste stream management and other fields. Appl Microbiol Biotechnol 103:4253–4268

Kawai F, Oda M, Tamashiro T, Waku T, Tanaka N, Yamamoto M, Mizushima H, Miyakawa T, Tanokura M (2014) A novel Ca^{2+}-activated, thermostabilized polyesterase capable of hydrolyzing polyethylene terephthalate from *Saccharomonospora viridis* AHK 190. Appl Microbiol Biotechnol 98:10,053–10,064

Kenny S-T, Runic J-N, Kaminsky W, Woods T, Babu R-P, Keely C-M, Blau W, O'Connor K-E (2008) Up-cycling of PET (polyethylene terephthalate) to the biodegradable plastic PHA (polyhydroxyalkanoate). Environ Sci Technol 42:7696–7701

Kershaw P, Katsuhiko S, Lee S, Samseth J, Woodring D (2011) Plastic debris in the ocean in UNEP Year Book 2011: emerging issues in our global environment. United Nations Environment Programme, Nairobi
Kim D-Y, Kim H-W, Chung M-G, Rhee Y-H (2007) Biosynthesis, modification, and biodegradation of bacterial medium-chain-length polyhydroxyalkanoates. J Microbiol 45:87–97
Klein S, Worch E, Knepper T-P (2015) Occurrence and spatial distribution of microplastics in river shore sediments of the Rhine-main area in Germany. Environ Sci Technol 49:6070–6076
Kuhn S, van Franeker J-A (2012) Plastic ingestion by the northern fulmar (Fulmarus glacialis) in Iceland. Mar Pollut Bull 64(6):1252–1254
Kuo F-J, Huang H-W (2014) Strategy for mitigation of marine debris: analysis of sources and composition of marine debris in northern Taiwan. Mar Pollut Bull 83:70–78
Labo R-S, Erfle D-J, Santerre J-P (1996) Elastase-induced hydrolysis of synthetic solid substrates: poly(ester-urea-urethane) and poly(ether-urea-urethane). Biomaterials 17:2381–2388
Laglbauer B-J-L, Franco-Santos R-M, Andreu-Cazenave M, Brunelli L, Papadatou M, Palatinus A, Grego M, Deprez T (2014) Macrodebris and microplastics from beaches in Slovenia. Mar Pollut Bull 89:356–366
Lamba N-M-K (1998) Polyurethanes in biomedical applications. CRC Press, Boca Raton
Lawson TJ, Wilcox C, Johns K, Dann P, Hardesty B-D (2015) Characteristics of marine debris that entangle Australian fur seals (Arctocephalus pusillus doriferus) in southern Australia. Mar Pollut Bull 98(1-2):354–357
Lebreton L-C, Greer S-D, Borrero J-C (2012) Numerical modelling of floating debris in the world's oceans. Mar Pollut Bull 64(3):653–661
Lebreton L, Slat B, Ferrari F, Sainte-Rose B, Aitken J, Marthouse R, Hajbane S, Cunsolo S, Schwarz A, Levivier A, Noble K, Debeljak P, Maral H, Schoeneich-Argent R, Brambini R, Reisser J (2018) Evidence that the Great Pacific Garbage patch is rapidly accumulating plastic. Sci Rep 8(1):4666
Lee J, Hong S, Song Y-K, Jang Y-C, Jiang M, Heo N-W, Han G-M, Kang D, Shim W-J (2013) Relationships among the abundances of plastic debris in different size, classes on beaches in South Korea. Mar Pollut Bull 77:349–354
Li W-C, Tse H-F, Fok L (2016) Plastic waste in the marine environment: a review of sources, occurrence and effects. Sci Total Environ 566-567:333–349
Lindborg V-A, Ledbetter J-F, Walat J-M, Moffett C (2012) Plastic consumption and diet of Glaucous-winged Gulls (Larus glaucescens). Mar Pollut Bull 64(11):2351–2356
Liu B, He L, Wang L, Li T, Li C, Liu H, Luo Y, Bao R (2018) Protein crystallography and site-direct mutagenesis analysis of the poly(ethylene terephthalate) hydrolase PETase from Ideonella sakaiensis. Chembiochem 19(14):1471–1475
Lobelle D, Cunlife M (2011) Early microbial biofilm formation on marine plastic debris. Mar Pollut Bull 62(1):197–200
Lusher A-L, Burke A, O'Connor I, Officer R (2014) Microplastic pollution in the Northeast Atlantic Ocean: validated and opportunistic sampling. Mar Pollut Bull 88(1-2):325–333
Lusher A-L, Tirelli V, Connor I, Officer R (2015) Microplastics in Arctic polar waters: the first reported values of particles in surface and sub-surface samples. Sci Rep 5:14,947
Lusher A-L, McHugh M, Thompson R-C (2013) Occurrence of microplastics in the gastrointestinal tract of pelagic and demersal fish from the English channel. Mar Pollut Bull 67:94–99
Macfayden G, Huntington T, Cappell R (2009) Abandoned, lost or otherwise discarded fishing gear. UNEP Regional Seas Reports and Studies 185, FAO Fisheries and Aquaculture Technical Paper 523, United Nations Environment Programme, Food and Agriculture Organisation of the United Nations, Rome
Magnusson K (2014) Microlitter and other microscopic anthropogenic particles in the sea area off Raunma and Turku. IVL Swedish Environmental Research Institute, Stockholm, p 18
Mandich A, Bottero S, Benfenati E, Cevasco A, Erratico C, Maggioni S, Massari A, Pedemonte F, Vigano L (2007) In vivo exposure of carp to graded concentrations of bisphenol A. Gen Comp Endocrinol 153:15–24

Marion G-M, Millero F-J, Camoes M-F, Spitzer P, Feistel R, Chen C-T-A (2011) pH of seawater. Mar Chem 126:89–96

Matsumura S, Harai S, Toshima K (2001) Lipase-catalyzed transformation of poly(trimethylene carbonate) into cyclic monomer, trimethylene carbonate: A new strategy for sustainable polymer recycling using an enzyme. Macromol Rapid Commun 22(3):215–218

Mathieu-Denoncourt J, Wallace S-J, de Solla S-R, Langlois V-S (2015) Plasticizer endocrine disruption: highlighting developmental and reproductive effects in mammals and non-mammalian aquatic species. Gen Comp Endocrinol 219:74–88

Matiddi M, Hochsheid S, Camedda A (2017) Loggerhead sea turtles (Caretta caretta): a target species for monitoring litter ingested by marine organisms in the Mediterranean Sea. Environ Pollut 230:199–209

Matsumiya Y, Murata N, Tanabe E, Kubota K, Kubo M (2010) Isolation and characterization of an ether-type polyurethane-degrading micro-organism and analysis of degradation mechanism by Alternaria sp. J Appl Microbiol 108:1946–1953

McIntosh R-R, Kirkwood R, Sutherland D-R, Dann P (2015) Drivers and annual estimates of marine wildlife entanglement rates: a long-term case study with Australian fur seals. Mar Pollut Bull 101:716–725

Min K, Cuiff J-D, Mathers R-T (2020) Ranking environmental degradation trends of plastic marine debris based on physical properties and molecular structure. Nat Commun 11(1):727

Montazer Z, Habibi Najafi M-B, Levin D-B (2020) Challenges with verifying microbial degradation of polyethylene. Polymers 12(1):123

Muhamad W, Othman R, Shaharuddin R-I, Irani M-S (2015) Microorganism as plastic biodegradation agent towards sustainable environment. Adv Environ Biol 9:8–14

Müller R-J, Kleeberg I, Deckwer W-D (2001) Biodegradation of polyesters containing aromatic constituents. J Biotechnol 86:87–95

Munn S, Goumenou M (2013) Thresholds for endocrine disrupters and related uncertainties. Report of the Endocrine Disrupters Expert Advisory Group (ED EAG). European Commission, Joint Research Centre. Institute for Health and Consumer Protection.

Murphy C-A, Cameron J-A, Huang S-J, Vinopal R-T (1996) Fusarium polycaprolactone depolymerase is cutinase. Appl Environ Microbiol J 62:456–460

Muthu S-S (2014) Roadmap to sustainable textiles and clothing: environmental and social aspects of textiles and clothing supply chain. Springer, New York

Naidoo T (2015) Plastic pollution in five urban estuaries of KwaZulu-Natal, South Africa (MSc Thesis) University of KwaZulu-Natal 1–69

Nakamiya K, Sakasita G, Ooi T, Kinoshita S (1997) Enzymatic degradation of polystyrene by hydroquinone peroxidase of Azotobacter beijerinckii HM121. J Ferment Bioeng 84:480–482

Negoro S, Shibata N, Tanaka Y, Yasuhira K, Shibata H, Hashimoto H, Lee Y-H, Oshima S, Santa R, Oshima S, Mochiji K, Goto Y, Ikegami T, Nagai K, Kato D, Takeo M, Higuchi Y (2012) Three-dimensional structure of nylon hydrolase and mechanism of nylon-6 hydrolysis. J Biol Chem 287:5079–5090

Nel H, Hean J, Noundou X, Froneman P (2017) Do microplastic loads reflect the population demographics along the southern African coastline? Mar Pollut Bull 115(1-2):115–119

Nelms S-E, Duncan E-M, Broderick A-C, Galloway T-S, Godfrey M-H, Hamann M, Lindeque P-K, Godley B-J (2015) Plastic and marine turtles: a review and call for research. ICES J Mar Sci:1–17

Nor N-H-M, Obbard J-P (2014) Microplastics in Singapore's coastal mangrove system. Mar Pollut Bull 79:278–283

O'Brine T, Thompson R-C (2010) Degradation of plastic carrier bags in the marine environment. Mar Pollut Bull 60(12):2279–2283

Oberbeckmann S, Loder M-G-J, Labrenz M (2015) Marine microplastic-associated biofilms—a review. Environ Chem 12:551–562

Oehlmann J-R, Schulte-Oehlmann U, Kloas W, Jagnytsch O, Lutz I, Kusk K-O, Wollenberger L, Santos E-M, Paull G-C, Van Look K-J-W, Tyler C-R (2009) A critical analysis of the biological impacts of plasticizers on wildlife. Philos Trans R Soc B 364:2047–2062

Orr I-G, Hadar Y, Sivan A (2006) Colonization, biofilm formation and biodegradation of polyethylene by a strain of Rhodococcus ruber. Appl Microbiol Biotechnol 65(1):97–104

Pickett J-E, Coyle D-J (2013) Hydrolysis kinetics of condensation polymers under humidity aging conditions. Polym Degrad Stab 98:1311–1320

Plastics Europe (2019). Plastics—the facts 2019: an analysis of European plastics production, demand and waste data

Plotkin P, Amos A-F (1990) Effects of anthropogenic debris on sea turtles in the northwestern Gulf of Mexico. Proceedings, Second International Conference on Marine Debris, 2–7 April 1989, Honolulu, HI, USA

Premraj R, Doble M (2005) Biodegradation of polymers. Indian J Biotechnol 4(2):186–193

Provencher J-F, Bond A, Avery-Gomm S, Borrelle S, Bravo R, Hammer S, Khn S, Lavers J, Mallory M, Trevail A, Van Franeker J (2017) Quantifying ingested debris in marine megafauna: a review and recommendations for standardization. Anal Methods 9(9):1454–1469

Provencher J-F, Bond A-L, Mallory M-L (2014) Marine birds and plastic debris in Canada: a national synthesis, and a way forward. Environ Rev 23(1):1–13

Provencher J-F, Borrelle S-B, Bond A-L, Lavers J-L, van Franeker J-A, Kuhn S, Hammer S, Avery-Gomm S, Mallory M-L (2019) Recommended best practices for plastic and litter ingestion studies in marine birds: collection, processing, and reporting. FACETS 4:111–130

Rabek J-F (1995) Polymer photodegradation mechanisms and experimental methods. Springer, Dordrecht

Rånby B (1989) Photodegradation and photo-oxidation of synthetic polymers. J Anal Appl Pyrolysis 15:237–247

Raum-Suryana K-L, Jemisonb L-A, Pitcherc K-W (2009) Entanglement of Steller sea lions (Eumetopias jubatus) in marine debris: identifying causes and finding solutions. Mar Pollut Bull 58(10):1487–1495

Reisser J, Shaw J, Hallegraeff G, Proietti M, Barnes D-K-A, Thums M (2014) Millimeter-sized marine plastics: a new pelagic habitat for microorganisms and invertebrates. PLoS One 9: e100289

Reisser J, Shaw J, Wilcox C, Hardesty B-D, Proiett M, Thums M, Pattiaratchi C (2013) Marine plastic pollution in waters around Australia: characteristics, concentrations, and pathways. PLoS One 8(11):e80466

Restrepo-Florez J-M, Bassi A, Thompson M-R (2014) Microbial degradation and deterioration of polyethylene—a review. Int Biodeter Biodegr 88:83–90

Rizzi M, Rodrigues F-L, Medeiros L, Ortega I, Rodrigues L, Monteiro D-S, Kessler F, Proietti M-C (2019) Ingestion of plastic marine litter by sea turtles in southern Brazil: abundance, characteristics and potential selectivity. Mar Pollut Bull 140:536–548

Roch S, Friedrich C, Brinker A (2020) Uptake routes of microplastics in fishes: practical and theoretical approaches to test existing theories. Sci Rep 10:3896

Rochman C-M, Hoh E, Hentschel B-T, Kaye S (2013) Long-term field measurement of sorption of organic contaminants to five types of plastic pellets: implications for plastic marine debris. Environ Sci Technol 47:1646–1654

Rodríguez B, Bécares J, Rodríguez A, Arcos J-M (2013) Incidence of entanglements with marine debris by northern gannets (Morus bassanus) in the non-breeding grounds. Mar Pollut Bull 75: 259–263

Romeo T, Pietro B, Peda C, Consoli P, Andaloro F, Fossi M-C (2015) First evidence of presence of plastic debris in stomach of large pelagic fish in the Mediterranean Sea. Mar Pollut Bull 95:358–361

Ronkvist A, Xie W, Lu W, Gross R-A (2009) Cutinase-catalyzed hydrolysis of poly(ethylene terephthalate). Macromolecules 42:5128–5138

Ruiz C, Main T, Hilliard N-P, Howard G-T (1999) Purification and characterization of two polyurethanase enzymes from Pseudomonas chlororaphis. Int Biodeter Biodegr 43:43–47
Ruiz-Dueñas F-J, Martínez A-T (2009) Microbial degradation of lignin: how a bulky recalcitrant polymer is efficiently recycled in nature and how we can take advantage of this. J Microbial Biotechnol 2(2):164–177
Rummel C-D, Jahnke A, Gorokhova E, Kuhnel D, Schmitt-Jansen M (2017) Impacts of biofilm formation on the fate and potential effects of microplastic in the aquatic environment. Environ Sci Technol Lett 4:258–267
Rummel C-D, Loder M-G, Fricke N-F, Lang T, Griebeler E-M, Janke M, Gerdts G (2016) Plastic ingestion by pelagic and demersal fish from the north sea and baltic sea. Mar Pollut Bull 102: 134–141
Santo M, Weitsman R, Sivan A (2013) The role of the copper-binding enzyme—laccase—in the biodegradation of polyethylene by the actinomycete Rhodococcus ruber. Int Biodeter Biodegr 84:204–210
Santos R-G, Andrades R, Fardim L-M, Martins A-S (2016) Marine debris ingestion and Thayer's law—the importance of plastic color. Environ Pollut 214:585–588
Schmidt J, Wei R, Oeser T, Dedavide Silva L, Breite D, Schulze A, Zimmermann W (2017) Degradation of polyester polyurethane by bacterial polyester hydrolases. Polymer 9:65
Schuyler Q, Hardesty B-D, Wilcox C, Kathy T (2012) To eat or not to eat? Debris selectivity by marine turtles. PLoS One 7(7):e40884
Shahul Hamid F, Bhatti M-S, Anuar N, Anuar N, Mohan P, Periathamby A (2018) Worldwide distribution and abundance of microplastic: how dire is the situation? Waste Manag Res 36(10): 873–897
Sheth M-U, Kwartler S-K, Schmaltz E-R, Hoskinson S-M, Martz E-J, Dunphy-Daly M-M, Schultz T-F, Read A-J, Eward W-C, Somarelli J-A (2019) Bioengineering a future free of marine plastic waste. Front Mar Sci 6:624
Shinozaki Y, Morita T, Cao X-H, Yoshida S, Koitabashi M, Watanabe T, Suzuki K, Sameshima-Yamashita Y, Nakajima-Kambe T, Fujii T, Kitamoto H-K (2013) Biodegradable plastic-degrading enzyme from Pseudozyma antarctica: cloning, sequencing, and characterization. Appl Microbiol Biotechnol 97:2951–2959
Shirke A-N, White C, Englaender J-A, Zwarycz A, Butterfoss G-L, Linhardt R-J (2018) Stabilizing leaf and branch compost cutinase (LCC) with glycosylation: mechanism and effect on PET hydrolysis. Biochemistry 57:1190–1200
Singh B, Sharma N (2008) Mechanistic implications of plastic degradation. Polym Degrad Stab 93: 561–584
Singh S, Rawat P-S (2019) Biodegradation of Plastic: an innovative solution to safe the human health and environment. In: Handbook of research on environmental and human health impacts of plastic pollution. Pennsylvania, USA, IGI Global, pp 435–461
Song Y-K, Hong S-H, Jang M, ShimW-J (2015) Occurrence and distribution of microplastics in the sea surface microlayer in Jinhae Bay, South Korea. Arch Environ Contam Toxicol 69(3):121
Summers J-W, Rabinovitch E-B (1999) Weatherability of vinyl and other plastics. In: Wypych G (ed) Weathering of plastics testing to mirror real life performance. William Andrew, Norwich, pp 61–68
Suyama T, Tokiwa Y (1997) Enzymatic degradation of an aliphatic polycarbonate, poly (tetramethylene carbonate). Enzyme Microb Technol 20:122–126
Szycher M (2013) Szycher's handbook of polyurethanes, Taylor & Francis, 2nd edn, Boca Raton, FL
Talsness C-E, Andrade A-J-M, Kuriyama S-N, Taylor J-A, vom Saal F-S (2009) Components of plastic: experimental studies in animals and relevance for human health. Philos Trans R Soc B 364:2079–2096
Tanaka K, Takada H, Yamashita R, Mizukawa K, Fukuwaka M-A, Watanuki Y (2013) Accumulation of plastic-derived chemicals in tissues of seabirds ingesting marine plastics. Mar Pollut Bull 69(1–2):219–222

Tanigaki N, Fujinaga Y, Kajiyama H, Ishida Y (2013) Operating and environmental performances of commercial-scale waste gasification and melting technology. Waste Manag Res 31:1118–1124

Terepocki A-K, Brush A-T, Kleine L-U, Shugart G-W, Hodum P (2017) Size and dynamics of microplastic in gastrointestinal tracts of Northern Fulmars (Fulmarus glacialis) and Sooty Shearwaters (Ardenna grisea). Mar Pollut Bull 116(1-2):143–150

Teuten E-L, Rowland S-J, Galloway T-S, Thompson R-C (2007) Potential for plastics to transport hydrophobic contaminants. Environ Sci Technol 41:7759–7764

Thiel M, Bravo M, Hinojosa I-A, Luna G, Miranda L, Núñez P (2011) Anthropogenic litter in the SE Pacific: an overview of the problem and possible solutions. Integr Coast Zone Manage 11: 115–134

Thiel M, Luna-Jorquera G, Álvarez-Varas R, Gallardo C, Hinojosa I-A, Luna N, Miranda-Urbina D, Morales N, Ory N, Pacheco A-S, Portflitt-Toro M, Zavalaga C (2018) Impacts of marine plastic pollution from continental coasts to subtropical gyres—fish, seabirds, and other vertebrates in the SE Pacific. Front Mar Sci 5:238

Thirunavukarasu K, Purushothaman S, Sridevi J, Aarthy M, Gowthaman M-K, Nakajima-Kambe T, Kamini N-R (2016) Degradation of poly(butylene succinate) and poly(butylene succinate-co-butylene adipate) by a lipase from yeast Cryptococcus sp. grown on agro-industrial residues. Int Biodeter Biodegr 110:99–107

Thompson R-C (2015) Microplastics in the marine environment: sources, consequences and solutions. In: Bergmann M, Gutow L, Klages M (eds) Marine anthropogenic litter. Springer International, New York, NY, pp 185–200

Tokiwa Y, Calabia B-P, Ugwu C-U, Aiba S (2009) Biodegradability of plastics. Int J Mol Sci 10(9): 3722–3742

Tolinski M (2009) Additives for polyolefins getting the most out of polypropylene, polyethylene and TPO. William Andrew Pub, Oxford

Townsend A-K, Barker C-M (2014) Plastic and the nest entanglement of urban and agricultural crows. PLoS One 9(1):e88006

Trevail A-M, Gabrielsen G-W, Kuhn S, van Franeker J-A (2015) Elevated levels of ingested plastic in a high Arctic seabird, the northern fulmar (Fulmarus glacialis). Polar Biol 38:975–981

Urbanek A-K, Rymowicz W, Mirończuk A-M (2018) Degradation of plastics and plastic-degrading bacteria in cold marine habitats. Appl Microbiol Biotechnol 102(18):7669–7678

Van Cauwenberghe L, Vanreusel A, Mees J, Janssen C-R (2013) Microplastic pollution in deep-sea sediments. Environ Pollut 182:495–499

van Franeker J-A, Blaize C, Danielse J, Fairclough K, Gollan J, Guse N, Hansen P-L, Heubeck M, Jensen J-K, Le Guillou G, Olsen O, K-O PJ, Stienen E-W, Turner D-M (2011) Monitoring plastic ingestion by the northern fulmar Fulmarus glacialis in the North Sea. Environ Pollut 159: 2609–2615

Van Houtan K, Francke D, Alessi S, Jones T, Martin S, Kurpita L, King C, Baird R (2016) The developmental biogeography of hawksbill sea turtles in the North Pacific. Ecol Evol 6(8): 2378–2389

Vasile C, Pascu M (2005) Practical guide to polyethylene. Rapra Technology Limited, Shrewsburg

Veerasingam S, Mugilarasanb M, Venkatachalapathyc R, Vethamonya P (2016) Influence of 2015 flood on the distribution and occurrence of microplastic pellets along the Chennai coast, India. Mar Pollut Bull 109(1):196–204

Venkatachalam S, Nayak S-G, Labde J-V, Gharal P-R, Rao K, Kelkar A-K (2012) Degradation and recyclability of poly (ethylene terephthalate), InTech

Verlis K-M-M, Campbell M-L-L, Wilson S-P-P (2013) Ingestion of marine debris plastic by the wedge-tailed shearwater, Ardenna pacifica, in the Great Barrier Reef, Australia. Mar Pollut Bull 72:244–249

Waluda C-M, Staniland I-J (2013) Entanglement of Antarctic fur seals at Bird Island, South Georgia. Mar Pollut Bull 74(1):244–252

Wardrop P, Shimeta J, Nugegoda D, Morrison P, Miranda A, Tang M, Clarke B (2016) Chemical pollutants sorbed to ingested microbeads from personal care products accumulate in fish. Environ Sci Technol 50(7):4037–4044

Wei R, Oeser T, Barth M, Weigl N, Lübs A, Schulz-Siegmund M, Hacker M-C, Zimmermann W (2014a) Turbidimetric analysis of the enzymatic hydrolysis of polyethylene terephthalate nanoparticles. J Mol Catal B: Enzym 103:72–78

Wei R, Oeser T, Schmidt J, Meier R, Barth M, Then J, Zimmermann W (2016) Engineered bacterial polyester hydrolases efficiently degrade polyethylene terephthalate due to relieved product inhibition. Biotechnol Bioeng 113:1658–1665

Wei R, Oeser T, Then J, Kühn N, Barth M, Schmidt J, Zimmermann W (2014b) Functional characterization and structural modeling of synthetic polyester-degrading hydrolases from Thermomonospora curvata. AMB Exp 4:44

Weinstein J, Crocker B, Gray A (2016) From macroplastic to microplastic: degradation of high-density polyethylene, polypropylene, and polystyrene in a salt marsh habitat. Environ Toxicol Chem 35(7):1632–1640

White G-F, Russell N-J, Tidswell E-C (1996) Bacterial scission of ether bonds. Microbiol Rev 60: 216–232

Wierckx N, Narancic T, Eberlein C, Wei R, Drzyzga O, Magnin A, Ballerstedt H, Kenny S-T, Pollet E, Avérous L (2018) Plastic biodegradation: challenges and opportunities. In: Consequences of microbial interactions with hydrocarbons, oils, and lipids: biodegradation and bioremediation. Springer International Publishing, Cham, Switzerland

Wilcox C, Heathcote G, Goldberg J, Gunn R, Peel D, Hardesty B-D (2015) Understanding the sources and effects of abandoned, lost and discarded fishing gear on marine turtles in Northern Australia. Conserv Biol 29(1):198–206

Wilkes R-A, Aristilde L (2017) Degradation and metabolism of synthetic plastics and associated products by *Pseudomonas* sp.: capabilities and challenges. J Appl Microbiol 123(3):582–593

Winton D-J, Anderson L-G, Rocliffe S, Loiselle S (2020) Macroplastic pollution in freshwater environments: focusing public and policy action. Sci Total Environ 704:135,242

Wright S-L, Thompson R-C, Galloway T-S (2013) The physical impacts of microplastics on marine organisms: a review. Environ Pollut 178:483–492

Wu C, Zhang K, Xiong X (2018) Microplastic pollution in inland waters focusing on Asia. In: Wagner M, Lambert S (eds) Freshwater microplastics, The handbook of environmental chemistry, vol 58. Springer, Cham

Yamashita R, Tanimura A (2007) Floating plastic in the Kuroshio Current area, western North Pacific Ocean. Mar Pollut Bull 54(4):485–488

Yoshida S, Hiraga K, Takehana T, Taniguchi I, Yamaji H, Maeda Y, Toyohara K, Miyamoto K, Kimura Y, Oda K (2016) A bacterium that degrades and assimilates poly(ethylene terephthalate). Science 351(6278):1196–1199

Zacharia J-T (2019) Degradation pathways of persistent organic pollutants (POPs) in the environment, persistent organic pollutants, Stephen Kudom Donyinah, IntechOpen, https://www.intechopen.com/books/persistent-organic-pollutants/degradation-pathways-of-persistent-organic-pollutants-pops-in-the-environment

Zettler E-R, Mincer T-J, Amaral-Zettler L-A (2013) Life in the plastisphere: microbial communities on plastic marine debris. Environ Sci Technol 47:7137–7146

Zhao S, Zhu L, Wang T, Li D (2014) Suspended microplastics in the surface water of the Yangtze Estuary System, China: first observations on occurrence, distribution. Mar Pollut Bull 86(1-2): 562–568

Zimmermann W, Billig S (2011) Enzymes for the biofunctionalization of poly(ethylene terephthalate). Adv Biochem Eng Biotechnol 125:97–120

Mitigation of the Plastic Pollution in the Marine Environment to Conserve the Marine Biota: An Overview

9

Sarika N. Kanade and Avinash B. Ade

Abstract

Plastics constitute the major portion of the solid waste entering the marine environment. Various environmental forces lead to the generation of small particles of the plastic which are referred to as microplastic, mesoplastic and nanoplastic. These plastic particles are more harmful, and due to extremely small size, they enter into food chain, disrupt the metabolic process and lead to malfunction of the tissues and organs and ultimately the behaviour of the organisms. To avoid the plastic pollution, several strategies are followed. The processes like landfilling, incineration and recycling are good for the macroplastics where the plastic waste can be easily collected and utilized in these processes. These processes have their own benefits and harms; however, these may not be adequate for the management of small- and minute-sized plastic waste. The minute-sized plastics are more harmful as these are ingested by the marine organisms and hence require specific strategies to overcome this accumulation. In the present chapter, the impact of micro- and nanoplastics on marine organisms is discussed.

Keywords

Microplastics · Nanoplastic · Recycling · Terrestrial · Aquatic · Ecosystems · Marine · Organisms

S. N. Kanade
Centre for Science Education and Communication, Savitribai Phule Pune University, Pune, Maharashtra, India

A. B. Ade (✉)
Department of Botany, Savitribai Phule Pune University, Pune, Maharashtra, India
e-mail: avinashade@unipune.ac.in

M. Shahnawaz et al. (eds.), *Impact of Plastic Waste on the Marine Biota*,
https://doi.org/10.1007/978-981-16-5403-9_9

9.1 Introduction

Plastic is the polymer derived from petrochemical which is unique in certain characteristics (Andrady and Neal 2009). It is used in different manners by modifications to certain extent. It occurs basically with two types, thermoplastic and thermoset plastic (Deng et al. 2015). Thermoplastic type is mouldable on supply of heat; therefore, it can regain the shape like that of the earlier (Mallick 2021). The examples of thermosetting polymers are epoxides and phenolic. This is further differentiated into two subtypes, crystalline or amorphous (Masuelli 2013). The examples of the crystalline forms are polyethylene, polyethylene terephthalates and nylon (Wang et al. 2014). The amorphous can be observed in the form of polystyrene and polycarbonate. Overall the plastic material is versatile and can be considered as novel alternative to the metals. Plastic substances are light weighted, highly flexible, moisture resistant, strong in tensile strength and relatively cheap as compared to any other material used for making goods of daily requirements (Asthana et al. 2006). Due to these qualities, plastic material is widely used all over the world. The plastic is highly inert and can withstand in the environment for longer time, and every year billion tonnes of plastic gets generated. Due to different routes and pathways, the plastic waste finally enters the oceans and leads to the death of billions of marine birds, animals, reptiles, fishes, etc (British Plastics Federation 2018). So, in the present chapter, it was attempted to overview the mitigation process of the plastic litter in the oceans to save the marine flora and fauna from the detrimental impact of plastic particles on the inhabitants of the marine environment.

9.2 Existing Methods of Tackling Plastic Pollution

Many strategies are employed to combat plastic pollution. These are recycling, landfilling and incineration. The larger plastic can be processed for recycling which is transformed into several products. Landfilling is not perfect if the plastic is not homogenized with the soil and vulnerable for exposure in case of soil erosion. Incineration results in release of toxic gases like chlorine from PVC material. Biodegradation is the best solution as it involves the microbes from soil. It is a comparatively cheap and eco-friendly method.

9.3 Fractionation of Plastic

In the plastic garbage disposal sites, the plastic is broken down into smaller pieces because of human or natural activities. These pieces are ultimately mixed with the soil and become part of runoff water. These pieces can be categorized as microplastic and nanoplastic depending on their size.

9.3.1 Microplastic

Microplastics are the fragments of plastic with maximum 5 mm in size. These are further fractionated into nanoplastics (Blair and Quinn 2016). Like any other plastic material, microplastics and nanoplastics also enter in the ecosystems through the cosmetics and clothing industries. Microplastic beads are used for making of toiletry products such as soaps, shampoo, deodorants, toothpastes, wrinkle creams, moisturizers, shaving creams, sunscreen lotions, facial masks, lipsticks, eye shadows, children's bubble bath, etc. (Bhattacharya 2016; Shahnawaz et al. 2019). The cleansers are updated as the major sources of the microplastic as per the recent figures. These contain polyethylene beads. These are therefore released in the river with the runoff water and later in ocean directly (Fendell and Sewell 2009). The actual studies regarding the effect of microplastics are rather difficult because of the unavailability of the information regarding plastic content in the cosmetics as the policies of so many manufacturers do not permit it.

Microplastics are also incorporated into the ecosystem in the form of fine fibres obtained from clothing according to the US National Oceanic and Atmospheric Administration (NOAA) and the European Chemicals Agency (2020). These are the synthetic fibres made of acrylic and polyester material (Arthur et al. 2009; Collignon et al. 2014).

Recent classifications of microplastics suggest two categories, primary microplastics and secondary. Primary microplastics are obtained from clothing and other allied products, microbeads or the pellets (Cole et al. 2013; Boucher and Friot 2017). Secondary microplastics are formed after degradation of larger plastic products through natural weathering processes after entering the environment. The secondary microplastics are obtained from the water and soda bottles, fishing nets, plastic carry bags and microwave containers (Boucher and Friot 2017; Conkle et al. 2018). Microplastics persisted in the environment at considerable levels in the soil and entered the terrestrial and aquatic ecosystems including fresh and marine ecosystems. Although the size is less, it takes much time for decomposition like any other plastics, and on the contrary, the risk of ingestion by the living system is increased. The microplastics are detected in the bodies of several organisms (Chamas et al. 2020). The toxic chemicals released from the microplastics from the sources of ocean and runoff are biomagnified in the food chain (Grossman 2015a; b; Nex 2021).

The marine environment is found to be contaminated with the microplastics from all the corners of the world. These are reported from both tropical and temperate conditions.

9.3.2 Nanoplastics

Nanoplastics are used in domestic and industrial applications. Some of these particles are very tiny that these may be carried for longer distances along with the wind currents and mixed with the raindrops or the snow during precipitation.

Nanoplastics are thought to be dangerous to the environment and the human health too as these are extremely tiny and can move across cell membranes and affect the normal functioning of cells. The lipophilic nature of the nanoplastics is evidenced by the predicted models which show that polyethylene nanoplastics can be incorporated into the hydrophobic lipid bilayers (Hollóczki and Gehrke 2020). Nanoplastics are reported to cross the epithelial membrane of fish and accumulate in the organs like the gall bladder, the pancreas and even the brain (Skjolding et al. 2017; Pitt et al. 2018). In zebra fish, polystyrene type of nanoplastics was found to induce a stress response pathway altering glucose and cortisol levels which showed behavioural changes in stress conditions (Brun et al. 2019). In *Daphnia*, polystyrene nanoplastic can be ingested by the freshwater species, *D. pulex*, and it can be found disturbances in growth and reproduction (Liu et al. 2018, 2019, 2020).

9.4 Entry of Plastic in Food Chain

The micro- and nanoplastics present in the sea enter the marine food chain. Consumption of these plastics by commercially important organisms with respect to physiology and toxicity studies has been carried out in the laboratories. In most of the parts of the world, only flesh of the fishes is preferred rather than the viscera due to the fear of plastics. The animals belonging to molluscs such as mussels, clams, scallops and oysters are eaten whole as these are smaller in size; therefore, there is a great risk for incorporating these animals in the diet.

The plastic was reported from the gastrointestinal tract of the annelid species, *Arenicola marina*, and the respiratory system of the crustacean *Carcinus maenas* (Grossman 2015a; b; Watts et al. 2014; Thompson et al. 2004). Due to the mimicking nature of the plastic particles like food material, it is consumed by fish which leads to blockage in the digestive tracts which creates incorrect feeding signals to these animals' brains. It takes longer duration for digestion of plastic rather than normal digestion (Watts et al. 2014). The accumulated plastic in the digestive tract of lantern fish is consumed by the commercial fishes, tuna and swordfish (Cozar et al. 2014). The role of plastics for absorbing chemical pollutants has been depicted by the researchers. Along with the plastics, these pollutants are also incorporated into the organism's tissues (Wardrop et al. 2016).

The benthic sea cucumbers feed on debris at the bottom of the ocean containing plastic (Wright et al. 2013). Upon exposure of microplastic and nanoplastic, Bivalvia's filtration capacity was found to be decreased (Tallec et al. 2018; Oliveira et al. 2018). Marine biologists in 2017 discovered microplastic fibres in underwater seagrass. The seagrass is an important component of the reef ecosystem which serves as food for parrotfish which is an important commercial fish (Hall et al. 2015; Risk and Edinger 2011).

In Europe, the first record of microplastic content was found in 2019 in the stomach of the amphibian European newt (*Triturus carnifex*) which was the first evidence for Caudata group in the amphibians (McAlpine 2019).

Zooplanktons were also reported to consume microplastic beads which was evidenced from their faecal matter containing microplastics. During ingestion the microplastics were found stuck with the appendages and exoskeleton (Boots et al. 2019). The production of dimethyl sulphide was also reported in the excreta of zooplanktons which is the digestion product of polypropylene (Iannella et al. 2019; Cole et al. 2013). The polypropylene is commonly found in plastic containers used for food packaging (Savoca et al. 2016). The microplastic filaments were reported in the planktonic organisms as well as in seaweeds (Dacey and Wakeham 1986).

The plastic is also being consumed by the microbes as these are found on the surface of microplastics. A layer of adsorbed microbes gives appearance of the biofilms to the plastic. Thus plastic sheets are also considered as habitats for the pathogenic microbes. Here exchange of genetic material occurs to develop highly effective pathogenic strains (Wu et al. 2019).

9.5 Hazards of Plastic Waste

Due to involvement of the plastic in our daily life, we have created hazardous consequences to the fellow organisms in the ecosystems, and still we are continuing the same. If this is not controlled in the future or if there are no sustainable strategies to overcome the plastic pollution, then predictions said that by 2050 there will be more plastic than fish in the world's oceans. Around 80% of marine pollution comes from land-based activities. Microplastic waste runs or is dumped into drains and rivers and hence the seas (Plastic Oceans - Future Agenda n.d.). Therefore it is essential to design the strategies from the land to take care the microplastic pollution of the oceans.

9.6 Strategies for Combating Plastic Pollution

For this purpose the following strategies can be implemented.

1. Conservation of water. If the usage of water is reduced, the excess runoff and wastewater will not flow into the ocean which restricts plastic pollution.
2. Reduction in the single-use plastic. The plastic material should be used at very minimum level. The recyclable plastic should be used over the single-use plastic.
3. Proper recycling. As far as possible, the recycling should be carried out immediately after the use of plastic so that its further disposal will not be deposited into the water current and hence into the ocean. The recycling should not be violating the global directives about the plastic recycling.
4. Using biodegradable plastic. It will minimize the time for degradation of the plastic material; hence during the journey from land to the ocean, it will have enough period for degradation.
5. Beach or river clean-up. The beaches and the rivers are to be cleaned regularly which will reduce the flow of plastic to the oceans.

6. Supporting plastic bans. Many countries as well as the local authorities have imposed ban on the plastic usage which is for the benefit of the masses. Therefore it should be supported unanimously.
7. Avoid products containing microbeads. Microbeads are the growing source of ocean plastic pollution in recent years. Microbeads are found in some face scrubs, toothpastes and body washes, and they readily enter the oceans and waterways through sewer systems.
8. Supporting organizations engaged with plastic pollution abatement. Many organizations from local to international as well as government and non-government levels should be supported for their clean environment drives.

Thus, if the management of the microplastic is to be planned and started, this is to be started right now to avoid the future complications in the ocean. The earth should be protected from the generation of fourth layer of plastic in addition to lithosphere, hydrosphere and atmosphere.

9.7 Conclusion

The plastic waste is created due to human activities which goes to the soil and then water streams, and via rivers it takes entry into the oceans where it may be settled down at the bottom or suspended in the water; hence, it causes harms to the living organisms in the oceanic water. To design the control strategies, the appropriate actions are needed on lands.

Acknowledgement The authors are thankful to the funding agencies who promoted the research work in plastic degradation.

References

Andrady AL, Neal MA (2009) Applications and societal benefits of plastics. Philos Trans R Soc Lond Ser B Biol Sci 364(1526):1977–1984

Arthur C, Baker J, Bamford H. (2009) Proceedings of the international research workshop on the occurrence, effects and fate of microplastic marine debris (PDF). NOAA Technical Memorandum

Asthana R, Kumar A, Dahotre NB (2006) Materials processing and manufacturing science. Elsevier

Bhattacharya P (2016) A review on the impacts of microplastic beads used in cosmetics. Acta Biomedica Scientia 3(1):47–52

Blair CC, Quinn B (2016) Microplastic pollutants, 1st edn. Elsevier Science

Boots B, Russell CW, Green DS (2019) Effects of microplastics in soil ecosystems. Environ Sci Technol 53:19. https://doi.org/10.1021/acs.est.9b03304

Boucher J, Friot D (2017) Primary microplastics in the oceans: a global evaluation of sources. IUCN, Gland, Switzerland

British Plastics Federation (2018) Where does marine litter come from? Marine Litter Facts. Retrieved on 5-5-2021

Brun NR, van Hage P, Hunting ER, Haramis AG, Vink SC, Vijver MG, Schaaf MJM, Tudorache C (2019) Polystyrene nanoplastics disrupt glucose metabolism and cortisol levels with a possible link to behavioural changes in larval zebrafish. Commun Biol 2(1):382
Chamas A, Moon H, Zheng J, Qiu Y, Tabassum T, Jang JH, Abu-Omar M, Scott SL, Suh S (2020) Degradation rates of plastics in the environment. ACS Sustain Chem Eng 8(9):3494–3511
Cole M, Lindeque P, Fileman E, Halsband C, Goodhead R, Moger J, Galloway TS (2013) Microplastic ingestion by zooplankton. Environ Sci Technol 47(12):6646–6655
Collignon A, Hecq JH, Galgani F, Collard F, Goffart A (2014) Annual variation in neustonic micro- and meso-plastic particles and zooplankton in the bay of Calvi (Mediterranean–Corsica) (PDF). Mar Pollut Bull 79(1–2):293–298
Conkle JL, Bez Del Valle CD, Turner JW (2018) Are we underestimating microplastic contamination in aquatic environments? Environ Manag 61(1):1–8
Cozar A, Echevarria F, Gonzalez-Gordillo JI, Irigoien X, Ubeda B, Hernandez-Leon S, Palma AT, Navarro S, Garcia-De-Lomas J, Ruiz A, Fernandez-De-Puelles ML, Duarte CM (2014) Plastic debris in the open ocean. Proc Natl Acad Sci 111(28):10239–10244
Dacey JWH, Wakeham SG (1986) Oceanic dimethylsulfide: production during zooplankton grazing on phytoplankton. Science 233(4770):1314–1316
Deng S, Djukic L, Paton R, Ye L (2015) Thermoplastic–epoxy interactions and their potential applications in joining composite structures–a review. Compos A: Appl Sci Manuf 68:121–132
European Chemicals Agency (2020). Restricting the use of intentionally added microplastic particles to consumer or professional use products of any kind. ECHA. European Commission. Retrieved 5-5-2020
Fendell LS, Sewell MA (2009) Contributing to marine pollution by washing your face: microplastics in facial cleansers. Mar Pollut Bull 58(8):1225–1228
Grossman E (2015a). How microplastics from your fleece could end up on your plate, Civil Eats, January 15, 2015
Grossman E (2015b). How plastics from your clothes can end up in your fish. Time
Hall NM, Berry KLE, Rintoul L, Hoogenboom MO (2015) Microplastic ingestion by scleractinian corals. Mar Biol 162(3):725–732
Hollóczki O, Gehrke S (2020) Can nanoplastics alter cell membranes? Chem Phys Chem 21 (1):9–12
Iannella M, Console G, D'Alessandro P (2019) Preliminary analysis of the diet of *Triturus carnifex* and pollution in mountain karst ponds in central Apennines. Water 44(129):11496–11506
Liu Z, Cai M, Yu P, Chen M, Wu D, Zhang M, Zhao Y (2018) Age-dependent survival, stress defense and AMPK in *Daphnia pulex* after short-term exposure to a polystyrene nanoplastic. Aquat Toxicol 204:1–8
Liu Z, Huang Y, Jiao Y, Chen Q, Wu D, Yu P, Li Y, Cai M, Zhao Y (2020) Polystyrene nanoplastic induces ROS production and affects the MAPK-HIF-1/NFkB-mediated antioxidant system in *Daphnia pulex*. Aquat Toxicol 220:105420
Liu Z, Yu P, Cai M, Wu D, Zhang M, Huang Y, Zhao Y (2019) Polystyrene nanoplastic exposure induces immobilization, reproduction, and stress defense in the freshwater cladoceran Daphnia pulex. Chemosphere 215:74–81
Mallick PK (2021) Thermoplastics and thermoplastic–matrix composites for lightweight automotive structures. In: Mallick PK (ed) Materials, design and manufacturing for lightweight vehicles. Woodhead Publishing, Cambridge, pp 187–228
Masuelli MA (2013) Introduction of fibre-reinforced polymers– polymers and composites: concepts, properties and processes. In: Masuelli MA (ed) Fiber reinforced polymers-the technology applied for concrete repair. IntechOpen, Rijeka
McAlpine KJ (2019) Have your plastic and eat it too. Bostonia (Boston University Alumni):36–37
Nex S (2021) How to garden the low carbon way: the steps you can take to help combat climate change (First American ed.). DK, New York
Oliveira P, Barboza LGA, Branco V, Figueiredo N, Carvalho C, Guilhermino L (2018) Effects of microplastics and mercury in the freshwater bivalve Corbicula fluminea (Müller, 1774):

filtration rate, biochemical biomarkers and mercury bioconcentration. Ecotoxicol Environ Saf 164:155–163

Pitt JA, Kozal JS, Jayasundara N, Massarsky A, Trevisan R, Geitner N, Wiesner M, Levin ED, Di Giulio RT (2018) Uptake, tissue distribution, and toxicity of polystyrene nanoparticles in developing zebrafish (*Danio rerio*). Aquat Toxicol 194:185–194

Plastic Oceans - Future Agenda (n.d.) www.futureagenda.org

Risk MJ, Edinger E (2011) Impacts of sediment on coral reefs. Encyclopedia of Modern Coral Reefs. Encyclopedia of Earth Sciences Series, pp 575–586

Savoca MS, Wohlfeil ME, Ebeler SE, Nevitt GA (2016) Marine plastic debris emits a keystone infochemical for olfactory foraging seabirds. Sci Adv 2(11):e1600395

Shahnawaz M, Sangale MK, Ade AB (2019) Bioremediation technology for plastic waste. Springer Nature Singapore Pte Ltd., Singapore

Skjolding LM, Ašmonaitė G, Jølck RI, Andresen TL, Selck H, Baun A, Sturve J (2017) An assessment of the importance of exposure routes to the uptake and internal localisation of fluorescent nanoparticles in zebrafish (*Danio rerio*), using light sheet microscopy. Nanotoxicology 11(3):351–359

Tallec K, Huvet A, Di Poi C, González-Fernández C, Lambert C, Petton B, Le Goïc N, Berchel M, Soudant P, Paul-Pont I (2018) Nanoplastics impaired oyster free living stages, gametes and embryos. Environ Pollut 242(Pt B):1226–1235

Thompson RC, Olsen Y, Mitchell RP, Davis A, Rowland SJ, John AW, McGonigle D, Russell AE (2004) Lost at sea: where is all the plastic? Science 304(5672):838

Wang RY, Chen XD, Xu QJ, Wang YJ, Zhang Q (2014) Study on crystallization performance of polyethylene terephthalate/polybutylene terephthalate alloys. J Polym Eng 34(8):747–754

Wardrop P, Shimeta J, Nugegoda D, Morrison PD, Miranda A, Tang M, Clarke BO (2016) Chemical pollutants sorbed to ingested microbeads from personal care products accumulate in fish. Environ Sci Technol 50(7):4037–4044

Watts AJR, Lewis C, Goodhead RM, Beckett SJ, Moger J, Tyler CR, Galloway TS (2014) Uptake and retention of microplastics by the shore crab *Carcinus maenas*. Environ Sci Technol 48 (15):8823–8830

Wright SL, Thompson RC, Galloway TS (2013) The physical impacts of microplastics on marine organisms: a review. Environ Pollut 178:483–492

Wu X, Pan J, Li M, Li Y, Bartlam M, Wang Y (2019) Selective enrichment of bacterial pathogens by microplastic biofilm. Water Res 165:114979

Mitigation of the Micro- and Nanoplastic Using Phycoremediation Technology

10

Rashidi Othman, Razanah Ramya, Nur Hanie Mohd Latif, Wan Syibrah Hanisah Wan Sulaiman, Farah Ayuni Mohd Hatta, Qurratu Aini Mat Ali, and Nor Hafizana Mat Jusoh

Abstract

Asia is the largest global plastic consumer, with about 35% of the world's plastic consumption. Considering that Malaysia is a part of Asia, it is evident that plastic use is extensive. Unfortunately, discarding plastic causes several environmental hazards and affects human wellbeing. The environmental authorities and the government have been organising campaigns that focus on propagating the reduce, recycling, and reuse concept among the Malaysian public. Nevertheless, after considering the extensive presence of microorganisms in the environment and their affinity towards degrading plastic, the use of such microorganisms and enzymes appears an efficacious approach. Environmental degradation of plastic typically happens through five processes: photodegradation, thermo-oxidative breakdown, hydrolytic degradation, mechanical degradation, and microbial degradation. Microbial degradation comprises plastic breakdown by microorganisms, which produce enzymes that can split long-chain polymers. Microbial enzymes are interesting since they are cost-effective and require

R. Othman (✉) · N. H. M. Jusoh
Herbarium Unit, Department of Landscape Architecture, Kulliyyah of Architecture and Environmental Design, International Islamic University Malaysia, Kuala Lumpur, Malaysia
e-mail: rashidi@iium.edu.my

R. Ramya
Institute of the Malay World and Civilization, The National University of Malaysia, Bangi, Selangor, Malaysia

N. H. M. Latif · W. S. H. W. Sulaiman
International Institute for Halal Research and Training, International Islamic University Malaysia, Kuala Lumpur, Malaysia

F. A. M. Hatta · Q. A. M. Ali
Institute of Islam Hadhari, The National University of Malaysia, Bangi, Selangor, Malaysia

M. Shahnawaz et al. (eds.), *Impact of Plastic Waste on the Marine Biota*,
https://doi.org/10.1007/978-981-16-5403-9_10

minimal maintenance; at the same time, they are easy to manipulate. *Rhizopus delemar*, *R. arrhizus*, *Pseudomonas* sp., *Penicillium funiculosum*, and *Aspergillus flavus* are the five microbes that have been cited extensively regarding their ability to break down specific plastics. Moreover, fungal, bacterial, cyanobacteria, and actinomycetes capabilities for plastic degradation are among the environmentally friendly techniques that can help the environment. This chapter discussed how cyanobacteria could be used to break down plastics. The projected research outcome is the identification of potent microbial agents that can rapidly degrade plastics with minimal environmental impact.

Keywords

Biodegradation mechanism · Cyanobacteria · Plastics · Phycoremediation

10.1 Introduction

Micro- and nanoplastics are pollutants that are becoming a cause for significant concern (Campanale et al. 2020). These particles are microscopic, and many living organisms consume them and which leads to nanoscale physiological effects (Ng et al. 2018). Presently, 95% of the packing material worth about USD 80 billion is discarded annually after a single use. Recycling is conducted as per the widely known three-cycle indicator. The world produces several tonnes of plastic waste every year. Over the last six decades, plastic waste has grown 300 times. The European Union, China, and nations under the North American Free Trade Agreement (NAFTA) were the significant plastic waste producers in 2017 (World Economic Forum 2016; Jiang 2018; Oliveira et al. 2019; Sobhani et al. 2020).

The adverse biological impact of macroplastic pieces on wildlife is widely recognised; plastic leads to starvation and suffocation. On the other hand, finer particles (1 mm) have been researched recently. Plastic can act as a conduit for contaminants and enhance environmental presence (Teuten et al. 2009). Plastic junk is getting accumulated in terrestrial and aquatic habitats. Plastic debris has been discovered in rivers, lagoons, estuaries, and oceans. In this regard, there are two critical challenges: regulation at source and addressing the extensive environmental passive aspects due to six decades of plastic use (plastics have steadily become costlier). Muddy and sandy sediment, plankton, ingestion by invertebrates and vertebrates, microplastics, and chemical effluents are typically researched (Sobhani et al. 2020; Teuten et al. 2009; Enfrin et al. 2020).

Marine species are at significant risk due to the effect of microplastics (Kögel et al. 2020). The challenges are because of two aspects: (i) microplastics are minute and can quickly accumulate in the food chain, thereby getting polluted on the surface, increasing the number of such particles, and (ii) plastic particles can absorb pollutants (Sharma and Chatterjee 2017; Arias-Andres et al. 2018; Deng et al. 2020;). The presence of microplastics in groundwater leads to adverse impacts. Beaches are affected from physical and other perspectives. The critical challenge is to ascertain if the extensive surface of the soil (organic and mineral) can take up these particles and create lumps in the soil, thereby restricting microplastic particles from getting to biota in the soil aggregate hierarchy (Rillig 2012; Wahl et al. 2021).

Microplastic particles have the potential to cause soil toxicity; consequently, cytotoxicity and oxidative stress can lead to enhanced uptake, influence energy equilibrium, or particle migration until soil organisms break down the particles. Numerous terrestrial habitats (soil, water, atmosphere, and sediment) are loaded with microplastic. Moreover, the use of sludge as composting mixtures for agriculture further pollutes soil with microplastics (Xu et al. 2020a, 2020b).

Additionally, research interest in microplastics is growing. There are often misunderstandings concerning the scientific consequences of microplastics; moreover, individuals might overestimate or understand these aspects poorly. Recovery from plastic packaging materials is minimal. 72% of all packaging is discarded: 40% goes into landfills, while 32% disperses through leaks in the collection system, thereby leading to the reduced collection. Such leaks are often dumped or illegally handled (World Economic Forum 2016). Building, packaging, healthcare, medical, and transportation are some sectors that use plastic extensively because of desirable aspects like manufacturing ease, toughness, thermal, corrosion, chemical, and shock resistance. Moreover, plastics can be shaped into almost any desired shape, size, or colour (Da Costa et al. 2016; Wan et al. 2018).

Most manufactured plastic is consumed by the packaging industry, which commonly uses polypropylene (PP), polyethylene (PE) (Crawford and Quinn 2017), low-density polyethylene (LDPE) and high-density polyethylene (HDPE), polymethylmethacrylate, and polystyrene. These substances are subjected to misuse, which causes them to be discarded in aquatic ecosystems in different sizes like nanoplastics (NPs) and microplastics (MPs) (Venâncio et al. 2019). Regardless of the extensive environmental presence, information concerning the microplastic impact on marine biota remains limited, explicitly for particles having less than 100 nm size (NPs).

There is no proof that plastic quantity being dumped in the oceans is being reduced. Recent research suggests that the production of plastic particles might exceed nanoscale (Oliveira et al. 2019). The marine ecosystem gets loaded with microplastics when macro- and mesoplastics are weathered due to abiotic and biotic processes. In processes such as mechanical abrasion, biological breakdown, disintegration, and solar ultraviolet radiation, plastic inside marine environments might break down macroplastic into NPs and MPs (Al-Thawadi 2020). Marine habitats are typically found with a plastic particle density of 0.01–2.3 g/cm^3. Such nanoparticles (particles less than 100 nm diameter) accumulate in household and industrial waste, which subsequently moves to wastewater processing facilities and finally manages to enter aquatic environments.

Several researchers suggest that microplastics typically accumulate inside marine and freshwater systems as specified below:

1. Use or breakdown of minute plastic used for abrasive scrubbing in personal care material (Browne et al. 2007).

2. Domestic and industrial cleaning materials such as the face, hand, body, and toilet cleaners (Xia et al. 2020).
3. The adverse effects of microplastic contaminants due to non-human aspects like rain. Prior research suggests rain as a significant pollution contributor (Xu et al. 2020a, 2020b).

Textile washing-created microfibre shares approximately 35% of the microplastic found in aquatic environments (Deng et al. 2020). Additionally, goods like agricultural plastic sheets, cosmetic and personal grooming products, synthetic turf, tyres, and road coatings contribute to microplastic pollution (Sharma and Chatterjee 2017; Deng et al. 2020).

It is possible to use plastic packaging in several ways to reduce greenhouse gas emissions. About 6% of world oil consumption is attributed to plastic production; about 25% of this is required for packaging plastic production (Ncube et al. 2021). Hence, the production and post-use plastic pathways lead to significant Greenhouse gas emissions (World Economic Forum 2016). Plastic use continues to grow at a substantial rate today; plastic production will account for about 15% of worldwide carbon footprint by 2050, from about 1% presently. More plastic is getting discharged into lakes because of climate change. Sedimental resuspension rate and frequency are increasing (Zhang et al. 2020a, 2020b).

Microplastics have a direct adverse impact on fish, zooplankton, and macroinvertebrates and vice versa. Laboratory investigations about several invertebrate categories indicated plastic ingestion; however, there is a lack of research concerning the effects of invertebrates on microplastics present in marine ecosystems. One probable cause for this shortage is the time- and technology-intensive nature of conducting such research (Barría et al. 2020; Kögel et al. 2020). Microplastics have physical characteristics that threaten living beings; moreover, microplastic ingestion might lead to intestinal blockage and pseudo-satiety, thereby causing stunted growth, reduced energy levels, oxidative stress, or even death. Plastic transformation conducted to obtain the benefits mentioned above leads to waste that causes severe environmental issues because of decomposition resistance, environmental concentration, and persistence (Chang et al. 2020; Xia et al. 2020).

Several organic materials and metals can employ the characteristics and evolving contaminant hazard of plastic-based polymers for microplastic adsorption (Fu et al. 2020a, 2020b). Particles of synthetic rubber disintegrating from car tyres cause maximum microplastic pollution (Kole et al. 2017). Stormwater discharge and runoff from roads lead such particles to enter the aquatic ecosystems (Ziajahromi et al. 2020). Nevertheless, evaluating atmosphere-based microplastic is challenging because there is little clarity concerning the dynamic particle chemical characteristics that change due to environmental aspects. The amount of plastic entering the oceans is approximately 8 million tons, which is equivalent to one truckload of garbage being dumped into every minute (Van Cauwenberghe et al. 2013). If mitigation measures are not deployed, this number is expected to reach two and four truckloads by 2030 and 2050, respectively (Jambeck et al. 2015). Moreover, it is estimated that by 2025, plastic quantity is expected to increase to 250 million tons, corresponding to an ocean plastic-to-fish ratio of 1:3 (Jennings et al. 2008).

Bisphenol A (BPA) and other phthalates utilised as plasticisers for polyvinyl chloride (PVC) production are also concerning substances because of potential adverse impact on health and the environment, thereby leading regulators and business organisations to act on this issue (Swan et al. 2015). Considering that microplastics on land could be about 4–23 times more concentrated compared to marine ecosystems, there are possibilities that plastics affect human health by entering the food chain through routes like taps, poultry, bivalves, and sea salt (Jin et al. 2019). Fu et al. (2020a, 2020b) asserted that microplastic pollution impacts marine creatures and can lead to harmful chemicals entering the food chain. Thirty-three studies conducted in China suggested higher microplastic ingestion by marine creatures (230 species); moreover, it is also true for 26 freshwater aquatic animal species. About 90% of the creatures are affected by microplastic pollution.

Plastics are the mainstay of the manufacturing industry; they offer distinct material characteristics at a minimal cost. Plastic use has increased by about 20 times in the last 50 years. It is projected that plastic use will witness 100% growth in the next 20 years. Presently, most people are regularly exposed to plastic (Wang et al. 2021). Plastic comprises long-chain polymers of organic or inorganic origin like hydrogen, oxygen, carbon, silicone, and chloride (Ivar Do Sul and Costa 2014).

The UNEP estimates that by 2050 that the weight of plastic in the oceans could exceed that of fish (World Economic Forum 2016). Plastic comprises long-chain polymers of organic or inorganic origin like hydrogen, oxygen, carbon, silicone, and chloride (Enfrin et al. 2020; World Economic Forum 2016).

In the present chapter, an attempt was made to overview the potential role of algae and cyanobacteria to mitigate plastic waste in the marine environment. The objectives of this chapter is to identify potential microbial agent that capable to degrade plastics with low environmental damage.

10.2 Microplastics

Microplastics are a reference to plastic particles with less than 5 mm diameter and are of interest for both experimental studies (Xia et al. 2020). Microplastic diameter ranges from 1 μm to 5 mm, and these particles comprise a heterogeneous set of various formed parts (fibres, granules, fragments, flakes, pellets, or beads) (EFSA Panel on Contaminants in the Food 2016; Frias and Nash 2019). The intensity of microplastic is increasing and has an extensive atmospheric presence; consequently, animals are affected. Therefore, microplastic pollution is a concerning global issue. Microplastic particles less than 5 mm size typically enter, move, and collect in natural ecosystems and are present across the poles, seabed, ocean floor, urban beaches, and sediments (Ivar Do Sul and Costa 2014).

Put differently, primary microplastic typically originates as plastic; on the other hand, fragmentation leads to the generation of microplastic (EFSA Panel on Contaminants in the Food 2016). Materials like cleaning substances, industrial abrasives, and cosmetics comprise large plastic parts that degrade to microplastic form (Auta et al. 2017; Rillig et al. 2019).

Microplastic particles are tiny (≤5 mm) and are found in terrestrial and marine ecosystems. Broad surface and volume characteristics cause microplastic particles to adsorb harmful substances from the environment (Ivar Do Sul and Costa 2014; Auta et al. 2017). Presently, over 96% of all studies concerning microplastics focus on marine ecosystems (Xu et al. 2020a, 2020b). Microplastic particle bioavailability might impact human wellbeing and exposed biota (Ziajahromi et al. 2020) because microplastics might exist in primary or secondary forms (Rillig 2012). Almost all aquatic systems like lakes, rivers, marine environments, and reservoirs are loaded with microplastics. Sedimentation and microplastic accumulation lead to increased eutrophication and lead to additional substances, primarily organic, being introduced into the food chain (Zhang et al. 2020a, 2020b).

Hence, high- and low-density microplastics present in marine ecosystems float on the sea surface and are ingested. Numerous organisms may ingest microplastics directly or indirectly. There are two classifications for marine microplastics: primary, produced by the mechanical disintegration of particles from larger plastics or due to biological and chemical degradation; the other category is secondary microplastics (Ivar Do Sul and Costa 2014; Piccardo et al. 2020).

Particle properties, extrinsic factors, physical site aspects, anthropogenic activity considering spatial and temporal aspects, and biofoliation magnitude affect microplastics; however, such data is often unavailable. The impact of biota is estimated to project the marine ecosystem (Meng et al. 2020).Nanoplastics are nanoscale materials having a broad range of external dimensions, nano-level surface characteristics (0.001–0.1 Lm), or internal structure (Meng et al. 2020). Breakdown of microplastic debris can lead to microplastic formation and might arise in the substances used for fabrication processes. When microplastic particles undergo fragmentation, nanoplastics are produced from engineered substrates.

Nanoplastics have a size range of about 1 nm to 100 nm (0.001–0.1 Lm), while microplastics have a size range of about 0.1–5000 Lm. The breakdown and destruction of macro plastics typically cause the formation of microplastic particles less than 1 mm; particle amount increases as size decreases (Zhang et al. 2020a, 2020b). Small plastics are introduced directly into the environment; on the other hand, relatively bigger particles steadily break down. Microplastics are emitted directly by primary sources; these particles reach the atmosphere as minute (mm) pellets and are employed for abrasive industrial application (Barnes et al. 2009). There is extensive plastic particle presence in the environment, and this is a grave concern considering that microparticles (ranging from 100 nm to 5 mm) and nanoparticles (100 nm in a comprehensive evaluation) are very minute that organisms take up such particles (EFSA Panel on Contaminants in the Food 2016).

Plastic may be produced from synthetic or semi-synthetic substances comprising polymers as the primary constituent. These polymers gain mobility when heated; hence, they can be extruded, moulded, or pressed into numerous shapes and dimensions. Plastic is a word derived from the Greek word *plastikos*; it indicates an ability to be shaped or transformed using heat. Plastic is durable and has a prolonged degradation rate, thereby causing pollution and adverse environmental effects. It is incredibly challenging to understand the significance of plastic in human life. Plastic affects human life significantly by allowing access to different varieties of products. It is infeasible to completely stop plastic use even once.

'Polymer' typically refers to plastic substrates, indicating the organic, carbon-derived substances having long connected chains. Chemically, plastics are extended hydrocarbon chain-based polymers with significant molecular weight. Petrochemicals are a prerequisite for plastic production. Chemical phenomena are utilised for the formation of long-chain polymers. Xenophobic components of plastics comprise about 5–8% of municipal solid waste (by dry weight) and are resistant to degradation. These polymers are responsible for environmental effects such as agricultural toxicology, damage to marine ecosystems, ozone depletion, and other adverse effects (Kaur 2014).

Typically, plastics are segmented into seven categories:

1. Natural plastics—these comprise polymers of naturally present substances that are prevalent in nature. Amber is one such resin derived from pine tree fossils Khan and Majeed 2020).
2. Semi-synthetic plastics—these comprise natural substances subjected to chemical processing. Rubber, cellulose acetate, and casein are some examples (Vroman and Tighzert 2009).
3. Synthetic plastics—these are human-made substances obtained by the degradation of carbon-derived material like gas, crude oil, and coal. Modification of molecular structure is typically affected using pressure and heat in petrochemical factories. These factories implement the initial processes required to transform these substances into standard household plastic (Boyle and Ormeci 2020).
4. Thermoplastics—this category comprises plastics that soften upon heating and are formed when soft; upon cooling, they retain the shape given during heat treatment. However, reapplication of heat leads to repeated softening. Styrene and acrylic are thermoplastics (Lambert et al. 2014).
5. Thermosetting plastics—these substances also soften using the thermal treatment and may be shaped into several designs. The shape is retained upon cooling; however, reheating does not soften these plastics; they retain their initial shape. Resin, polyester, and melamine formaldehyde are thermosetting plastics (Kausar et al. 2016).
6. Microplastics—these are derived from industrial and domestic waste; breakdown of waste material leads to plastic debris that accumulates in the environment. Any plastic particle under 5 mm is classified as microplastic (Horton et al. 2017).
7. Nanoplastics—these particles are in size range of 0.001–1 μm; nanoplastics exhibit colloidal characteristics and are formed by the degradation of plastic parts (Ng et al. 2018).

10.3 Society of Plastic Industry (SPI) Resin Identification Code of Plastics Classification

In 1998, the Society of Plastic Industry formulated classification criteria to facilitate users and recyclers to understand the various plastic types used to create widely used household products like packaging and water bottles. Under this system, plastic products have a number engraved on the bottom. The SPI number conveys essential information about the kind of plastic used in the object. SPI codes are described in Table 10.1.

10.4 Plastic Biodegradation Mechanism in the Marine Environment

Distinct marine ecosystems like sea surface, water column, beach, and seafloor lead to marine plastic debris (MPD) exposure to varying environmental conditions; consequently, MPD breakdown rate increases or decreases. Plastic breakdown in various marine ecosystems is affected by environmental aspects like UV intensity, exposure duration, temperature, physical abrasion, and biological breakdown. Plastic breakdown happens due to photo-oxidation resulting from ultraviolet radiation; hence, plastic degrades faster on land than water (Bhuyan et al. 2021). Plastic breakdown in the marine environment is primarily through five processes (Table 10.2) (Zheng et al. 2005):

10.4.1 Photodegradation

Polymers naturally break down through photodegradation; this process is regulated by the amount of light and light-initiated chemical changes because carbonyl groups comprising the polymer chain absorb ultraviolet radiation; consequently, thermo-oxidative degradation results. Solar radiation comprises ultraviolet rays that have the necessary activation energy to start integrating oxygen with polymer molecules (Zheng et al. 2005; Webb et al. 2013). Consequently, plastic particles gradually break into smaller particles because the plastic is brittle. The breakdown process continues until the molecular weight of the polymer chain reaches a level that microorganisms can metabolise (Brandon et al. 2016). The microbes either break the carbon off the polymer to produce carbon dioxide or integrate it with the biomolecules (Muthukumar et al. 2011).

10.4.2 Thermo-Oxidative Degradation

Polymers naturally break down through photodegradation; this process is regulated by the amount Oxidative degradation involves the disintegration of macromolecules by the action of oxygen on the substrate (oxidation). Free radicals are formed which

Table 10.1 SPI resin identification code for plastic classification

RIC code	Material		References
1	PET (polyethylene terephthalate)	It is the most widely used thermoplastic resin polymer; it falls in the polyester classification and utilised for fibre, which is ultimately used for manufacturing clothes, containers for food, thermoforming in industry. Moreover, it is used with glass fibre for resin production	Käppler et al. (2016)
2	HDPE (high-density polyethylene)	Ethylene is used as a prerequisite chemical for producing thermoplastic polymer. Pipes with corrosion resistance, plastic lumber, plastic bottles, and geomembranes are some applications of HDPE	Kothari (2008)
3	PVC (polyvinyl chloride)	It is an artificial plastic polymer that is produced in flexible and rigid forms. The rigid variant is applied for pipe manufacturing and window and door production. Non-food packaging material, bottles, sheets for covering food, and cards are some other applications. On the other hand, the flexible variant is employed for fabrication insulation for electrical wires, plumbing material, signage, floor material, imitation leather, and phonograph cartridges [8], and inflatable objects; it is used as a replacement material for rubber [9]. Canvas is fabricated by using PVC with linen or cotton	Bhuyan et al. (2021)
4	LDPE (low-density polyethylene)	LDPE is an ethylene-derived thermoplastic that is extensively utilised for producing dispensing containers, bottles, tubes, plastic objects for computers, lab equipment; however, plastic bags are its most common application	Bhuyan et al. (2021)
5	PP (polypropylene)	PP is a thermoplastic that has extensive applications. This polymer is fabricated using the chain growth technique; propylene monomer leads to polymer formation. PP has excellent fatigue resistance, and plastic hinges constitute a significant application area	Bhuyan et al. (2021)
6	PS (polystyrene)	PS is a synthetic polymer of the aromatic hydrocarbon category; styrene is its monomer form [5]. It is typically used for producing packaging material since it can be produced in foamed or solid variants	Bhuyan et al. (2021)
7	Others	Nylon, polylactic acid (PLA), polycarbonate, and multilayer varieties of numerous plastics fall under this category	Bhuyan et al. (2021)

react with oxygen-producing oxy- and peroxy-radicals (Chamas et al. 2020). These, on the other hand, can participate in many reactions: they may react with each other or remove hydrogen from polymer chains. Oxidative degradation can be initiated by three factors such as UV light, heat, or mechanical stress in the presence of oxygen atmosphere (Brandon et al. 2016).

Oxidative degradation can proceed according to two mechanisms: photo-oxidation and thermal oxidation. Photo-oxidation is caused by the action of UV

Table 10.2 Modes of plastics degradation and their contributing factors

	Degradation mechanism	Contributing factors	Reference
1.	Thermal degradation	Exposure to heat	Kopinke et al. (1996), McNeill and Leiper (1985a, 1985b)
2.	Thermo-oxidative degradation	Exposure to heat and oxygen	Kopinke et al. (1996), McNeill and Leiper (1985a, 1985b)
3.	Photodegradation	Exposure to visible light and ultraviolet (UV) light	Sakai et al. (2001), Yasuda et al. (2010)
4.	Mechanical degradation	Exposure to mechanical stress	Briassoulis (2006, 2007)
5.	Chemical degradation	Exposure to chemical attack such as hydrolysis, Ozonolysis, catalytic degradation	Moore and Saunders (1997), Södergård and Stolt (2002), Lunt (1998), Hu et al. (2011)
6.	Biodegradation	Exposure to aerobic and anaerobic environment	Williams and Peoples (1996), de Wilde (2005)

light in the presence of oxygen, and due to its limited penetration capability, it takes place only on the surface and subsurface layers of polymer (Iannuzzi et al. 2013). Thermal oxidation, on the other hand, can extend throughout the material. It takes place as a result of the simultaneous interaction of oxygen with the polymer and high temperature. Oxidation leads to formation of, inter alia, hydroxyl, carbonyl, aldehyde groups, or peroxides, along the polymer chain or at its ends. As a result of oxidation, the mechanical properties and utility of the polymer are significantly reduced (Brzozowska-Stanuch et al. 2009).

Most polymers will undergo significant changes over time when exposed to heat, light, or oxygen (Pielichowski and Njuguna 2005). The degradation of polymers can be induced by heat (thermal degradation), oxygen (oxidative and thermal-oxidative degradation), light (photodegradation), or weathering (generally UV/ozone degradation). The deterioration due to oxidation and heat is greatly accelerated by stress and exposure to other reactive compounds like ozone (Vahabi et al. 2014).

All polymers will undergo some degradation during service life. The result will be a steady decline in their (mechanical) properties caused by changes to the molecular weight and molecular weight distribution and composition of the polymer (Rabek 1996; Razza and Innocenti 2012). Other possible changes include:

- Embrittlement (chain hardening)
- Softening (chain scission)
- Colour changes
- Cracking and charring (weight loss)

In general, the resistance to degradation will depend on the chemical composition of the polymer. For example, polymers such as polypropylene (PP), polyvinylchloride (PVC), and polybutadiene (PBD) are very susceptible to degradation and can only withstand UV, oxidative, and thermal degradation/decomposition when formulated with UV stabilisers and antioxidants, whereas polymers such as polysulfone (PES, PSU), polyetherketone (PEEK), and polysiloxanes (silicones) possess excellent resistance to thermal and oxidative degradation due to the strong bonds in the long-chain backbone and in the side groups (Iannuzzi et al. 2013; Siracusa 2019; Chamas et al. 2020).

10.4.3 Hydrolytic Degradation

This mechanism can be defined as the breaking of chemical bonds in the polymer backbone by the water to form oligomers and finally monomers. The rate of degradation is modulated by hydrophilic characteristics of the plastics or having heteroatoms in the main or side chain (polyesters, polyamides, and polyurethanes show higher susceptibility to hydrolysis), temperature, exposure time, and stress levels. In general, hydrolysis is usually not a significant mechanism in seawater for the degradation of most commercial fossil fuel-derived plastics (Andrady 2011), polymers with pure carbon backbones, and vinyl polymers carrying aromatic carbocyclic rings such as polystyrene (Zheng et al. 2005).

10.4.4 Mechanical Degradation

Mechanical degradation may take place through the combined efforts of wave and tide action and abrasion from sediment particles, which can scratch the surface of plastics and increase its rate of fragmentation. Surface alterations in plastic fragments, resulting from environmental erosion, increase the overall surface area and polarity and can facilitate the sorption of persistent organic pollutants (Fotopoulou and Karapanagioti 2012; Fotopoulou and Karapanagioti 2015).

10.4.5 Microorganism Biodegradation

Mechanical microbial degradation takes place through the action of enzymes or by-products generated by microorganisms such as bacteria, yeasts, and fungi. Two key steps happen in the microbial polymer degradation process: (1) depolymerisation and (2) mineralisation. Depolymerisation is a free-radical mechanism in which the polymer is degraded to monomers, oligomers, and dimers which when exposed to an aerobic environment will yield carbon dioxide and water. However, when the same depolymerised oligomers, dimers, and monomers are exposed to anaerobic environment, the end products are methane, carbon dioxide, and water. Depolymerisation occurs outside the organism due to the size of the polymer chain (Siracusa 2019). Extracellular enzymes are responsible for this step (Cho 2005).

Abiotic hydrolysis is the most important reaction for initiating the environmental degradation of synthetic polymers such as polycarboxylates, poly(ethylene terephthalate) (PET), polylactic acid (PLA), and their copolymers. For all of these materials, hydrolysis acts as the initial step of splitting the polymer into its monomers, after which the monomers can be biodegraded. Abiotic oxidation can also initiate the degradation of some polymers. As an example, polyethylene undergoes an auto-oxidation, which gradually reduces its molecular weight to the point where biodegradation can proceed (Müller et al. 2001).

Aerobic biodegradation is the process of breaking down organic contaminants by microorganisms during the presence of oxygen. Organic contaminants are rapidly degraded under aerobic conditions by aerobic bacteria called aerobes. Anaerobic degradation generally occurs when anaerobic microbes are dominant over aerobic microbes such as *Clostridia* and *Eubacterium* spp. Typically, there are four key stages of anaerobic degradation (Andrady 2011):

- Hydrolysis
- Acidogenesis
- Acetogenesis
- Methanogenesis

10.5 Mitigation of Micro and Nanoplastics Through Phycoremediation Technology

The dynamics of plastic accumulation have undergone immense change since the 1950s, leading to a boom in the industrial production of plastic. Polymer-based plastic has a global production of about 8300 million metric tons (MT) (Al-Thawadi 2020). In 2015, approximately 6300 MT of waste plastic was accumulated; 4.8–12.7 MT of this waste reached the oceans (Al-Thawadi 2020). Land-based plastic waste that might reach the oceans is estimated to increase to 88 million metric tons in 2025 from about 8.8 million metric tons in 2010 if waste mitigation techniques are not deployed significantly (Barone et al. 2020).

Plastics exhibit bulk and significant strength; hence, discarded plastic degrades over extended durations, leading to volumetric accumulation. Weather and environmental microflora (e.g., bacteria and microalgae) characteristics lead to increasing accumulation in freshwater and marine environments, thereby leading to continuous structural damage to the plastic. Marine basins and several large rivers have an abundance of plastic particles; these ecosystems adversely impact global food chains, as researchers recently asserted.

Cyanobacteria like *Synechocystis* sp. PCC 6803 and *Synechococcus elongatus* PCC 7942, also called blue-green algae or photosynthetic organisms, can use solar energy and carbon dioxide (a greenhouse gas) to break down plastic (Barone et al. 2020). In this regard, using these microorganisms for high-value products has received immense attention. Such microbes are used for numerous applications like antioxidants, fertilisers, biofuel, and 'superfood' processing. Consequently, microalgae *Phaeodactylum tricornutum*, *Chlamydomonas reinhardtii*, and others are being researched for biotechnological and artificial use of solar light (Barone et al. 2020). Phycoremediation refers to using micro- and macro-algae and cyanobacteria to eliminate inorganic and organic effluents from the atmosphere and edaphic environment (Barone et al. 2020). This environmentally friendly technology is considered cost-effective, efficient, and sustainable (Kamani et al. 2019).

Phycoremediation process efficiency is regulated by factors like strain, genotype, and species used for the process; moreover, abiotic environmental factors affect the process immensely (Shanmugam et al. 2021). Microalgae are single-celled marine- or freshwater-based organisms that perform photosynthesis, i.e. they use water, carbon dioxide, and sunlight to produce algal biomass. Additionally, microalgae have cell walls with carbohydrates that can capture harmful substances like hydrocarbons, pesticides, endocrine disrupting agents, and cyanides containing nitrogen and carbon. These photosynthetic aspects, along with metabolic versatility and capability to conduct photoautotrophic, heterotrophic, and mixotrophic metabolism, are of unprecedented benefits for mitigating micro- and nanoplastic pollution (Lusher et al. 2017).

Microalgae are considered model photosynthetic species in stress research; these organisms are at the bottom of the trophic chain (Déniel et al. 2020). Algae possess favourable characteristics like rapid growth and response, cost-effectiveness, and stress sensitivity that are considered for a majority of the typical toxicity tests conducted in Europe. Consequently, they fall in the emission bioindicator category. For instance, *Chlamydomonas reinhardtii* is a single-celled biflagellate that develops rapidly. Hence, this species comprises an abundant ecotoxicological framework pertaining to aquatic environments (Déniel et al. 2020).

Microalgae have numerous species with varying size and shape; cell dimensions range from 0.5 to 200 micrometres (Roy and Mohanty 2019). The significant effect of algae in different environments and the distinct characteristics of species is an argument in favour of additional research on this subject, explicitly when these microorganisms are exposed to several pollutants like microplastics prevalent in marine ecosystems (Almeida et al. 2019). Considering that the harmful impact on microalgae will disrupt species higher in the food chain, the effects of microplastics require additional research (Wan et al. 2018).

Microalgae have been the subject of extensive research for biotechnological use in recent years; biofuel production is an attractive research area (Shanmugam et al. 2021; Correa et al. 2021). Microalgae are at the lowest point in the food chain. These are primary producers and have a significant role in the aquatic food chain in marine ecosystems (Zhu et al. 2020). Genetic engineering, nanotechnology, synthetic genetics, and novel biotreatment methods are promising fields that can help society realise critical objectives (Vallero et al. 2020).

Plastics may be regarded as carbon sources for specific microorganisms living in a specific habitat. Phenomena that require microbial metabolism require nutrients for such processes. Using the sorption phenomenon for eliminating pollutants from wastewater, it is feasible to use living organisms or non-living cells (Barone et al. 2020). Prokaryotic organisms and cyanobacteria that employ oxygen-based photosynthesis have been researched for potential uses considering their photosynthetic system. Moreover, such Gram-negative bacteria can convert CO_2, water, and solar energy into chemical form and produce oxygen during the process; oxygen is released into the atmosphere (Barone et al. 2020).

These organisms were considered good vitamin and protein sources; therefore, these are employed for producing healthy consumable products. Additionally, they are used as human food sources and also as fertilisers for paddy cultivation (Correa et al. 2021). Such photosynthetic microorganisms may manufacture active substances having activity against fungus, bacteria, virus, and cancer (Shanmugam et al. 2021). Barone et al. (2020) considered cyanobacteria as green cell-producing agents. Such micro-photosynthetic factories have immense potential as vital catalysts for biochemical processes and sources of carbon-neutral replenishable substances like biofuels and bio fertilisers. Moreover, these organisms can use solar energy to process water and carbon dioxide into hydrogen. The study led to cyanobacteria being considered photosynthetic microorganisms that might be utilised as green cell producers to produce carbon-neutral and renewable fuels and other substances (Shanmugam et al. 2021).

Microorganisms like cyanobacteria (e.g. *Synechocystis* sp. PCC 6803 and *Synechococcus elongatus* PCC 7942), which have photosynthetic capabilities, were earlier called blue-green algae and are now researched for its greenhouse gas (CO_2) and solar energy absorption capability for manufacturing high-value entities (Barone et al. 2020). Moreover, these organisms possess the capability to formulate ligninolytic and oxidative proteins that facilitate the effective breakdown of polyethylene (PE) (Barone et al. 2020). Plastic bioremediation can be transformed to address more social challenges by utilising synthetic biology and sophisticated genetic concurrently (Barone et al. 2020). Microalgal areas are critical indicators of the stability of the aquatic environment considering that these organisms are primary producers (Prata et al. 2019). The absence of cell walls makes single-celled freshwater organism *Euglena gracilis* specifically susceptible to environmental strain; it is typically employed for modelling freshwater microalgae to ascertain contaminant harmfulness (Xiao et al. 2020; Déniel et al. 2020). Consequently, MP presence can reduce microalgae production and photosynthesis and lead to the formation of reactive oxidative species (ROS), causing oxidative in oxidative impact (Prata et al. 2019; Choi et al. 2020).

Metal phycoremediation is another potential area where charophytes such as *Nitella opaca* and *Chara aculeolate* might be used for isolating zinc, lead, and cadmium (Jasrotia et al. 2017). Jasrotia et al. (2017) discovered that the chemical processing of aquatic brown algae (*Cystoseira indica*) is feasible for isolation of chromium. Barring physical adsorption, research suggests that metal chemisorption (chemical adsorption) for copper and chromium is the phenomenon used by blue-green algae *Spirulina* sp.

10.6 Potential of Microalgae and Cyanobacteria for Micro and Nanoplastic Biodegradation

Biosorption phenomena require metabolism-based processes; Barone et al. (2020) suggest that there are two stages. The steps are (i) attachment of the contaminant to the surface of the cell, or vice versa, based on relative sizes, and (ii) the active or passive movement of the contaminants to enter the cell (Wenten et al. 2020). Consequently, it is feasible to use the movement of biomass or integrating membrane isolation to enhance the uptake of plastic particles. The beginning of the biodegradation process is called biodeterioration; long carbon chain instability generation is called bio-fragmentation; the movement of particular carries across the cytoplasmic boundary is called assimilation by microorganisms; the last step is mineralisation, whereby biodegradation stops. These four steps comprise the vital stages (Jacquin et al. 2019; Barone et al. 2020). Access to oxygen and the metabolic phenomena of the species involved are determinants of plastic as a potential nutrient source for microorganisms.

Cyanobacteria have a critical role in the natural phenomena that occur because of the microbial layer on plastics. Nevertheless, there is research scarcity concerning cyanobacteria's potential for the plastic breakdown; moreover, the effectiveness of NP remediation is not considered. Cyanobacteria possess the highest potential for absorbing solar energy than other photosynthetic oxygenic organisms; additionally, cyanobacteria also have adequate CO_2 concentrating and fixation abilities. Traditional plastic has a prolonged rate of microorganism-based degeneration even at ideal laboratory settings; however, polymers with chemical sensitivity provide a better environment for microbial development (Barone et al. 2020). It is understood that cyanobacterial action enhances the hydrophilicity of PE surfaces by producing carbonyl, a critical biodegradation indicator. Considering that light is critical for cyanobacteria to survive, it is typically understood that degradation caused by such bacteria would be feasible only for surface-specific particles. Damage to the cell membrane can occur if plastic particles remain attached to this microalgae surface (Sarmah and Rout 2018).

In contrast, it might indicate a cell-specific technique for reducing polystyrene (PS) concentration in water. The unearthing of lipase-creating cold-adapted bacteria highlighted a critical property concerning the biodegradation phenomenon (Tokiwa and Calabia 2004; Yu et al. 2009). Moreover, different enzymes produced by bacteria captured from cold settings are likely to have biodegradability potential. Microbial strains gathered from Arctic Ocean ice obtained from the Canada Basin demonstrate extracellular lipase with plastic degradation potential (Barone et al. 2020).

As per Barone et al. (2020), enzymes typically act when the conditions are mild; they have substantial selectivity for natural target substances. Degradable polymers may be verified using changes in biodegradability demonstrated to ascertain the targets based on enzymes. Considering that cyanobacteria may be utilised as cellular factories, such microorganisms could be utilised for producing enzymes that can cause the breakdown of polymeric structures directly, thereby decreasing the number of MPs and NPs in the environment. Considering its expression specific to the wild-type enzyme PETase (polyethylene terephthalate) in primary microorganisms and rapid growth, two variants of C. *reinhardtii* (CC-124 and CC-503) were studied (Barone et al. 2020). Additional information concerning plastic degradation through the action of wild-type enzymes produced by cyanobacteria and microalgae is needed; moreover, new phenomena specific to this behaviour might uncover promising results and uses of such interesting microorganisms (Jacquin et al. 2019; Oberbeckmann et al. 2014).

Plastic contamination is a critical threat to the marine ecosystem. NPs create more challenges as compared to MPs, and additional information is needed to perform a detailed seafood safety risk evaluation (Barone et al. 2020). Research suggests that the digestive tracts in marine organisms have a significant quantity of MPs (Lusher et al. 2017; Barone et al. 2020). Pollution elimination using biosorption is widely utilised as a cost-effective technique to decontaminate the environment since long back (Singh et al. 2020). *Rivularia* and *Phormidium* cyanobacteria have a high presence likelihood in biofilms in proximity to the water.

Additionally, *Proteobacteria* and *Bacteroidetes* (*Flavobacteriaceae*) are typically abundant as the seafloor microbiome (*Alcanivoracaceae* and *Rhodobacteraceae*). Though numerous aspects affect biofilm characteristics, studies concerning the North Pacific indicate biogeographic sources having higher importance than polymer variety. To disintegrate plastic, there is a need for a complex microbial environment instead of one species. Considering the extensive types of synthetic polymers available in the market, the metabolic pathways concerning biodegradation are extensively diverse. Consequently, considering the challenges presented by natural environments and using data from existing studies, preliminary studies concerning metabolic engineered cyanobacteria specific to plastic bio-degeneration must be performed employing a culture-based technique (Kumar et al. 2017; Sarmah and Rout 2018).

10.7 Other Potential Bioremediation of Plastics

'Bioremediation' indicates the biological phenomena of pollution elimination from the environment through the metabolic capability of microorganisms to act on a broad variety of organic substances (Perelo 2010). Using microbial metabolic abilities (bioremediation) to remove contaminants from polluted ecosystems is typically understood as a green and cost-effective technique (Pandey et al. 2009). Bioremediation happens through live ecosystems comprising microorganisms that act as catalysts to the pollutant degradation process and reducing environmental impact (Daba and Ezeronye 2005). It is a technique for eliminating or converting numerous pollutants to a harmless form using natural biological processes (Shukla et al. 2010). Research indicates novel bioremediation techniques like bioaugmentation, biostimulation, and phytoremediation, which are used for sediments. Furthermore, plastic contaminants may be classified depending on the polluting medium. The soil ecosystem facilitates the bioremediation phenomenon based on the type of microorganisms like fungi, bacteria, and algae.

Additionally, invertebrates, plants, or insects must be immediately researched to observe the ecological effects on the overall impact created by plastic pollution on the comprehensive soil ecosystem (Chae and An 2018). Plastic polymer comprises, long chains of hydrocarbons having significant molecular height, a different technique to reverse the phenomenon, where the long chains disintegrate to simpler hydrocarbons, is understood as a greener approach. This phenomenon may be performed by specific microorganisms producing a chemical that disintegrate the long polymer chain present in plastics (Muhamad et al. 2015). The count of acknowledged enzymes and microbes acting on synthetic polymers remains minimal. Consequently, additional research should emphasise ascertaining species that act on an extensive range of polymers in use today.

The critical roadblocks include the initiation of the degradation of high-molecular-weight material, polymers with supreme characteristics, and crystalline characteristics (Danso et al. 2019). Shukla et al. (2010) suggest that living organisms are employed to treat or remediate pollution using biological phenomenon to remove, reduce, or convert pollutants and contaminants. Critical bioremediation aspects include soil structure, pollutants, moisture levels, pH, nutritional characteristics, hydrogeology, site microbial diversity, oxidation-reduction (redox) potential, and temperature.

Perelo (2010) suggests several bioremediation techniques, as specified below:

10.7.1 Monitored Natural Recovery (MNR)

MNR refers to regulated natural recovery, which integrates the natural microbial population's 'self-healing' natural abilities with environmentally happening chemical and physical phenomena.

10.7.2 Biostimulation

It refers to the techniques for changing the determinants that affect microbial growth to stimulate the indigenous population.

10.7.3 Bioaugmentation

It refers to the method where the required organisms are used for catalysing the breakdown of specific contaminants.

10.7.4 Phytoremediation

It refers to the phenomena where pollutants are degraded and eliminated from the atmosphere by utilising algae and plants.

Biostimulation and phytoremediation are bioremediation techniques that may be utilised concurrently. Plants may stimulate the process of decaying microorganisms at the root levels; this phenomenon is referred to as rhizoremediation (Shukla et al. 2010). The activity comprising plastic breakdown has also been assessed to improve plastic biodegradation. There are two process categories: direct action, where plastic degeneration produces trophic material for microbial formation, and indirect action, where plastic characteristics are affected by the metabolic substances produced by microorganisms (Caruso 2015).

10.8 In Situ Bioremediation

Bioremediation in situ comprises techniques that treat pollutants in their natural surroundings. Bioremediation can be utilised on sites contaminated with numerous chemical pollutants considering that microbes possess a significant capacity for breaking down synthetic substances (Iwamoto and Nasu 2001). In situ bioremediation focuses on increasing the rate of contaminant degradation kinetics through an increase in the natural reduction techniques (biostimulation) or through concurrently used techniques by adding specific potent pollutant-degrading strain(s) externally (bioaugmentation) (Pandey et al. 2009). Both techniques (bioaugmentation and biostimulation) may be classified as in situ bioremediation. The inclusion of elements like nitrogen, carbon, potassium, and phosphorus has proven to be a powerful biostimulation technique. Adding electron donors like nitrate, acetate, sulphate, or glutamate or introducing gaseous mixtures to a contaminated environment is another powerful biostimulation technique (Perelo 2010). Theoretically, in situ bioremediation is performed in order to eliminate or reduce pollutant concentration in natural environments by utilising microbial metabolic activities without excavating contaminated samples (Pandey et al. 2009). In situ bioremediation provides several benefits:

1. On-site execution reduces transportation cost, permanent waste reduction, applicable for low concentration and extensively diffused pollutants, and affordable (Iwamoto and Nasu 2001).
2. In-place treatment of groundwater and soil without needing excavation (Shukla et al. 2010).

10.9 Ex Situ Bioremediation

On the other hand, ex situ approaches comprise methods that remove toxins from ecosystems and move them to a treatment facility (Iwamoto and Nasu 2001). Excavation is performed until ex situ processes are implemented to control it (Shukla et al. 2010). Conventional management technique may provide a less-than-expected effect but higher cost compared to contemporary methods. It is a lasting solution that may facilitate complete pollutant mineralisation while being economical. Moreover, this is a non-invasive environmentally friendly technique. Bioremediation can handle lesser pollutant amount that might be challenging to remediate using a chemical or physical technique (Perelo 2010). Ex situ bioremediation techniques consist of processes undertaken to break down chemical contaminants inside excavated samples; conversely, in situ bioremediation techniques consist of interventions to break down chemical effluents from excavated samples. It is easy to regulate ex situ bioremediation using physicochemical processes aimed at a specific pollutant before and during breakdown because it happens in non-natural settings (Pandey et al. 2009). Consequently, ex situ bioremediation techniques are less economical compared to in situ techniques. Another advantage associated with ex situ bioremediation techniques is that environmental aspects do not affects such processes, thereby making them insignificant at affecting effectiveness (Pandey et al. 2009).

10.10 Bioremediation Potential of Marine Animals and Aquatic Plants

Considering their significant retention rates, cephalopods and decapods comprise marine invertebrates that may be used for bioremediation (Masiá et al. 2020). Filter-feeding organisms might possess minute microplastic (MP) retention ability. The circulatory system of these organisms might retain MP particles for up to 2 days; however, most MP fibres are copious and excreted after 24 h, thereby reducing removal efficacy. Rhizoremediation is the most sophisticated form of bioremediation; it comprises the interaction of plant roots with suitable microbial flora to eliminate specific contaminants from waste substances in polluted environments (Shukla et al. 2010).

Microorganisms utilise contaminants as energy sources or nutrients during bioremediation. Adding nutrients (phosphorus and nitrogen), substrates (methane, phenol, and toluene), and electron acceptors (oxygen) and adding microorganisms having the required catalytic abilities increase the bioremediation process facilitated by microbes (Pramila et al. 2012). Plants interact with soil microbes to create a rhizospheric area (a system of extensively dynamic synergistic and symbiotic associations) that may be employed for enhancing degradation rate or removing pollutants (Masiá et al. 2020). Some studies were conducted concerning the capabilities of specific plant types to react to pollutants; for instance, seaweed such as *Fucus vesiculosus* can retain suspended MP on its surface and eliminate such pollutants from the soil by taking up metals and transform them to harvestable form is known as phytoremediation (Masiá et al. 2020; Castro-Castellon et al. 2021; Shukla et al. 2010).

Rhizofiltration is another concept related to phytoremediation. It is associated with cleaning polluted groundwater than cleaning affected soil (Shukla et al. 2010). Plant roots uptake the pollutants, or these substances undergo adsorption by the surface of the root.

Plants utilised for rhizofiltration are initially introduced to the contaminant before planting. The hydroponic technique is used for plant growth where clean water is used instead of soil till the roots have developed significantly. Also, the seagrass plant species have a promising potential for processing pollutants in proximity to the ocean (Anawar et al. 2008). Masiá et al. (2020) suggested that initial research indicated MP attachment to seagrass through the encrustation process and correlated to the macrophyte. There is adhesion with the polysaccharide-based mucus film. Hence, seagrasses may capture or sink MPs, indicating promising sludge processing potential considering that they grow on sludge.

Consequently, seagrass meadows might behave as long-term plastic sinks. Considering that terrestrial plants and seagrass share common physiologies. These aspects might create a significant effect on seagrass health and development in seagrass soil (Diaz-almela et al. 2021). Seagrass leaves were another area where microplastic was discovered. Several herbivorous organisms can consume the leaves, creating another pathway for microplastics to reach the food chain.

10.11 Conclusion

This chapter discussed how bioremediation and phycoremediation technologies could be utilised to degrade plastics. The five plastic biodegradation mechanism in the marine environment that have been extensively referenced for their ability to degrade certain plastics are photodegradation, thermo-oxidative degradation, hydrolytic degradation, mechanical degradation, and microorganism biodegradation. Microbial enzymes are intriguing since they are inexpensive and require little care while yet being simple to modify. Furthermore, the ability of fungi, bacteria, cyanobacteria, and actinomycetes to degrade plastic is one of the environmentally beneficial strategies that can aid the environment. Microbial degradation is the breakdown of plastic by microbes, which produce enzymes capable of splitting long-chain polymers. Given the abundance of microorganisms in the environment and their propensity for decomposing plastic, the employment of such microbes and enzymes appears to be an effective strategy to reduce the environmental risks caused by excessive plastics waste.

Acknowledgements The research was supported by the Ministry of Higher Education (Malaysia) and International Islamic University Malaysia under RPDF19-003-0013 grant.

References

Almeida AC, Gomes T, Habuda-Stanić M, Lomba JAB, Romić Ž, Turkalj JV, Lillicrap A (2019) Characterization of multiple biomarker responses using flow cytometry to improve environmental hazard assessment with the green microalgae Raphidocelis subcapitata. Sci Total Environ 687:827–838

Al-Thawadi S (2020) Microplastics and nanoplastics in aquatic environments: challenges and threats to aquatic organisms. Arab J Sci Eng 45:4419–4440

Anawar HM, Garcia-Sanchez A, Alam TKM, Rahman MM (2008) Phytofiltration of water polluted with arsenic and heavy metals. Int J Environ Pollut 33(2/3):292–312

Andrady AL (2011) Microplastics in the marine environment. Mar Pollut Bull 62:1596–1605

Arias-Andres M, Klümper U, Rojas-Jimenez K, Grossart HP (2018) Microplastic pollution increases gene exchange in aquatic ecosystems. Environ Pollut 237:253–261

Barnes DKA, Galgani F, Thompson RC, Barlaz M (2009) Accumulation and fragmentation of plastic debris in global environments. Philos Trans R Soc B Biol Sci 364:1985–1998

Barone GD, Ferizović D, Biundo A, Lindblad P (2020) Hints at the applicability of microalgae and cyanobacteria for the biodegradation of plastics. Sustain 12:1–15

Barría C, Brandts I, Tort L, Oliveira M, Teles M (2020) Effect of nanoplastics on fish health and performance: a review. Mar Pollut Bull 151:1–7

Bhuyan MS, Venkatramanan S, Selvam S, Szabo S, Hossain MM, Rashed-Un-Nabi M, Paramasivam CR, Jonathan MP, Islam MS (2021) Plastics in marine ecosystem: a review of their sources and pollution conduits. Reg Stud Mar Sci 41:101539

Boyle K, Örmeci B (2020) Microplastics and nanoplastics in the freshwater and terrestrial environment: a review. Water 12:2633. https://doi.org/10.3390/w12092633

Brandon J, Goldstein M, Ohman MD (2016) Long-term aging and degradation of microplastic particles: comparing in situ oceanic and experimental weathering patterns. Mar Pollut Bull 110 (1):299–308

Briassoulis D (2006) Mechanical behaviour of biodegradable agricultural films under real field conditions. Polym Degrad Stab 91:1256–1272

Briassoulis D (2007) Analysis of the mechanical and degradation performances of optimised agricultural biodegradable films. Polym Degrad Stab 92:1115e32

Browne MA, Galloway T, Thompson R (2007) Microplastic--an emerging contaminant of potential concern? Integr Environ Assess Manag 3(4):559–561

Brzozowska-Stanuch A, Rabiej S, Stanuch W (2009) The influence of accelerated weathering conditions-UV radiation and temperature on polyamides and polypropylene. Tech Trans 1 (3):43–49

Campanale C, Massarelli C, Savino I, Locaputo V, Uricchio VF (2020) A detailed review study on potential effects of microplastics and additives of concern on human health. Int J Environ Res Public Health 17(4):1–26

Caruso G (2015) Plastic degrading microorganisms as a tool for bioremediation of plastic contamination in aquatic environments. J Pollut Eff Control 3(3):1000e112

Castro-Castellon AT, Hughes JMR, Read DS, Azimi Y, Chipps MJ, Hankins NP (2021) The role of rhizofiltration and allelopathy on the removal of cyanobacteria in a continuous flow system. Environ Sci Pollut Res 28:27,731–27,741

Chae Y, An YJ (2018) Current research trends on plastic pollution and ecological impacts on the soil ecosystem: a review. Environ Pollut 240:387–395. https://doi.org/10.1016/j.envpol.2018.05.008. Epub 2018 May 9

Chamas A, Moon H, Zheng J, Qiu Y, Tabassum T, Jang JH, Abu-Omar M, Scott SL, Suh S (2020) Degradation rates of plastics in the environment. ACS Sustain Chem Eng 8:3494–3511

Chang X, Xue Y, Li J, Zou L, Tang M (2020) Potential health impact of environmental micro- and nanoplastics pollution. J Appl Toxicol 40:4–15

Cho DO (2005) Challenges to marine debris management in Korea. Coast Manag 33:389–409

Choi JS, Hong SH, Park JW (2020) Evaluation of microplastic toxicity in accordance with different sizes and exposure times in the marine copepod Tigriopus japonicus. Mar Environ Res 153: 104838

Correa DF, Beyer HL, Possingham HP, Fargione JE, Hill JD, Schenk PM (2021) Microalgal biofuel production at national scales: Reducing conflicts with agricultural lands and biodiversity within countries. Energy 215:119033

Crawford CB, Quinn B (2017) Plastic production, waste and legislation. Microplast Pollut 30:39–56

Da Costa JP, Santos PSM, Duarte AC, Rocha-Santos T (2016) (Nano)plastics in the environment—sources, fates and effects. Sci Total Environ 566–567:15–26

Daba AS, Ezeronye OU (2005) Bioremediation of textile effluent using Phanerochaete chrysosporium. Afr J Biotechnol 4:1548–1553

Danso D, Chow J, Streita WR (2019) Plastics: environmental and biotechnological perspectives on microbial degradation. Appl Environ Microbiol 85:1–14

Deng H, Wei R, Luo W, Hu L, Li B, Di Y, Shi H (2020) Microplastic pollution in water and sediment in a textile industrial area. Environ Pollut 258:113658

Déniel M, Lagarde F, Caruso A, Errien N (2020) Infrared spectroscopy as a tool to monitor interactions between nanoplastics and microalgae. Anal Bioanal Chem 412:4413–4422

Diaz-almela E, Granberg M, Dahl M, Bergman S, Bj M, Magnusson K, Marco-Méndez C, Piñeiro-Juncal N, Mateo MÁ (2021) A temporal record of microplastic pollution in Mediterranean seagrass. Environ Pollut 273:116451. https://doi.org/10.1016/j.envpol.2021.116451

EFSA Panel on Contaminants in the Food Chain (2016) Presence of microplastics and nanoplastics in food, with particular focus on seafood. EFSA J 14:1–30

Enfrin M, Lee J, Gibert Y, Basheer F, Kong L, Dumée LF (2020) Release of hazardous nanoplastic contaminants due to microplastics fragmentation under shear stress forces. J Hazard Mater 384: 121393

Fotopoulou KN, Karapanagioti HK (2012) Surface properties of beached plastic pellets. Mar Environ Res 81:70–77

Fotopoulou K, Karapanagioti H (2015) Surface properties of beached plastics. Environ Sci Pollut Res Int 14:11,022–11,032

Frias JPGL, Nash R (2019) Microplastics: finding a consensus on the definition. Mar Pollut Bull 138:145–147

Fu D, Chen CM, Qi H, Fan Z, Wang Z, Peng L, Li B (2020a) Occurrences and distribution of microplastic pollution and the control measures in China. Mar Pollut Bull 153:1–14

Fu Z, Chen G, Wang W, Wang J (2020b) Microplastic pollution research methodologies, abundance, characteristics and risk assessments for aquatic biota in China. Environ Pollut 266: 115,098

Horton AA, Walton A, Spurgeon DJ, Lahive E, Svendsen C (2017) Microplastics in freshwater and terrestrial environments: evaluating the current understanding to identify the knowledge gaps and future research priorities. Sci Total Environ 586:127–141

Hu HT, Shin TC, Lee SY, Chen CC, Yang JC (2011) Influence of hydrolytic degradation on the surface properties of poly-5d/95l-lactide resorbable bone plates. Polym Degrad Stab 96:1522–1529

Iannuzzi G, Mattsson B, Rigdahl M (2013) Color changes due to thermal ageing and artificial weathering of pigmented and textured ABS. Polym Eng Sci 53:1687–1695

Ivar Do Sul JA, Costa MF (2014) The present and future of microplastic pollution in the marine environment. Environ Pollut 185:352–364

Iwamoto T, Nasu M (2001) Current bioremediation practice and perspective. J Biosci Bioeng 92:1–8

Jacquin J, Cheng J, Odobel C, Pandin C, Conan P, Pujo-Pay M, Barbe V, Meistertzheim A-L, Ghiglione J-F (2019) Microbial ecotoxicology of marine plastic debris: a review on colonization and biodegradation by the "Plastisphere". Front Microbiol 10:865

Jambeck JR, Geyer R, Wilcox C, Siegler TR, Perryman M, Andrady A, Narayan R, Law KL (2015) Plastic waste inputs from land into the ocean. Science (80-) 347:768 LP–768771

Jasrotia S, Kansal A, Mehra A (2017) Performance of aquatic plant species for phytoremediation of arsenic-contaminated water. Appl Water Sci 7:889–896

Jennings S, Mélin F, Blanchard JL, Forster RM, Dulvy NK, Wilson RW (2008) Global-scale predictions of community and ecosystem properties from simple ecological theory. Proc R Soc B 275:1375–1383

Jiang JQ (2018) Occurrence of microplastics and its pollution in the environment: a review. Sustain Prod Consum 13:16–23

Jin Y, Lu L, Tu W, Luo T, Fu Z (2019) Impacts of polystyrene microplastic on the gut barrier, microbiota and metabolism of mice. Sci Total Environ 649:308–317

Kamani MH, Eş I, Lorenzo JM, Remize F, Roselló-Soto E, Barba FJ, Clark J, Mousavi Khaneghah A (2019) Advances in plant materials, food by-products, and algae conversion into biofuels: use of environmentally friendly technologies. Green Chem 21:3213–3231

Käppler A, Fischer D, Oberbeckmann S, Schernewski G, Labrenz M, Eichhorn K-J, Voit B (2016) Analysis of environmental microplastics by vibrational microspectroscopy: FTIR, Raman or both? Anal Bioanal Chem. https://doi.org/10.1007/s00216-016-9956-3

Kaur S (2014) Mechanism of aerobic and anaerobic biodegradation. MSc Thesis

Kausar A, Anwar Z, Muhammad B (2016) Recent developments in epoxy/graphite, epoxy/graphene and epoxy/graphene nanoplatelet composites: a comparative review. Polym Plast Technol Eng 55:1192–1210

Khan AK, Majeed T (2020) Biodegradation of synthetic and natural plastics by microorganisms: a mini review. JNASP 1(2):180–184

Kögel T, Bjorøy Ø, Toto B, Bienfait AM, Sanden M (2020) Micro- and nanoplastic toxicity on aquatic life: determining factors. Sci Total Environ 709:1–16

Kole PJ, Löhr AJ, Van Belleghem FGAJ, Ragas AMJ (2017) Wear and tear of tyres: a stealthy source of microplastics in the environment. Int J Environ Res Public Health 14(10):1265

Kopinke FD, Remmler M, Mackenzie K, Moder M, Wachsen O (1996) Thermal decomposition of biodegradable polyestersdII Poly (lactic acid). Polym Degrad Stab 53:329–342

Kothari VK (2008) Polyester and polyamide fibres—apparel applications. Polyesters and Polyamides Elsevier Ltd, pp 419–440

Kumar RV, Kanna GR, Elumalai S (2017) Biodegradation of polyethylene by green photosynthetic microalgae. J Bioremed Biodegr 8(381):2

Lambert S, Sinclair C, Boxall A (2014) Occurrence, degradation, and effect of polymer-based materials in the environment. Rev Environ Contam Toxicol 227:1–53

Lunt J (1998) Large-scale production, properties and commercial applications of polylactic acid polymers. Polym Degrad Stab 59:145–152

Lusher A, Hollman P, Mendoza-Hill J (2017) Microplastics in fisheries and aquaculture: status of knowledge on their occurrence and implications for aquatic organisms and food safety. FAO

Masiá P, Sol D, Ardura A, Laca A, Borrell YJ, Dopico E, Laca A, Machado-Schiaffino G, Díaz M, Garcia-Vazquez E (2020) Bioremediation as a promising strategy for microplastics removal in wastewater treatment plants. Mar Pollut Bull 156:1–7

McNeill IC, Leiper HA (1985a) Degradation studies of some polyesters and polycarbonates-2 Polylactide: degradation under isothermal conditions, thermal degradation mechanism and photolysis of the polymer. Polym Degrad Stab 11:309–326

McNeill IC, Leiper HA (1985b) Degradation studies of some polyesters and polycarbonates-1 Polylactide: general features of the degradation under programmed heating conditions. Polym Degrad Stab 11:267–285

Meng Y, Kelly FJ, Wright SL (2020) Advances and challenges of microplastic pollution in freshwater ecosystems: a UK perspective. Environ Pollut 256:113445

Moore GF, Saunders SM (1997) Advances in biodegradable polymers. Rapra Technology Limited

Muhamad WNAW, Othman R, Shaharuddin RI, Hasni MSI (2015) Microorganism as plastic biodegradation agent towards sustainable environment. Adv Environ Biol 9:8–13

Müller RJ, Kleeberg I, Deckwer WD (2001) Biodegradation of polyesters containing aromatic constituents. J Biotechnol 86:87–95
Muthukumar T, Aravinthan A, Lakshmi K, Venkatesan R, Vedaprakash L, Doble M (2011) Fouling and stability of polymers and composites in marine environment. Int Biodeterior Biodegrad 65: 276–284
Ncube LK, Ude AU, Ogunmuyiwa EN, Zulkifli R, Beas IN (2021) An overview of plasticwaste generation and management in food packaging industries. Recycling 6:1–25
Ng EL, Huerta Lwanga E, Eldridge SM, Johnston P, Hu HW, Geissen V, Chen D (2018) An overview of microplastic and nanoplastic pollution in agroecosystems. Sci Total Environ 627: 1377–1388
Oberbeckmann S, Loeder MG, Gerdts G, Osborn AM (2014) Spatial and seasonal variation in diversity and structure of microbial biofilms on marine plastics in Northern European waters. FEMS Microbiol Ecol 90:478–492
Oliveira M, Almeida M, Miguel I (2019) A micro(nano)plastic boomerang tale: a never ending story? TrAC—Trends Anal Chem 112:196–200
Pandey J, Chauhan A, Jain RK (2009) Integrative approaches for assessing the ecological sustainability of in situ bioremediation. FEMS Microbiol Rev 33:324–375
Perelo LW (2010) Review: in situ and bioremediation of organic pollutants in aquatic sediments. J Hazard Mater 177:81–89
Piccardo M, Renzi M, Terlizzi A (2020) Nanoplastics in the oceans: theory, experimental evidence and real world. Mar Pollut Bull 157:111317
Pielichowski K, Njuguna J (2005) Thermal degradation of polymeric materials Shawbury: Rapra Technology Limited
Pramila R, Padmavathy K, Ramesh KV, Mahalakshmi K (2012) Brevibacillus parabrevis, Acinetobacter baumannii and Pseudomonas citronellolis—potential candidates for biodegradation of low-density polyethylene (LDPE). J Bacteriol Res 4:9–14
Prata JC, da Costa JP, Lopes I, Duarte AC, Rocha-Santos T (2019) Effects of microplastics on microalgae populations: a critical review. Sci Total Environ 665:400–405
Rabek JF (1996) Photodegradation of polymers physical characteristics and applications. Springer, Berlin
Razza F, Innocenti FD (2012) Bioplastics from renewable resources: the benefits of biodegradability. Asia Pac J Chem Eng 7(3):301–309
Rillig MC (2012) Microplastic in terrestrial ecosystems and the soil? Environ Sci Technol:6453–6454
Rillig MC, Lehmann A, de Souza Machado AA, Yang G (2019) Microplastic effects on plants. New Phytol 223:1066–1070
Roy M, Mohanty K (2019) A comprehensive review on microalgal harvesting strategies: current status and future prospects. Algal Research 44:101683
Sakai W, Kinoshita M, Nagata M, Tsutsumi N (2001) ESR studies of photosensitized degradation of poly(L-lactic acid) via photoionization of dopant. J Poly Sci A Polym Chem 39:706–714
Sarmah P, Rout J (2018) Efficient biodegradation of low-density polyethylene by cyanobacteria isolated from submerged polyethylene surface in domestic sewage water. Environ Sci Pollut Res 25:33,508–33,520
Shanmugam S, Hari A, Kumar D, Rajendran K, Mathimani T, Atabani AE, Brindhadevi K, Pugazhendhi A (2021) Recent developments and strategies in genome engineering and integrated fermentation approaches for biobutanol production from microalgae. Fuel 285: 119052
Sharma S, Chatterjee S (2017) Microplastic pollution, a threat to marine ecosystem and human health: a short review. Environ Sci Pollut Res 24:21,530–21,547
Shukla KP, Singh NK, Sharma S (2010) Bioremediation: developments, current practices and perspectives. Genet Eng Biotechnol J 2010:1–20
Singh S, Kumar V, Datta S, Dhanjal DS, Sharma K, Samuel J, Singh J (2020) Current advancement and future prospect of biosorbents for bioremediation. Sci Total Environ 709:135895

Siracusa V (2019) Microbial degradation of synthetic biopolymers waste. Polymers 11(6):1066
Sobhani Z, Zhang X, Gibson C, Naidu R, Megharaj M, Fang C (2020) Identification and visualisation of microplastics/nanoplastics by Raman imaging (i): down to 100 nm. Water Res 174:1–10
Södergård A, Stolt M (2002) Properties of lactic acid based polymers and their correlation with composition. Prog Polym Sci 27:1123–1163
Swan SH, Sathyanarayana S, Barrett ES, Janssen S, Liu F, Nguyen RH, Redmon JB (2015) First trimester phthalate exposure and anogenital distance in newborns. Hum Reprod 30(4):963–972
Teuten EL, Saquing JM, Knappe DRU, Barlaz MA, Jonsson S, Björn A, Rowland SJ, Thompson RC, Galloway TS, Yamashita R, Ochi D, Watanuki Y, Moore C, Viet PH, Tana TS, Prudente M, Boonyatumanond R, Zakaria MP, Akkhavong K, Ogata Y, Hirai H, Iwasa S, Mizukawa K, Hagino Y, Imamura A, Saha M, Takada H (2009) Transport and release of chemicals from plastics to the environment and to wildlife. Philos Trans R Soc B Biol Sci 364:2027–2045
Tokiwa Y, Calabia BP (2004) Review degradation of microbial polyesters. Biotechnol Lett 26: 1181–1189
Vahabi H, Sonnier R, Ferry L (2014) Effects of ageing on the fire behaviour of flame-retarded polymers: a review Polym Int 64: 313–328
Vallero D, Daniel AV, Gunsch KC (2020) Applications and implications of emerging biotechnologies in environmental engineering. J Environ Eng 146:3120005
Van Cauwenberghe L, Vanreusel A, Mees J, Janssen CR (2013) Microplastic pollution in deep-sea sediments. Environ Pollut 182:495–499
Venâncio C, Ferreira I, Martins MA, Soares AMVM, Lopes I, Oliveira M (2019) The effects of nanoplastics on marine plankton: a case study with polymethylmethacrylate. Ecotoxicol Environ Saf 184:109632
Vroman I, Tighzert L (2009) Biodegradable polymers. Materials 2:307–344
Wahl A, Le Juge C, Davranche M, El Hadri H, Grassl B, Reynaud S, Gigault J (2021) Nanoplastic occurrence in a soil amended with plastic debris. Chemosphere 262:1–7
Wan JK, Chu W-L, Kok Y, Lee C (2018) Distribution of microplastics and nanoplastics in aquatic ecosystems and their impacts on aquatic organisms, with emphasis on microalgae. In: de Voogt P (ed) Reviews of environmental contamination and toxicology. Springer, Cham
Wang L, Wu WM, Bolan NS, Tsang DCW, Li Y, Qin M, Hou D (2021) Environmental fate, toxicity and risk management strategies of nanoplastics in the environment: current status and future perspectives. J Hazard Mater 401:123415
Webb HK, Arnott J, Crawford RJ, Ivanova EP (2013) Plastic degradation and its environmental implications with special reference to poly(ethylene terephthalate). Polymers (Basel) 5:1–18
Wenten IG, Khoiruddin K, Harimawan A, Ting YP, Boopathy R (2020) Membrane biosorption: recent advances and challenges. Curr Pollut Reports 6:152–172
de Wilde B (2005) International norms on biodegradability and certification procedures. In: Bastioli C (ed) Handbook of biodegradable polymers. Rapra Technology Limited, Sawbury, pp 145–182
Williams S, Peoples O (1996) Biodegradable plastics from plants. CHEMTECH 26:38–44
World Economic Forum (2016) The new plastics economy—rethinking the future of plastics. Ellen MacArthur Foundation and McKinsey & Company
Xia W, Rao Q, Deng X, Chen J, Xie P (2020) Rainfall is a significant environmental factor of microplastic pollution in inland waters. Sci Total Environ 732:139065
Xiao Y, Jiang X, Liao Y, Zhao W, Zhao P, Li M (2020) Adverse physiological and molecular level effects of polystyrene microplastics on freshwater microalgae. Chemosphere 255:126914
Xu B, Liu F, Cryder Z, Huang D, Lu Z, He Y, Wang H, Lu Z, Brookes PC, Tang C, Gan J, Xu J (2020a) Microplastics in the soil environment: occurrence, risks, interactions and fate–a review. Crit Rev Environ Sci Technol 50:2175–2222
Xu C, Zhang B, Gu C, Shen C, Yin S, Aamir M, Li F (2020b) Are we underestimating the sources of microplastic pollution in terrestrial environment? J Hazard Mater 400:123228

Yasuda N, Wang Y, Tsukegi T, Shirai Y, Nishida H (2010) Quantitative evaluation of photodegradation and racemization of poly(l-lactic acid) under UV-C irradiation. Polym Degrad Stab 95:1238–1243

Yu Y, Li H, Zeng Y, Chen B (2009) Extracellular enzymes of cold-adapted bacteria from Arctic Sea ice, Canada Basin. Polar Biol 32:1539–1547

Zhang D, Cui Y, Zhou H, Jin C, Yu X, Xu Y, Li Y, Zhang C (2020a) Microplastic pollution in water, sediment, and fish from artificial reefs around the Ma'an Archipelago, Shengsi, China. Sci Total Environ 703:134768

Zhang Y, Liang J, Zeng G, Tang W, Lu Y, Luo Y, Xing W, Tang N, Ye S, Li X, Huang W (2020b) How climate change and eutrophication interact with microplastic pollution and sediment resuspension in shallow lakes: a review. Sci Total Environ 705:135979

Zheng Y, Yanful EK, Bassi AS (2005) A review of plastic waste biodegradation. Crit Rev Biotechnol 25:243–250

Zhu X, Zhao W, Chen X, Zhao T, Tan L, Wang J (2020) Growth inhibition of the microalgae Skeletonema costatum under copper nanoparticles with microplastic exposure. Mar Environ Res 158:105005

Ziajahromi S, Drapper D, Hornbuckle A, Rintoul L, Leusch FDL (2020) Microplastic pollution in a stormwater floating treatment wetland: Detection of tyre particles in sediment. Sci Total Environ 713:136356

Part II

Impact of the Plastic Waste on the Marine Ecosystem

Entry of Macro, Micro, and Nanoplastic in the Food Chain and Their Impact on Marine Life (from Source to Sink)

11

Syed Abrar Ahmad and Varsha Wankhade

Abstract

Plastic is one of the and most widely used polymer in the world. It degrades very slow rate in the environment and took around 1000 years for complete degradation. Due to its enormous properties, it's demand is increasing exponentially, and each year huge amount of plastic waste is getting generated at an alarming rate. All plastic waste finally enters the marine environment and is reported to impacts a wide range of animals. Due to various environmental forces, the plastic gets deteriorated into small particles, viz., micro, macro, and nanoplastic. These plastic particles are ingested by marine animals and get transferred to other trophic levels through the food chain and lead to the death of billions of marine animals annually. In the present chapter, an attempt was made to overview the entry of these small-sized plastic in the food chain and their impact on the consumer organisms. This study also emphasizes plastic distribution in the marine environment and its consumption and different entry points in the marine ecosystem.

Keywords

Persistent organic pollutants · Marine ecosystem · Marine debris · Plastic waste · Bioplastics

S. A. Ahmad (✉) · V. Wankhade
Department of Zoology, Savitribai Phule Pune University, Pune, Maharashtra, India
e-mail: varsha3w@unipune.ac.in

M. Shahnawaz et al. (eds.), *Impact of Plastic Waste on the Marine Biota*,
https://doi.org/10.1007/978-981-16-5403-9_11

11.1 Introduction

The word plastics has been derived from the Greek word *plastikos* meaning "fit for molding" which refers to the flexibility of these materials at the time of manufacturing (Liddell et al. 1968). Plastics are mostly used materials in the world due to their durability and flexibility. The downside of plastic use is they persist in the environment over long periods thus affecting the ecosystem (UNEP 2005). Plastics are biochemically inert due to their non-reactiveness and non-penetration into the cell membrane of an organism due to their macromolecular structure. Plastic pollution also felicitates the transport of alien species floating on debris which eventually invade new ecosystems and leads to the extinction of other species (Aliani and Molcard 2003).

Microplastics are fragmented parts of plastic with a size of <5 mm. Due to different shapes, sizes, and density, microplastics are distributed in an ecosystem at different levels further taken by marine biota (Sharma and Chatterjee 2017). Microplastics are leached into terrestrial and aquatic ecosystem mainly by three forms of activities: domestic (microbeads and microplastic fragments), industrial (pellets and resin powders), and coastal activities (fishing practices, aqua tourism, and marine industries) (Anthony and Andrady 2011; Michiel et al. 2011). Macroplastics are >25 mm in size. Macroplastics are also known as "macro litter" (Andrady 2011), "anthropogenic litter" (Chin and Fung 2019), "plastic litter" (Bond et al. 2018), "marine litter" (Hengstmann et al. 2017), "marine plastic" (Barnes et al. 2018), and "plastic debris (Derraik 2002). Macroplastics have a profound impact on aquatic ecosystems, with the entanglement of aquatic animals leading to suffocation and severe wounds (Fischer et al. 2016; Campana et al. 2018).

Nanoplastic size ranges from 1 to 100 nm produced by the degradation of microplastics or released by industrial or domestic source (EFSA 2016). As nanoplastics are mainly produced from macro and microplastics, commercially these are produced from beauty products such as abrasive cleaning supplies, plastic powders, and air-blasting mechanisms (Claessens et al. 2011; Carr et al. 2016).

In the present chapter, an attempt was made to discuss the entry of the plastic particles in the food chain and their detrimental impact on marine animals under the following line: (i) to overview the total plastic consumed and waste generated at the global level, (ii) to enlist the routes followed by the plastic waste to reach the marine environment, (iii) to register different sources of plastics (macro, micro, and nano) in the marine ecosystem, (iv) to brief the accumulation and degradation of the plastic waste in the marine environment, and (v) to depict the mechanism of particle transfer across the tropic level and deaths of marine animals.

11.2 Global Plastic Production and Consumption

The global production of plastics is increasing exponentially since 1950. In the year 2016, 2.01 billion tonnes of municipal solid waste was produced which may go up to 3.40 billion tonnes by 2050. By the end of 2050, 26,000 million tonnes (Mt) of resins, 6000 Mt of PP&A fibers, and 2000 Mt of additives will be added. The share distribution of non-fiber plastics is polyester (70%), PE (36%), PP (21%), and PVC (12%), followed by PET, PUR, and PS (<10% each) (Geyer et al. 2017). Food and green waste comprise 50% of total waste in low and middle-income countries, and global terms it turns out 0.74 kg of waste per capita per day. The global plastic production rate in terms of waste composition scenario was 12%. As per income levels, high-income countries produce 13% of plastics, while low-income countries add only 6.4% (Kaza et al. 2018). Southern Asia produced the highest mismanaged plastic waste (MPW) 18.4 Mt/year in 2015 followed by Eastern Asia (17.4), Southeast Asia (11.7), Eastern Africa (5.89), South America (5.81), Western Africa (4.47), Western Asia (3.31), Northern Africa (3.40), Eastern Europe (2.66), and Middle Africa (2.39) (Lebreton and Andrady 2019). During the year 2015, 407 Mt of primary plastics were produced, and 302 Mt was left discarded. In just 13 years, 3900 Mt of plastic was produced worldwide (Geyer et al. 2017). By this pace, global production of plastic may reach up to 1124 Mt (World Economic Forum 2016).

11.3 Source of Different Plastic Types

The life cycle of plastics mainly passes through three phases: production from raw materials to convert plastic to manufacture of plastic products which are ultimately exported. In 2018 alone the world production of plastic reached 350 Mt as were produced 348 Mt in 2017. In Europe, the production of plastic reached 61.8 Mt in 2018 as compared to 64.4 Mt in 2017. As per plastic demand by segment in Europe, packing, building, and construction segment represents (39.9%); construction industry (19.8%); mechanical, furniture, and medicinal (16.7%); automotive (9.9%); electric and electronic (6.2%); household, leisure, and sports usage (4.1%); and agriculture (3.4%). Plastic distribution as per resin in 2018 was PP > PE-LD/PE-LLD > HDPE > PVC > PUR > PET > EPS > others (ABS, PBT, PC, PMMA, PTFE). Polypropylene (PP) represents highest share (19.3%), linear low-density polyethylene (17.5%), high-density polyethylene (12.2%), polyvinyl chloride (10%), polyurethane (7.9%), polyethylene terephthalate (7.7%), expanded polystyrene (6.4%), and others (19%) (Plastics Europe 2019).

Based on origin, microplastics are categorized as primary and secondary. Primary microplastics mostly come from microbeads used in cosmetic and healthcare products as exfoliates and from synthetic clothes. Secondary microplastics are fragmented particles of larger plastic parts including macro and mesoplastics due to biodegradation, photodegradation, thermo-oxidative degradation, thermal degradation, and hydrolysis (Sharma and Chatterjee 2017). Microplastics includes various products such as plastic pellets; personal care products including microbeads, toothpaste, sunscreen, shower gel, and hair dye, paint; washing waste water including household laundry; washing plant wastewater; sewage effluents including industrial sewage and runoff; sewage sludge; artificial turf; rubber asphalt roads in cities; vehicle tire ware; plastic bags; plastic bottles; disposable utensils; plastic packings; fishing wastes including buoys, floating boxes, fishing rods, fish tanks, fishing nets, fishing lines, and cables; and agricultural farming firms (An et al. 2020).

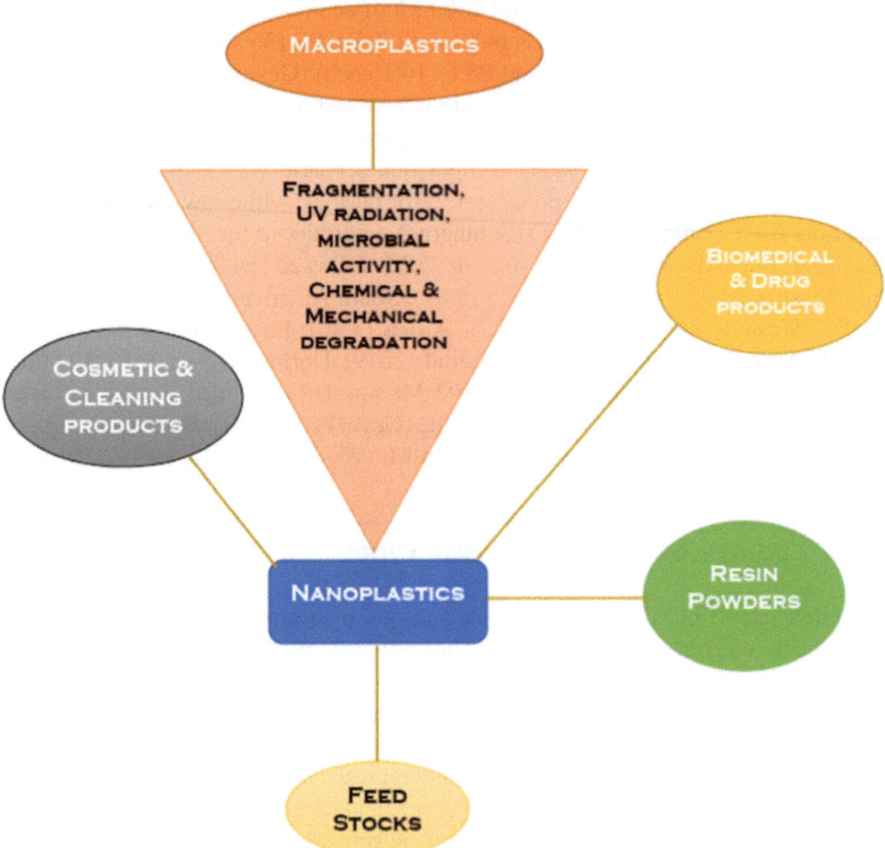

Fig. 11.1 Source of nanoplastics and degradation of large plastic by biotic and abiotic weathering

Nanoplastics are more deleterious due to their smaller size and penetration across cell membranes (Lambert et al. 2014). Nanoplastics represent a small size and higher surface area/unit mass, interaction between particles and surface energy have important role in toxicity (Lowry et al. 2012; Zoroddu et al. 2014). The primary source includes a wide range of products including paints, adhesives, biomedical products, diagnostics, magnetic and electronics, and drug delivery. The secondary source includes microplastic fragmentation by physical and breakage of energy bonds and degradation by biotic and abiotic forces (Koelmans et al. 2015) (Fig. 11.1).

11.4 Entry and Routes of Plastics in a Marine Ecosystem

Plastic can enter the environment at any stage of production due to poor management, accidental leakage, illegal dumping, and deliberate littering (Barnes et al. 2009). The first point of loss of plastics in a marine environment is resin pellets spillage; after conversion into plastic products, these may also enter in aquatic environment by improper management, such as disposal in open landfills, and littering (Law 2017). Four out of five plastic pieces originate from land-based use (Derraik 2002). Land- and ocean-based plastic inclusion into marine ecosystem occurs through both in situ and ex situ pathways (Thushari and Senevirathna 2020). Land-based sources account for 80% of plastic litter in the marine environment, while 18% is attributed to the fishing industry (Andrady 2011). Industrial and domestic sewage including hand and body cleaners are also a source of microplastics in water (Browne et al. 2007). On daily basis 5 million items are lost from ships and are overthrown. The main sources of ocean-based source of waste are merchant ships, ferries, cruiseliners, naval research vessels, offshore oil or gas platforms, and fishing vessels (accounts 50–90% of marine debris) (UNEP 2009). Rivers are key agents for the transportation of macro and microplastics besides wind and ice in the ocean ecosystem (Moore et al. 2011). Beach sands can be a source of microplastics transported by landward-directed surface currents (Chubarenko et al. 2018). 12% of waste in oceans which equals approximately 0.8 Mt/year originates from land mainly includes municipal landfills located on the coasts, discharge of untreated sewage and stormwater, industrial facilities, and pollution by tourists (Sheavly 2005; UNEP 2005).

11.5 Degradation and Accumulation of Plastics in a Marine Ecosystem

Around 49% of all plastic produced is buoyant which makes them to travel different locations (EPA 2008). Based on chemical composition, plastics undergo different processes such as thermal oxidation, biodegradation, hydrolysis, and photooxidative degradation. Sinkable plastic undergoes thermal oxidation as UV light is absorbed by water while as buoyant plastic undergoes photodegradation or photothermal oxidation. A "heat buildup" mechanism is associated with land-based plastics due to absorption of infrared radiation as compared to buoyant plastics on sea surfaces (Andrady 2003). As per UNEP 2001 report, of all plastics dumped in the North Sea, 15% are floating on the surface, 70% will sink at the bottom, and 15% washed to shore. An average of 110 pieces of debris per km^2 occurs in the North Sea seabed which equals 600,000 m^3 of the seabed. Due to high circular oceanic currents in the North Pacific central gyre results in accumulation of debris in the central part, this is now known as "Garbage Patch" or "Pacific Trash Vortex" (Allsopp et al. 2007). Between the 1986 and 2008 investigation in the North Atlantic gyre, 62 percent of all net tows contains plastics which equals 580,000 pieces/km (Law et al. 2010). Another observation in North Pacific gyre in the Kuroshio Current area shows plastic abundance ranging from 0 to 3,520,000 particles/km^2 and mass (0–153,000 g/km^2) (Yamashita and Tanimura 2007). The density of marine benthic debris in eastern Mediterranean coastal areas/1000 m^2 was 0–251 items with plastics (55.47%), metal (25.73%), rubber (5.16%), paper-cardboard (3.53%), clothing (2.71%), glass (2.09%), ceramic (0.59%), and others (4.73%) (Katsanevakis and

Katsarou 2004). Plastics are considered biochemically inert, but the presence of additives causes some toxic effects and can penetrate biochemical membranes. Phthalates and BPA can leach in the environment which affects the endocrine system, reproductive system, development impairment, and genetic aberrations (Teuten et al. 2009; Diamanti-Kandarakis 2009).

11.6 Entry and Impact of Plastics across Different Trophic Levels

Macroplastics are mainly released into the ecosystem through land-based and ocean-based pathways (Lechthaler et al. 2020). Despite the large size of macroplastics, a large number of reports suggest that macroplastics are ingested by many sea birds including fish and cetaceans (Derraik 2002; Teuten et al. 2007). Ingestion of macroplastics results in blockage of the intestinal tract, gastric enzyme inhibition, decrease in feeding stimuli, delay ovulation, and inhibit reproduction (Li et al. 2016). The presence of macro and microplastics in marine biota has a potential effect on endangered species such as swordfish and bluefin tuna. A positive correlation was present between levels of PBT compounds (PAH, PCB, PBDE, DDT) and alteration on the reproductive system. The levels of PBT compounds are biomagnified due to circulation in the food chain (Romeo et al. 2015). A negative correlation was present between body weight and the number of plastic particles ingested in seabirds due to digestive system impairment, presence of toxins in bird's body, and reduced foraging efficiency (Spear et al. 1995). Ryan (1988) also reported that ingested plastic reduces meal size and food consumption which in turn decreases fat deposits thus reduces fitness. Seabirds regurgitate undigested hard plastics which eventually accumulate in gastrointestinal blockage which in turn causes the problem in feeding stimuli and different activity levels (Derraik 2002). Marine-associated bird species dependent on the zooplankton diet may increase their plastic consumption if birds cannot distinguish between zooplankton (e.g., amphipods, copepods, or euphausiids) and small-sized neustonic plastic particles (Avery-Gomm et al. 2013). Some seabirds select specific plastic shapes and colors mistakenly taken as prey without any health issues reported (Moser and Lee 1992). Plastic debris is a trap for persistent organic pollutants (POPs) including polychlorinated biphenyl's (PCBs), DDT (dichloro-diphenyl-trichloroethane), polycyclic aromatic hydrocarbons (PAHs), and aliphatic hydrocarbons. Marine organisms ingest plastic debris mistakenly and POPs are also carried into the bodies which do not only affect the initial organism but organisms within the food web (Rios et al. 2007). Marine plastic debris act as inverse sediments for polychlorinated biphenyls (PCBs), chlorinated pesticides, and polycyclic aromatic hydrocarbons (PAHs) and are mistaken for food which provides a risk for all marine species and also humans as these pollutants circulate into the food web (Rios et al. 2010).

As per the report of the European Food Safety Authority (EFSA) (2016), microplastics have the potential to be transferred between various trophic levels, e.g., fish meat is used in poultry, and pig meal thus ends up in nonmarine foods.

Bivalves are eaten without removal of the digestive tract thus poses exposure of microplastics to fish and other seafoods, and particles <150 μm are transferred to gut epithelium which results in systemic exposure (EFSA 2016). Planktons (microalgae) are the carbon fixers from inorganic CO_2 and energy from sunlight. Polychaete larvae of the genus *Marenzelleria* have the highest proportion of microplastics ingestion. Penetration of microplastics results in the reduction of chlorophyll absorption (Nerland et al. 2014). Zooplanktons are heterotrophic organisms ranging from microscopic to large animals including jellyfish. Zooplanktons are the primary consumers of aquatic food and thus represent an integral part of the food chain. Food quality is essential than size selection as zooplanktons use chemoreceptors for prey selection (DeMott 1988). The interaction between microplastics and zooplanktons is high due to the presence of >333 μm in diameter size of debris in the marine environment. Microplastics at the range of $\geq$7000 microplastics mL^{-1} show a significant impact on algal ingestion by the copepod *Centropages typicus* in a dose-response relationship (Cole et al. 2013). BioMPP (polyhydroxybutyrate (PHB) and petroleum-based MPP (polymethylmethacrylate (PMMA) in *Gammarus fossarum* significantly decreases the assimilation efficiency, lowers weight gain, and causes digestive constraints on the amphipods (Straub et al. 2017). Polyethylene microplastic fibers have a significantly decreased reproduction and growth. Egestion time and growth were also significantly less on exposure to polypropylene microplastic fibers on the freshwater amphipod, *Hyalella azteca* (Au et al. 2015). Polystyrene microplastics reduce the feeding activity at a PS dose of 7.4% dry weight in lungworm *Arenicola marina* (Besseling et al. 2013). The immediate effects of microplastics ingestion are gastrointestinal injury and obstructions thus reducing nutrition and food consumption ultimately leading to starvation and death (Da Costa et al. 2017). Ingestion of native microplastics in sea bass fish results in swelling of villi, increased number of vacuoles, increased goblet cells, detachment of lamina propria, and swelling in mucosal layers in the distal part of the intestine (Peda et al. 2016). Microplastic exposure in fish and planktons, in turn, leaches plastic additives including plasticizers, persistent organic pollutants which are toxic to biota which may lead to the death of organisms (Da Costa et al. 2017). Induction of polystyrene microplastics in zebrafish results in inflammation and lipid accumulation in the liver. Higher levels of oxidative stress were detected in terms of increased activities of superoxide dismutase (SOD) and catalase (CAT) (Lu et al. 2016). Polystyrene microplastic beads also have a negative impact on higher mammals including mice, rats, and dogs. Ingestion of microplastics is accumulated in the liver, kidney, and gut which leads to inflammation and damage to organs. In dogs, microplastic is transported through the lymphatic system to different organs such as the gastrointestinal tract, liver, pancreas, and spleen (Volkheimer 1974).

Nanoplastics are the least known but most dangerous due to the formation of large particles by the process of nanofragmentation (Andrady 2011; Shim et al. 2014). Charged polystyrene nanoplastics increase ROS production and hinder photosynthesis due to CO_2 depletion, and decreased absorbance (Bhattacharya et al. 2010) population growth inhibition, decreased chlorophyll concentration in green algae, reduced body size, and reproduction alterations in Daphnia (Besseling et al. 2014).

Nanoplastics affect *Mytilus edulis* (mussels) by reducing the valve diameter thus decreasing the filtering activity by producing pseudofeces (Wegner et al. 2012). Bivalve *Mytilus* on exposure to cationic polystyrene nanoplastics decreases phagocytic activity and increases lysosome activity. An increase in ROS and NO production and induction of apoptotic processes were also reported (Canesi et al. 2015). Carboxylated polystyrene nanoplastics also induce hindering mortality, hampering the molting, and feeding process of brine shrimp larvae (Bergami et al. 2016).

Exposure of polystyrene and polycarbonate nanoplastics in fathead minnow fish showed higher oxidative stress in terms of neutrophil oxidative burst by increasing degranulation and higher innate immune response (Greven et al. 2016). Polystyrene nanoparticles (PS-NP, 52 nm) move from the external surface to internal organs which results in decreased survival and reproduction in terms of abnormal development and decreased hatching rate (Cui et al. 2017). Exposure of sea urchin embryos to polystyrene nanoparticles results in upregulation of key stress response and development genes including Pl-Hsp70, Pl-p38 Mapk, Pl-Univin, Pl-Cas8, and Pl-Univin at 24 and 48-hour postfertilization (hpf) (Pinsino et al. 2017). Transportation of nanoplastics from algae through zooplanktons to fish affects lipid metabolism as triglycerides/cholesterol ratio, cholesterol distribution, weight loss, and behavioral disturbance of the top consumer (Cedervall et al. 2012). Uptake of nanoparticles through the trophic level food chain (algae, zooplankton, and fish) affects hunting behavior, doubling feeding time, strong shoaling behavior, less explorative, and disorder of cellular functions (Mattsson et al. 2015). Trophic transfer of nanoplastics through algae, water flea, fish (second consumer), and fish (end consumer) results in a negative impact on fish activity, histopathological changes, damage to intestinal walls, liver tissue destruction, shoal behavior, and lipid metabolism disturbances (Chae et al. 2018). Silver nanowire (AgNW) uptake through the food chain (algae, water flea, zebrafish) has a negative effect on algal growth, higher necrosis and death, and damage to the digestive system of water flea which in turn may affect humans (Chae and An 2016). Nanoplastic ingestion retards root growth of the mung bean plant and trophic transfer to snails which decreases feeding and foraging speed and dysmorphism to digestive system, reduces the viability of gut microbiota to half, and induces behavioral changes (Chae and An 2020). Transfer of nanoplastics from zooplankton reduces survival, penetrates the blood-brain barrier of fish and induces behavioral changes, reduces feeding rate, lowers activity, and induces morphological changes in the brain which in turn may further get transferred to higher trophic levels (Mattsson et al. 2017).

11.7 Conclusion

Plastics have a wide range of effects on freshwater to terrestrial and in turn on human lives. It is increasing exponentially as high-income countries contribute more as compared to low-income countries. Macroplastics is a snare for many persistent organic pollutants which ultimately end up in the animal bodies and affect the whole food web. Being heavier of all types of plastics macroplastics may end up at the bottom of the sea which may result in entanglement or may become the hiding place for various predators. Microplastics being smaller in size are transferred to different trophic levels and may get shifted from marine food chain to nonmarine food chain. Nanoplastics of the smallest of all are easily transferred across cellular membranes and are taken up by cellular organisms which may be transported to humans through various primary and secondary consumers. To reduce plastic pollution and environmental sustainability, proper sampling, reuse, and effective recycling should be implemented.

Acknowledgments None.

Conflicts of Interest No conflicts to declare.

References

Aliani S, Molcard A (2003) Hitch-hiking on floating marine debris: macrobiotic species in the Western Mediterranean Sea. Hydrobiologia 503(1):59–67

Allsopp M, Walters A, Santillo D et al (2007) Plastic debris in the World's oceans, Greenpeace

An L, Liu Q, Deng Y et al (2020) Sources of microplastic in the environment. In: He D, Luo Y (eds) Microplastics in terrestrial environments: emerging contaminants and major challenges. Springer, Cham, pp 143–159

Andrady AL (ed) (2003) Plastics and the environment. John Wiley & Sons

Andrady AL (2011) Microplastics in the marine environment. Mar Pollut Bull 62:1596–1605

Au SY, Bruce TF, Bridges WC et al (2015) Responses of Hyalella azteca to acute and chronic microplastic exposures. Environ Toxicol Chem 34(11):2564–2572

Avery-Gomm S, Provencher JF, Morgan KH et al (2013) Plastic ingestion in marine-associated bird species from the eastern North Pacific. Mar Pollut Bull 72(1):257–259

Barnes DKA, Morley SA, Bell J et al (2018) Marine plastics threaten giant Atlantic marine protected areas. Curr Biol 28:R1137–R1138

Barnes DK, Galgani F, Thompson RC et al (2009) Accumulation and fragmentation of plastic debris in global environments. Philos Trans R Soc B Biol Sci 364(1526):1985–1998

Bergami E, Bocci E, Vannuccini ML et al (2016) Nano-sized polystyrene affects feeding, behavior and physiology of brine shrimp Artemia franciscana larvae. Ecotoxicol Environ Saf 123:18–25

Besseling E, Wang B, Lürling M, Koelmans AA (2014) Nanoplastic affects growth of S. obliquus and reproduction of D. magna. Environ Sci Technol 48(20):12,336–12,343

Besseling E, Wegner A, Foekema EM et al (2013) Effects of microplastic on fitness and PCB bioaccumulation by the lugworm Arenicola marina (L.). Environ Sci Technol 47(1):593–600

Bhattacharya P, Lin S, Turner JP, Ke PC (2010) Physical adsorption of charged plastic nanoparticles affects algal photosynthesis. J Phys Chem C 114(39):16,556–16,561

Bond T, Ferrandiz-Mas V, Felipe-Sotelo M et al (2018) The occurrence and degradation of aquatic plastic litter based on polymer physicochemical properties: a review. Crit Rev Environ Sci Technol 48:685–722

Browne MA, Galloway T, Thompson R (2007) Microplastic—an emerging contaminant of potential concern? Integr Environ Assess Manag Int J 3(4):559–561

Campana I, Angeletti D, Crosti R et al (2018) Seasonal patterns of floating macro-litter across the Western Mediterranean Sea: a potential threat for cetacean species. Rend Lincei-Sci Fis 29:453–467

Canesi L, Ciacci C, Bergami E et al (2015) Evidence for immunomodulation and apoptotic processes induced by cationic polystyrene nanoparticles in the hemocytes of the marine bivalve Mytilus. Mar Environ Res 111:34–40
Carr SA, Liu J, Tesoro AG (2016) Transport and fate of microplastic particles in wastewater treatment plants. Water Res 91:174–182
Cedervall T, Hansson LA, Lard M, Frohm B, Linse S et al (2012) Food chain transport of nanoparticles affects behaviour and fat metabolism in fish. PLoS One 7(2):e32254
Chae Y, An YJ (2016) Toxicity and transfer of polyvinylpyrrolidone-coated silver nanowires in an aquatic food chain consisting of algae, water fleas, and zebrafish. Aquat Toxicol 173:94–104
Chae Y, An YJ (2020) Nanoplastic ingestion induces behavioral disorders in terrestrial snails: trophic transfer effects via vascular plants. Environ Sci Nano 7(3):975–983
Chae Y, Kim D, Kim SW et al (2018) Trophic transfer and individual impact of nano-sized polystyrene in a four-species freshwater food chain. Sci Rep 8(1):1–1
Chin LW, Fung TH (2019) Plastic in marine litter. In: Plastics and the environment. Royal Society of Chemistry, Cambridge, pp 21–59
Chubarenko IP, Esiukova EE, Bagaev AV et al (2018) Three-dimensional distribution of anthropogenic microparticles in the body of sandy beaches. Sci Total Environ 628:1340–1351
Claessens M, De Meester S, Van Landuyt L et al (2011) Occurrence and distribution of microplastics in marine sediments along the Belgian coast. Mar Pollut Bull 62:2199–2204
Cole M, Lindeque P, Fileman E et al (2013) Microplastic ingestion by zooplankton. Environ Sci Technol 47(12):6646–6655
Cui R, Kim SW, An YJ (2017) Polystyrene nanoplastics inhibit reproduction and induce abnormal embryonic development in the freshwater crustacean Daphnia galeata. Sci Rep 7(1):1–0
Da Costa JP, Duarte AC, Rocha-Santos TA (2017) Microplastics–occurrence, fate and behaviour in the environment. Compr Anal Chem 75:1–24
DeMott WR (1988) Discrimination between algae and detritus by freshwater and marine zooplankton. Bull Mar Sci 43(3):486–499
Derraik JGB (2002) The pollution of the marine environment by plastic debris: a review. Mar Pollut Bull 44(9):842–852
Diamanti-Kandarakis E (2009) Endocrine-disrupting chemicals: an endocrine society scientific statement. Endocr Rev 30(4):293–342
EFSA European Food Safety Authority (2016) Presence of microplastics and nanoplastics in food, with particular focus on seafood. EFSA J 14(6):4501
EFSA Panel on Contaminants in the Food Chain (CONTAM) (2016) Presence of microplastics and nanoplastics in food, with particular focus on seafood. EFSA J 14(6):e04501
EPA (2008) Municipal solid waste generation, recycling and disposal in the United States: facts and figures for 2008, United States Environmental Protection Agency Agency
Fischer EK, Paglialonga L, Czech E et al (2016) Microplastic pollution in lakes and Lake shoreline sediments—a case study on Lake Bolsena and Lake Chiusi (Central Italy). Environ Poll 213: 648–657
Geyer R, Jambeck JR, Law KL (2017) Production, use, and fate of all plastics ever made. Sci Adv 3 (7):e1700782
Greven AC, Merk T, Karagöz F (2016) Polycarbonate and polystyrene nanoplastic particles act as stressors to the innate immune system of fathead minnow (Pimephales promelas). Environ Toxicol Chem 35(12):3093–3100
Hengstmann E, Gräwe D, Tamminga M et al (2017) Marine litter abundance and distribution on beaches on the isle of Rügen considering the influence of exposition, morphology and recreational activities. Mar Pollut Bull 115:297–306
Katsanevakis S, Katsarou A (2004) Influences on the distribution of marine debris on the seafloor of shallow coastal areas in Greece (eastern Mediterranean). Water Air Soil Pollut 159(1):325–337
Kaza S, Yao L, Bhada-Tata P et al (2018) What a waste 2.0: a global snapshot of solid waste management to 2050. World Bank Publications, Washington, DC

Koelmans AA, Besseling E, Shim WJ (2015) Nanoplastics in the aquatic environment, critical review. Mar Anth Litt:325–340

Lambert S, Sinclair C, Boxall A (2014) Occurrence, degradation, and effect of polymer-based materials in the environment. Rev Environ Contam Toxicol 227:1–54

Law KL (2017) Plastics in the marine environment. Annu Rev Mar Sci 9:205–229

Law KL, Morét-Ferguson S, Maximenko NA et al (2010) Plastic accumulation in the North Atlantic subtropical gyre. Science 329(5996):1185–1188

Lebreton L, Andrady A (2019) Future scenarios of global plastic waste generation and disposal. Palgrave Commun 5(1):1

Lechthaler S, Waldschläger K, Stauch G et al (2020) The way of macroplastic through the environment. Environments 7(10):73

Li W, Tse H, Fok L (2016) Plastic waste in the marine environment: a review of sources, occurrence and effects. Sci Total Environ 566:333–349

Liddell HG, Scott R, Jones HS (1968) A Greek-English lexicon compiled by Henry George Liddell and Robert Scotte, vol 2. The Clarendon, London

Lowry GV, Gregory KB, Apte SC et al (2012) Transformations of nanomaterials in the environment. Environ Sci Technol 46:6893–6899

Lu Y, Zhang Y, Deng Y et al (2016) Uptake and accumulation of polystyrene microplastics in zebrafish (Danio rerio) and toxic effects in liver. Environ Sci Technol 50(7):4054–4060

Mattsson K, Ekvall MT, Hansson LA et al (2015) Altered behavior, physiology, and metabolism in fish exposed to polystyrene nanoparticles. Environ Sci Technol 49(1):553–561

Mattsson K, Johnson EV, Malmendal A et al (2017) Brain damage and behavioral disorders in fish induced by plastic nanoparticles delivered through the food chain. Sci Rep 7(1):1–7

Michiel C, Steven DM, Lieve VL et al (2011) Occurrence and distribution of microplastics in marine sediments along the Belgian coast. Mar Pollut Bull 62(10):2199–2204

Moore CJ, Lattin GL, Zellers AF (2011) Quantity and type of plastic debris flowing from two urban rivers to coastal waters and beaches of Southern California. J Integr Coast Zone Manag 11:65–73

Moser ML, Lee DS (1992) A fourteen-year survey of plastic ingestion by western North Atlantic seabirds. Col Waterbirds:83–94

Nerland IL, Halsband C, Allan I et al (2014) Microplastics in marine environments: occurrence, distribution and effects. NIVA:6754–2014

Peda C, Caccamo L, Fossi MC et al (2016) Intestinal alterations in European sea bass Dicentrarchus labrax (Linnaeus, 1758) exposed to microplastics: preliminary results. Environ Pollut 212:251–256

Pinsino A, Bergami E, Della TC et al (2017) Amino-modified polystyrene nanoparticles affect signaling pathways of the sea urchin (Paracentrotus lividus) embryos. Nanotoxicology 11 (2):201–209

Plastics Europe (2019) Plastics. The facts. An analysis of European plastics production, demand and waste data. Plastics Europe, Brussels, Belgium

Rios LM, Moore C, Jones PR (2007) Persistent organic pollutants carried by synthetic polymers in the ocean environment. Mar Pollut Bull 54:1230–1237

Rios LM, Jones PR, Moore C et al (2010) Quantitation of persistent organic pollutants adsorbed on plastic debris from the northern Pacific Gyre's "eastern garbage patch". J Environ Monit 12: 2226–2236

Romeo T, Pietro B, Pedà C et al (2015) First evidence of presence of plastic debris in stomach of large pelagic fish in the Mediterranean Sea. Mar Pollut Bull 95(1):358–361

Ryan PG (1988) Effects of ingested plastic on seabird feeding: evidence from chickens. Mar Pollut Bull 19(3):125–128

Sharma S, Chatterjee S (2017) Microplastic pollution, a threat to marine ecosystem and human health: a short review. Environ Sci Pollut Res 24(27):21,530–21,547

Sheavly SB (2005) Marine debris-an overview of a critical issue for our oceans. Sixth meeting of the UN open-ended informal consultative processes on oceans & the Law of the sea

Shim WJ, Song YK, Hong SH et al (2014) Producing fragmented micro- and nano-sized expanded polystyrene particles with an accelerated mechanical abrasion experiment. SETAC annual meeting, Basel, Switzerland, p 11–15

Spear LB, Ainley DG, Ribic CA (1995) Incidence of plastic in seabirds from the tropical pacific, 1984–1991: relation with distribution of species, sex, age, season, year and body weight. Mar Environ Res 40(2):123–146

Straub S, Hirsch PE, Burkhardt-Holm P (2017) Biodegradable and petroleum-based microplastics do not differ in their ingestion and excretion but in their biological effects in a freshwater invertebrate Gammarus fossarum. Int J Environ Res Public Health 14(7):774

Teuten EL, Rowland SJ, Galloway TS et al (2007) Potential for plastics to transport hydrophobic contaminants. Environ Sci Technol 41:7759–7764

Teuten EL, Saquing JM, Knappe DRU et al (2009) Transport and release of chemicals from plastics to the environment and to wildlife. Philos Trans R Soc B Biol Sci 364(1526):2027–2045

Thushari GG, Senevirathna JD (2020) Plastic pollution in the marine environment. Heliyon 6(8): e04709

UNEP (2001) Marine litter—trash that kills, United Nations Environment Programme

UNEP (2005) Marine litter, an analytical overview. United Nations Environment Programme, Nairobi

UNEP (2009) Abandoned, lost or otherwise discarded fishing gear. United Nations Environment Programme, Food and Agriculture Organization of the United Nations, Rome

Volkheimer G (1974) Passage of particles through the wall of the gastrointestinal tract. Environ Health Perspect 9:215

Wegner A, Besseling E, Foekema EM et al (2012) Effects of nanopolystyrene on the feeding behavior of the blue mussel (Mytilus edulisL.). Environ Toxicol Chem 31(11):2490–2497

World Economic Forum, Ellen MacArthur Foundation, and McKinsey & Company (2016) The new plastics economy: rethinking the future of plastics. newplasticseconomy.org/about/publications

Yamashita R, Tanimura A (2007) Floating plastic in the kuroshio current area, Western North Pacific Ocean. Mar Pollut Bull 54(4):485–488

Zoroddu MA, Medici S, Ledda A et al (2014) Toxicity of nanoparticles. Curr Med Chem 21 (33):3837–3853

Effect of Meso-, Micro-, and Nano-Plastic Waste on the Benthos

12

Osikemekha Anthony Anani, Charles Oluwaseun Adetunji, Gloria Anwuli Anani, John Ovie Olomukoro, Tunde O. T. Imoobe, Alex Ajeh Enenuku, and Isioma Tongo

Abstract

Plastics are organic, synthetic, and semisynthetic polymer materials that have various uses. It can be pressed, reused, and molded into several solid materials and forms, making it an essential material in the global manufacturing and industrial sectors. Studies have shown that plastics are nonbiodegradable. The overuse of plastics and the noneffective recycling of the products have been documented to be linked to several probabilistic environmental and health risks to humans and animals in the environment. Fragmented debris of plastics such as meso-, micro-, and nano-plastics, which are termed plastic wastes residues, have been recently shown to have an organismal negative influence, especially on aquatic benthos when they reside via sedimentation in the aquatic ecosystem. Of recent, this characterized plastic debris has shown potential food chain transfer

O. A. Anani (✉)
Laboratory for Ecotoxicology and Forensic Biology, Department of Biological Science, Faculty of Science, Edo State University Uzairue, Auchi, Edo State, Nigeria
e-mail: osikemekha.anani@edouniversity.edu.ng

C. O. Adetunji
Applied Microbiology, Biotechnology and Nanotechnology Laboratory, Department of Microbiology, Edo State University Uzairue, PMB 04, Auchi, Edo State, Nigeria

G. A. Anani · J. O. Olomukoro · T. O. T. Imoobe · I. Tongo
Department of Animal and Environmental Biology, Faculty of Life Science, University of Benin, Benin City, Nigeria

J. O. Olomukoro
Department of Animal and Environmental Biology, Faculty of Life Science, University of Benin, Benin City, Nigeria

A. A. Enenuku
Department of Environmental Management and Toxicology, Faculty of Life Science, University of Benin, Benin City, Nigeria

M. Shahnawaz et al. (eds.), *Impact of Plastic Waste on the Marine Biota*,
https://doi.org/10.1007/978-981-16-5403-9_12

with possible health impacts like cancer and environmental distortion like clogging of the certain microecological niche and creating high pollution risk factor. It can also assemble or combine with the microbial consortia in the benthic region causing an ecological condition known as biofouling. This chapter intends to examine the effects of meso-, micro-, and nano-plastic waste on the benthos with the special scenario on the classification of meso-, micro-, and nano-plastics. The ecosystem impacts of meso-, micro-, and nano-plastics on the benthic environment are discussed. Highlights are made on the physical and chemical influence of the association and leaching of meso-, micro-, and nano-plastics on the benthic environment. Possible identification of the ecological and health damages of meso-, micro-, and nano-plastics on the benthic community is discussed. Highlights on various molecular and cellular mechanisms pathways of meso-, micro-, and nano-plastics absorbed by the benthos are discussed. The examination of the accumulation and effects of meso-, micro-, and nano-plastics on the sediment as well as their reaction with the microbial communities are highlighted. The investigation of the influence of biofouling in the benthic region is discussed. Several mitigation measures in the management of meso-, micro-, and nano-plastic pollution in the benthic region are proffered.

Keywords

Sedimentation · Benthos · Plastics · Biofouling · Nanoparticles · Food chain

12.1 Introduction

Mesoplastics, microplastics, and nano-plastics are usually defined based on their sizes; 5 mm–2.5 cm; 0.1 μm – 5 mm, and 0.001–0.1 μm (Moore 2008). In recent times, environmental pollution from plastic remains of smaller sizes has been known to cause deleterious effects on the aquatic ecosystem and associated water resources (Thompson et al. 2004; Rochman et al. 2016; Chae and An 2017; Haegerbaeumer et al. 2019).

Generally, plastics like polyethylene terephthalate (PET), polyvinyl chloride (PVC), and polystyrene (PS) have complicated charges, densities, and shapes that are continually changing with time (Galloway et al. 2017). However, the knowledge of the ecological fate of small- to large-scale plastics is essential in the valuation of the probable hazards they portend. PET plastics have specific gravity higher than H_2O which can cause the residual materials to reside in the benthic region of the aquatic environment. However, polypropylene (PP), high-density polyethylene (HDPE), and low-density polyethylene (LDPE) plastics are mainly floaters in different columns or layers of water because their density is lighter compared to the density of water (Auta et al. 2017; Haegerbaeumer et al. 2019). Based on several biological activities like biofouling and accumulation of organic matter, the gravity of plastics particles may become higher, by which they possibly reside in the benthic sediment layer of the water ecosystem (Lattin et al. 2004; Galloway et al. 2017).

It has been evaluated by Wright et al. (2013), Ivar do Sul and Costa (2014), and Scherer et al. (2017) that biofouling and accumulation of organic matters in benthic sediments can intensify the bioavailability of meso-, micro-, and nano-plastics for benthic inhabiting faunas to ingest during their feeding process because the sizes of the plastics are smaller or of the same size as the sediment granules. The ingestion of plastic debris by benthic faunas in freshwater and marine ecosystem are of immense concern because they donate up to 90 percent of the biomass of the fish food and to the stabilization of the food chain structure (Weber and Traunspurger 2015; Olomukoro and Anani 2019). Thus, it is crucial to evaluate the ecotoxicological properties of different sizes of plastics on the macrobenthic community because of the influence of trophic or energy interactions.

Studies have shown that the ecotoxicological properties of polymers and monomers of meso-, micro-, and nano-plastics might be through natural, chemical, and physical influence from plastic leaching which could result in endocrine disruptions and carcinogenic problems, etc. (Barnes et al. 2009; Talsness et al. 2009; Wright et al. 2013; Adetunji and Anani 2021a; Adetunji and Anani 2021b). Due to their scale to small sizes and volume to surface area ratio, plastics can be easily contaminated with an associated level of pollutants like hydrophobic persistent organic pollutants (POPs) in a magnitude higher than those of the surrounding environment (Hirai et al. 2011; Haegerbaeumer et al. 2019).

Possible accumulation by ingestion of the combination of plastics debris and POPs has been estimated to bioaccumulate and biomagnified into the aquatic food web structure thus affecting the transfer of energy from one trophic level to another (vom Saal et al. 2008; Adetunji and Anani 2021a; b). This chapter examined the effects of meso-, micro-, and nano-plastic wastes on the benthos with the special scenario on the classification of meso-, micro-, and nano-plastics.

12.2 Ecosystem Impacts: Ecological and Health Damages of Meso-, Micro-, and Nano-Plastics on the Benthic Environment

Haegerbaeumer et al. (2019) investigated the impacts of nano-, meso-, and micro-sized particles of plastics on benthic macroinvertebrates. The authors stated that plastic pollution is one of the recent environmental pollutions. They specified that one of the major issues of the understanding of nano- and micro-sized plastics materials is their ever-changing charges, shapes, and sizes. In addition, the biological activities of bacterial colonies such as biofouling and aggregation of organic matter with plastics can increase the densities of the debris of plastics in the benthic region of the aquatic ecosystem. Thus, the hazard posed by the pollution of plastics to benthic animals is very high. However, the evaluation of the influence of plastics on pelagic-benthic organisms has not yet been fully exhausted. The authors concluded that the influence of various sizes of plastics on benthic animals is based on the routes of exposure, the shapes of the debris, and fractional sizes. In addition, a total assessment of the risk hazards using an ecosystem model from the chemical effects of plastic debris (nano-, meso-, and micro-sizes) on the affected ecosystem should be carried out to prospect a sustainable environment for meiobenthic and microbenthic faunas.

Plastics have been known as one of the most used and highly demanded resources on the globe. However, this global utilization has led to plastic pollution as a result of poor management and has caused serious environmental pollution. Bellasi et al. (2020) did a review of the pollution of benthic organisms and the sediment layer of freshwater with microplastics. The authors stated that the breakdown of plastics in the ecosystem has led to the production of plastic debris of smaller sizes greater than 5 mm (microplastics). These microplastics have potential and uncleared ecological effects when dispersed into the aquatic environment. However, recent data has shown that the presence of microplastics in the freshwater ecosystem is still understudied. Freshwater and marine sediment are some of the residual sites of microplastics. Benthic organisms which play a pivotal role in the biotic food chain are major accumulators of microplastic debris and a clear cut of the ecotoxicological transfer of plastic toxins to the higher top food chain levels. In addition, plastics act as a carrier of noxious chemicals to living animals in the benthic region. The authors stated that the noxious effects of microplastics especially the desorption to adsorption stabilities in certain environmental settings need to be understudied. Nonetheless, investigation on the harmonization of the pretreatment procedures and sampling techniques for multifaceted environmental conditions for microplastics should also be put into place because of the possible threats to freshwater and marine ecosystems they portend.

Thushari and Senevirathna (2020) investigated the effects of pollution from plastic wastes on the marine ecosystem. Pollution from plastic is a serious human ecological issue that has affected the marine and coastal environments of the globe. The continuous and unprecedented accumulation of wastes from plastics especially from anthropogenic bases is worrisome. There have been incidences of indirect and direct interruption of the biotic ecosystem by these actions of humans. The authors stated that plastic pollution is mainly distributed via the land. They enter the water bodies in different sizes such as microplastics, mesoplastics, nano-plastic, macroplastics, and megaplastics. However, microplastics in their secondary and primary forms disclose a general distribution into coastal and marine habitats, biota, and sediment regions of the water environment. In marine and coastal environments, microplastics range from 0.2 to 8766 particles/m^3 and 0.001 to 140 particles/m^3 in sediment and water, respectively. Pollution from plastic creates different abnormal consequences in conjunction with socioeconomic and ecological effects. Moreover, threats to trophic and biodiversity relationships, offensive species introduction displacement of organisms from their ecological niche, dispersal, starvation, and suffocation are some toxicological and ecological influences of microplastics to benthic faunas. This could also lead to modification and degradation of values and services of the marine ecosystems and their associated bioresources. Therefore, these emerging plastic pollutants could also affect human health, shipping, fishery, and tourism. In conclusion, the authors recommended the application of the 3Rs: reuse, recycle, and reduce in conjunction with capacity building and awareness to combat plastic contaminants in the environment. The enforcement of regulations, legislations, and adopted and existing policies to reduce plastic wastes in the coastal and marine zones are tenable at this point to ensure a healthy and plastic-free blue aquatic ecosystem.

In recent times, plastics of different sizes have been found to habit the estuaries, rivers, and lake systems of the aquatic world via uncontrolled anthropogenic activities. Redondo-Hasselerharm et al. (2018) evaluated in a biological study the influence of microplastics on six freshwater benthos (*Tubifex* sp., *Sphaerium corneum, Asellus aquaticus, Hyalella azteca, Gammarus pulex, and Lumbriculus variegatus*) as well as the thresholds of plastics in the water environments. The biological experiment lasted for 28 days performed under controlled environmental conditions, utilized microplastics of size 20–500 μm (polystyrene), and then combined it with 0–40 percent of dry weight sediment concentration. The effects of the microplastics showed no positive impacts on *Tubifex* sp., *Sphaerium corneum, Asellus aquaticus, Hyalella Azteca, and Gammarus pulex* on their survival. In addition, there was no impact on the reproductive faculties of *Lumbriculus variegatus*. There were no significant differences at $P > 0.05$ in the growth and development pattern of *Tubifex* sp., *L. variegatus*, *S. Corneum*, *A. Aquaticus*, and *H. Azteca*. Nonetheless, there was a significant difference in the decrease in the growth rate of *G. pulex* at $1.07\% = EC_{10}$ of the sediment dry weight, and the uptake of the microplastics correlated with the sediment concentration. Findings from the study revealed that ecotoxicological risks of the microplastic concentration were low. However, there might be some level of toxicological impacts on the benthic fauna community because they are sensitive and dependable species of the ecosystem.

The impacts of plastics on the aquatic environment have been long identified as one of the major drivers of aquatic (inland and marine) health deterioration. The impacts of plastics on the benthic environment have been traditionally neglected for a while; however, macroinvertebrates which serve as a tool for water quality evaluation, are supposed to be assessed on their responses to pollutants such as microplastics. On this note, Gallitelli et al. (2021) in a review looked at the baseline pieces of evidence of the effects of microplastic: polyvinylidene fluoride (PVDF), PS, PP, and PET on freshwater macrobenthos such as *Ephemera danica* (mayfly) and *Odontocerum albicorne* (caddisflies). The results from the experimental study showed that the selected microplastics impacted both the mayflies and caddisflies negatively thus reducing their burrowing and case-building abilities. It was also observed that the mayflies did not burrow into their natural substrates instead in the microplastics an implication that macroinvertebrates use microplastics for their ecological function. However, microplastics did not show as a straight stressor to the microbenthic invertebrates. This finding is of particular note to plastic profile contamination in the aquatic environment. The authors suggested further investigations on the possible chronic impacts of microplastics on the community structure of the benthic organisms specifically on their larvae stages and the attraction of microbenthic to the microplastics; since it was founded, the organisms use blue plastics for the building of their cast (Fig. 12.1).

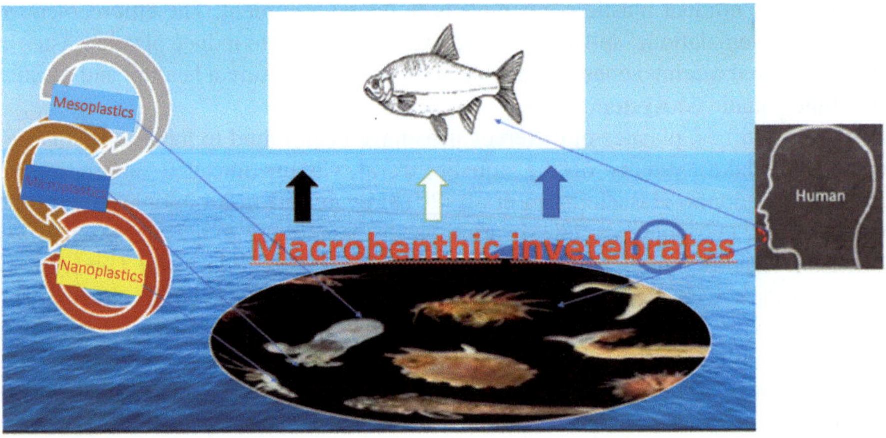

Fig. 12.1 Plastic impacts on marine macrobenthic invertebrates and possible food chain/trophic level transfers

12.3 Examination of the Accumulation and Effects of Meso-, Micro-, and Nano-Plastics on the Sediment As Well as their Reaction with the Benthic Communities

Amelia et al. (2021) examined the role of microplastics as carriers of main pollutants in the marine environment and possible hazards to humans and the ecosystem. The authors stated that groundwater, air, land, and water environments have been polluted by microplastics daily with serious ecological and biological impacts on humans, animals, and plants. Amelia et al. (2021) reported that microplastics have been known as carriers of pollutants by sorption, accumulation, and transferring them across the food chain. The properties of water, interactions of chemicals, and the characteristics of the microplastics play an immense role in the process of microplastic sorption in the environment by aquatic biota and humans along with the food web. However, the effects of direct health risks by microplastics are lacking; nonetheless, substantial health and ecological risks have been linked with pollutants as additives in their diets causing serious conditions like cancer and sexual disorders. Examples of such microplastics are PP and PE which are mainly contributed by activities from the marine, fishery, urban, and anthropogenic sources.

Kelsey et al. (2020) evaluated the fate of microplastics by the action of microbes in the marine environment. The authors reported that microbes are the major driver of the biotic, abiotic, and biogeochemical processes in the environment. Their presence in the aqua-ecosystem can influence the activities of microplastics in the sediment and water layers thus augmenting the elements pathways and food web cycles as well as the degradation process to reduce possible health and environmental risks. The fate of microplastics in the environment depends on the interrelationship between the microorganisms and the microplastics and the time of biodegradation. More so, understanding the levels and routes of microplastics in the ecosystem can also pave more ways in the understanding of the mechanisms of the hazards it portends. This will also enable the environmental researcher to capture the hazard characterization and exposure evaluation as well as set regulation on the thresholds or limits of microplastics debris or litters in benthic animals, humans, water, and sediment.

Sendra et al. (2020) investigated the impact of nano- and macroplastics on the immunology of *Mytilus galloprovincialis* hemocyte. The authors stated that *Mytilus galloprovincialis* is one of the well-known bivalves to be used to evaluate the toxicity of nano-plastics because their hemocytes (R1, R2, and R3) are specific and specialize for such task. Different sizes of plastics 1 μm, 100 nm, and 50 nm were proposed for the study and exposed to 1–10 mg $L^{-}1$ of nano- and microplastics. The results from the study showed the 1 μm plastics elicited greater immunological replies when compared to the 100 and 50 nm nano-plastics which was caused by the increase in the stability of the shape and size of the hemolymph blood serum. More so, the subpopulation of the R1 was most hit when compared to the R3 and R2 subpopulations. In addition, there was a sharp increase in the discharge of the trailing of the lysosomes, apoptotic messengers, noxious radicals, and a reduction in the action of the phagocytes in the R1 subpopulations. In conclusion, the authors reported that the inherent system of the *Mytilus galloprovincialis* is known to be a custodian of knowledge as regards the influence of nanomaterials. The inherent nature of nano-plastics and the immediate environment, the hemolymph blood serum, can be a second-degree identity in accordance to accumulation and the protein corona development, and the novel shape is developed. The use of confocal and cytometry images to verify the Fluoresbrite internalization of micro- and nano-plastics at 100 nm is needed to give a better picture of their nature before the utilization for immunotoxicity purposes.

In the deep sea and coastal benthic sediment layer, microplastics are inhabitants in this region from the activities of humans, a consequence of the ever-increasing population. There has been a paucity of literature on the effect of microplastics in this region. Based on this, Seeley et al. (2020) examined the influence of microplastics on benthic sediment and the effects on the nitrogen cycle and microbial consortia. The authors did a microcosm test by employing polylactic acid (PLA), polyurethane foam (PUF), PVC, and PE (plastics mixed with salt-mash sediment). The results showed alteration of the sediment-microbial community and the cycling of nitrogen in the ecosystem when compared to the control group with no microplastics. The amended PLA and PUF sediments indicated promotion of denitrification and nitrification in contrast to PVC which inhibit the two processes. The findings indicated that the cycling of nitrogen can be significantly influenced by various microplastics that can serve as a vector of organic carbon materials for the microbial consortia. In conclusion, the evaluation of the microplastic effects on the microorganisms' community and the cycling of nutrients are essential for a sustainable global environment.

Harrison et al. (2014) assessed the rapid microbial settlement of LDPE microplastics in the benthic sediment region of microcosms' coastal water bodies.

Harrison et al. (2014) opined that one of the emerging plastic pollutants is artificial microplastics of size $\leq$5 mm that could be found to reside in the benthic sediments of the deep sea and coastal regions of the world. The activities of these microplastics can elicit ecological and biological impacts on the microbial community and the benthic biota respectively if not properly managed. The authors stated that the association of plastics and microorganisms in the environment has not yet been exhausted. In this study, the authors employed a 2-week biological test on LDPE using a bacterial colony on three coastal sediment regions in the United Kingdom, the estuary of Humber, and Spurn Point. The association of the bacterial colonies and the LDPE at the sediment was done using catalyzed reporter deposition-fluorescence in situ hybridization and CARD-FISH and scanning electron microscopy (SEM). The outcome from the study revealed that there was a Log-fold upsurge in the richness of the genes; 16S rRNA from the linked LDPE bacterial consortia within 7 days which also differed significantly across the different sediments as indicated by the polymerase chain reaction (PCR). The terminal-restriction fragment length polymorphism (T-RFLP) analysis confirmed a swift assortment of the association between the LDPE bacterial colonies accumulations whose composition and structure varied significantly compared to the ones in the nearby sediments. Furthermore, the T-RFLP analysis showed progression conjunction of the association between the LDPE and bacterial consortia from the various sediments in the experimental period. The consortia were dominated by *Colwellia* sp. and *Arcobacter* sp. at a sequence sum of 84–93% on day 14 by employing the 16S rRNA genes' cloned sequencing. *Colwellia* sp. was found to be attached to the LDPE in the sediments as picked up by the catalyzed reporter deposition fluorescence in situ hybridization machine. The findings of this study showed that the bacterial colonies have been previously associated with areas with the high breakdown of hydrocarbon pollutants in lower-temperature aquatic ecosystems. However, the data derived from this study propose hydrocarbonoclastic recruitment of microbes on microplastics with shared characteristics between the pelagic and benthic macro- and mega-habitats.

In the biosphere, microplastics have been reported to have a serious ecological and biological concern to humans and the aquatic faunas by influencing toxicity and possible microbiota pathogenicity. Yang et al. (2020) examined in a review the ecological niche microplastics provide for microbes in the aquatic environment and the possible toxicity they portend. They reported that microplastics are generally found in the lakes, rivers, and oceans; they act as vectors for many microbes which form synthetic substrates and biofilms. This action of the microplastics enables them to provide novel ecological microorganism niches in both freshwater and marine environments. The microbes generate more biofilms compared to the particle-linked or natural microbial consortia in the adjacent water surroundings. Yang et al. (2020) suggested that biochemical, physiological, morphological, and multi-omic analyses should be collectively put together to understand the ecological and biological function and association of the microbial biofilms and the microplastics in aquatic environs. This will pave a better way to evaluate the pathway of animal and human contamination in the ecosystem specifically on the wide abundance and distribution of microplastic consequent of human population increase and great anthropogenic activities.

Neto et al. (2019) examined the activities of microplastics-microorganisms in the sediment layer of Vitória Bay and estuarine system in southeast Brazil. The authors stated that the accumulation of wastes from plastics in the environment poses a serious ecological threat to the aquatic biota in the water and sediment layer as well as humans along the food chains with consequent economic impacts. Plastics can accumulate for a longer time in oceans and rivers in the form of macro-nano-plastics. They reported that the concentration of microplastics in the sediment layer of the Bay ranged from 0 to 38 pps (particles per samples) with a sum of 247 particles. The number of artificial fibers microplastics sourced from fishing nets amounts to about 77% of plastic contents in marine habitats. The findings from the SEM showed that the spores of microorganisms, fungi, and bacteria colonized the microplastic debris (SVB) and formed biofilms in the sediment of the Bay region. This is an indication of a symbiotic relationship between the abiotic (microplastics) and biotic (microbial biofilm) in the Vitória bay estuarine system in southeast Brazil.

Oberbeckmann and Labrenz (2020) evaluated the degradation, adaptation, and diversity function of assemblages of marine microbes in influencing the activities of microplastics. The authors stated that for over 45 years microplastics have been found to habit the ocean in conjunction with assemblages of microbial communities. The role of microbes in the breakdown of plastics has not fully been investigated. However, the establishment of specific plastisphere and biofilms by pathogenic bacteria of the genus *Vibrio* has been observed to play an immense role in the breakdown of plastic in the marine ecosystem. The authors reported that the microbes under this genus have been established to be resourceful colonists that have the faculty to colonize both on artificial and natural surfaces. Nonetheless, the microplastics act as a vector of many recalcitrant organic pollutants and bacterial colonies, therefore posing a lesser risk to benthic faunas. The microplastics also represent to impede the metabolic pathway of the microbes thus reducing the limits to which they can degrade the debris. The authors suggested that since the elimination of microplastics from the coastal and marine ecosystem by microbes is not too efficient, a strategic alternative in preventing the entering of plastic into the ecosystem should be developed to forestall environmental sustainability.

12.4 Effects of Meso-, Micro-, and Nano-Plastic Waste on the Benthos with the Special Scenario on the Classification of Meso-, Micro-, and Nano-Plastics

Campanale et al. (2020) evaluated the effects and potentials of additives of microplastics on the health of humans. The authors stated that the abundance and distribution of microplastics in the aquatic environment pave way for it to be an indicator for the plasticene community. The ecological and biological effects of microplastics are not fully understood because of the complexity in their chemical and physical properties which empower them to look like a complicated stressor.

Microplastics are known as vectors of noxious pollutants which they transport to the benthic environment to impede the life processes of the faunas therein. The fate of plastic debris disposed of in the surrounding water environment could elicit various health-related sicknesses like cancer and hormonal alterations in humans when they come in contact with them via food chain or trophic transfer. The points of entry might be inhalation, dermal, and ingestion pathways. However, the effects and fate of microplastics in humans, which are still vague specifically of the ones between the sizes 10 and 20 μm which can infiltrate the muscles, placental, blood, liver, brain, organs, and cell membranes, should be considered important in the environmental valuation of plastic debris pollutants. The mode of action is the composition of the polymers, maturation of the particles, variations of the pH status, and the relation of the hydrophobic nature of the plastic debris. In addition, the process of desorption sorption and absorption should be evaluated to understand the extrinsic and intrinsic nature of the polymer materials in the immediate environment they are found as well as their resultant exposure to the process of weathering.

For the past century, the intensive utilization of plastics and their derivatives has increased with non-limited contamination into various macrohabitats globally. Espinosa et al. (2016) investigated the toxicological consequences of microplastics in aquatic fauna. The authors reported that apart from the possible negative impact plastic portend in the marine environment, they also could result in an incremental risk to faunas when exposed to related plastic chemicals.

More so, plastics serve as vector noxious pollutants and microorganisms that might be ingested by benthic animals thus influencing their behavior and physiological activities. Microplastics can results in ill health conditions like obstruction of the intestinal pathway, biological stress, and progression health and welfare alteration in aquatic animals specifically the benthos. This problem is not only related to marine animals but a translational or biomagnified issue in the food chain structure (Table 12.1).

12.5 Possible Ways for Abating Different Sizes of Plastics Debris in the Benthic Region around the Globe

Globally, it has been established and validated that the world's plastics deposit is mainly in sub-Saharan Africa (Loulad et al. 2017). A large portion of the ocean is accumulated with wastes of plastics of different shapes and sizes, proliferating the water ecosystem, causing ecosystem deterioration, faunas' death, and displacement. Most developing nations and some developed countries of the world cannot reduce, recycle, and reuse plastic wastes because of a lack of manpower and technical know-how (Xanthos and Walker 2017; Anonymous 2018). However, there is a need for the cleanup and prevention of the aquatic system from uncontrolled wastes from plastics debris produced by the developing nations by keying and domesticating proper technology use by the highly developed nations of the world.

In Netherlands, Australia, India, Mexico, EU countries, and America, they have developed wastewater treatment plants that can be used to manage and control

Table 12.1 Different plastic debris in surface and benthic faunas in marine and coastal ecosystem of the world

Sr. no.	Species	Level of microplastic ingestion	Location found	References
1	*Venerupis philippinarum*	–	Columbia (British colony)	Davidson and Dudas (2016)
2	*Perna perna*	75%	Brazil	Santana et al. (2016)
3	Nine different species of crabs	40%	China	Li et al. (2015)
4	*Mytilus edulis*	60%, 33–4700 μm, 20–90 μm, and 5%	China, Newfoundland, Netherlands, France, Belgium, and Germany	De Witte et al. (2014), Mathalon and Hill (2014), Li et al. (2015), and Van Cauwenberghe et al. (2015)
5	*Crassostrea gigas*	45%	Germany	Van Cauwenberghe et al. (2015)
6	Demersal and pelagic fish assortments	83%	The United Kingdom	Murray and Cowie (2011)
7	*Sciades hersbergii*, *Cathorops spixii*, and *Cathorops agassizii*	18%, 18%, and 33%	South America, Brazil, Goiniana estuary	Possatto et al. (2011)
8	*Patella* sp. (limpet), *Littorina* sp. (periwinkle), *Saccostrea cucullata* (rock oyster)	7.2–2.8 counts/g	Sri Lanka and southern coaster H_2O	Wijethunga et al. (2019)
9	*Diplodus vulgaris* and seabream	73%	Portugal (Mondego estuary)	Bessa et al. (2018)
10	*Temora longicornis* (copepods)	77%	Channel of Western English	Cole et al. (2014)
11	*Crangon crangon* (brown shrimp)	About 15,033 and 267 μ spheres/mL in the fluid medium (hemolymph)	The United Kingdom	Farrell and Nelson (2013)
12	*Lepas anatifera* (gooseneck barnacle)	33.5%	North Pacific	Goldstein and Goodwin (2013)
13	*Metapenaeus monoceros* and *Penaeus monodon*	3.87 and 3.40 items/g GT	Bangladesh and Northern Bay of Bengal	Hossaina et al. (2019)

(continued)

Table 12.1 (continued)

Sr. no.	Species	Level of microplastic ingestion	Location found	References
14	*A. marina* feces and tissue, *Mytilus edulis* feces and tissue	0.3, 1.2 parts/g and 0.1 and 0.2 parts/g	The coastal line of French-Dutch-Belgian	Van Van Cauwenberghe et al. (2015)
15	*Balarus amphitrite* (striped barnacle)	0.23–0.43 particles/g	The eastern coastal line of Thailand	Thushari et al. (2017)

macroplastics, microplastic, and nano-plastics from source and non-source points (Mason et al. 2016; Murphy et al. 2016; Ziajahromi et al. 2017). The use of garbage stops and banning of plastic bottles, plastic jars, plastic cups and straws, and plastic bags were also enforced to ensure a sustainable and clean plastic-free environment. In the United Kingdom and Mexico, the use of red algae and fish skin as natural materials was developed to produce plastic alternates and cactus plant to produce non-noxious comestible plastics (Student Designs Sustainable Plastic Alternative Made of Fish Skin and Algae 2020, and Scientist in Mexico Creates Biodegradable Plastic from Prickly Pear Cactus 2020).

It was also suggested by Brandon et al. (2018) and Urbanek et al. (2018) that biological breakdown of plastic wastes using engineered microorganisms to convert them to water and carbon (IV) oxide is also tenable. In addition, the use of the mealworms specifically the yellow strain and fungi can also be employed to break down PLA, PHB, PVC, and PET microplastics using specific enzymes in the conversion process (Ghosh et al. 2013; Papadopoulou et al. 2019).

12.6 Conclusion, Recommendations, and Future Outlook

This chapter review evaluated the effect of meso-, micro-, and nano-plastic wastes on the benthic community with the special scenario on the classification of meso-, micro-, and nano-plastics. The ecosystem impacts, ecological and health damages of meso-, micro-, and nano-plastics on the benthic environment, were documented. It was observed that the harmonization of the pretreatment procedures and sampling techniques for multifaceted environmental conditions for microplastics should put into place because of the possible threats to freshwater and marine ecosystems they portend. The enforcement of regulations, legislations, and adopted and existing policies to reduce plastic wastes in the coastal and marine zones are tenable at this point to ensure a healthy and plastic-free blue aquatic ecosystem. The examination of the accumulation and effects of meso-, micro-, and nano-plastics on the sediment and their reaction with the microbial communities were discussed. It was established that environmental researchers should capture the hazard characterization and exposure evaluation as well as set regulation on the thresholds or limits of microplastic debris or litters in benthic animals, humans, water, and sediment to ensure a more sustainable ecosystem. The influence of meso-, micro-, and nano-plastic waste on the benthos with the special scenario on the classification of meso-, micro-, and nano-plastics was highlighted. It was noted that macro-, micro-, and nano-plastics can result in ill health conditions like obstruction of the intestinal pathway, biological stress, and progression health and welfare alteration in aquatic animals specifically to the benthos. This problem is not only related to marine animals but a translational or biomagnified issue in the food chain structure. It was recommended that different integrated ways for abating different sizes of plastic debris in the benthic region should be adhered to forestall efficient plastic waste management, control, recycling, and reuse for the present and future generations.

Acknowledgments We appreciate the support of Edo State University, Uzairue, and University of Benin, both in Edo State Nigeria for providing the enabling environment and materials for conducting this extensive book chapter review.

References

Adetunji CO, Anani OA (2021b) Bioremediation of polythene and plastics using beneficial microorganisms. In: Panpatte DG, Jhala YK (eds) Microbial rejuvenation of polluted environment, vol 1. Springer, Singapore, pp 353–372

Adetunji CO, Anani OA (2021a) Plastic-eating microorganisms: recent biotechnological techniques for recycling of plastic. In: Panpatte DG, Jhala YK (eds) Microbial rejuvenation of polluted environment, vol 1. Springer, Singapore, pp 373–398

Anonymous (2018) Africa is on the right path to eradicate plastics. Ecosystem and Biodiversity. UN Environment Programme (online link: https://www.unenvironment.org/news-and-stories/story/africa-right-path-eradicate-plastics, accessed on 4 February 2020)

Amelia TSM, Khalik WMAWM, Ong MC et al (2021) Marine microplastics as vectors of major ocean pollutants and its hazards to the marine ecosystem and humans. Prog Earth Planet Sci 8 (12):1–26. https://doi.org/10.1186/s40645-020-00405-4

Auta HS, Emenike CU, Fauziah SH (2017) Distribution and importance of microplastics in the marine environment: a review of the sources, fate, effects, and potential solutions. Environ Int 102:165–176. https://doi.org/10.1016/j.envint.2017.02.013

Barnes DK, Galgani F, Thompson RC, Barlaz M (2009) Accumulation and fragmentation of plastic debris in global environments. Philos Trans R Soc B Biol Sci 364:1985–1998. https://doi.org/10.1098/rstb.2008.0205

Bellasi A, Binda G, Pozzi A, Galafassi S, Volta P, Bettinetti R (2020) Microplastic contamination in freshwater environments: a review, focusing on interactions with sediments and benthic organisms. Environments 7(4):30. https://doi.org/10.3390/environments7040030

Bessa F, Barría P, Neto JM, Frias JPGL, Otero V, Sobral P, Marques JC (2018) Occurrence of microplastics in commercial fish from a natural estuarine environment. Mar Pollut Bull 128: 575–584

Brandon AM, Gao SH, Tian R, Ning D, Yang SS, Zhou J, Wu WM, Criddle CS (2018) Biodegradation of polyethylene and plastic mixtures in mealworms (larvae of *Tenebrio molitor*) and effects on the gut microbiome. Environ Sci Technol 52:6526–6533. https://doi.org/10.1021/acs.est.8b02301

Campanale C, Massarelli C, Savino I, Locaputo V, Uricchio VF (2020) A detailed review study on potential effects of microplastics and additives of concern on human health. Int J Environ Res Pub Health 17:1212. https://doi.org/10.3390/ijerph17041212

Cole M, Webb H, Lindeque PK, Fileman ES, Halsband C, Galloway TS (2014) Isolation of microplastics in biota-rich sea water samples and marine organisms. Sci Rep 4:4528

Chae Y, An YJ (2017) Effects of micro- and nanoplastics on aquatic ecosystems: current research trends and perspectives. Mar Pollut Bull 124:624–632. https://doi.org/10.1016/j.marpolbul.2017.01.070

Davidson K, Dudas SE (2016) Microplastic ingestion by wild and cultured Manila clams (*Venerupis philippinarum*) from Baynes sound, British Columbia. Arch Environ Contam Toxicol 71:147–156

De Witte B, Devriese L, Bekaert K, Hoffman S, Vandermeersch G, Cooreman K, Robbens J (2014) Quality assessment of the blue mussel (*Mytilus edulis*): comparison between commercial and wild types. Mar Pollut Bull 85(1):146–155

Espinosa CM, Esteban A, Cuesta A (2016) Microplastics in aquatic environments and their toxicological implications for fish. In: Toxicology: new aspects to this scientific conundrum. IntechOpen, Rijeka, pp 114–145

Farrell P, Nelson K (2013) Trophic level transfer of microplastic: *Mytilus edulis* (L.) to *Carcinus maenas* (L.). Environ Pollut 177:1–3

Gallitelli L, Cera A, Cesarini G, Pietrelli L, Scalici M (2021) Preliminary indoor evidences of microplastic effects on freshwater benthic macroinvertebrates. Sci Rep 11:720. https://doi.org/10.1038/s41598-020-80606-5

Galloway TS, Cole M, Lewis C (2017) Interactions of microplastic debris throughout the marine ecosystem. Nat Ecol Evol 1:116. https://doi.org/10.1038/s4s1559-017-0116

Ghosh SK, Pal S, Ray S (2013) Study of microbes having potentiality for biodegradation of plastics. Environ Sci Pollut Res Int 20:4339–4355. https://doi.org/10.1007/s11356-013-1706-x

Goldstein MC, Goodwin DS (2013) Gooseneck barnacles (*Lepas* spp.) ingest microplastic debris in the north pacific subtropical gyre. Peer J:1–184

Haegerbaeumer A, Mueller MT, Fueser H, Traunspurger W (2019) Impacts of micro- and nano-sized plastic particles on benthic invertebrates: a literature review and gap analysis. Front Environ Sci 7:17. https://doi.org/10.3389/fenvs.2019.00017

Harrison JP, Schratzberger M, Sapp M et al (2014) Rapid bacterial colonization of low-density polyethylene microplastics in coastal sediment microcosms. BMC Microbiol 14:232. https://doi.org/10.1186/s12866-014-0232-4

Hirai H, Takada H, Ogata Y, Yamashita R, Mizukawa K, Saha M, et al. (2011) Organic micropollutants in marine plastics debris from the open ocean and remote and urban beaches. Mar Pollut Bull 62:1683–1692. doi: https://doi.org/10.1016/j.marpolbul.2011.06.004

Hossaina MS, Shajjadur MR, Mohammad NU, Sharifuzzaman SM, Sayedur RC, Subrata S, Shah MNC (2019) Microplastic contamination in Penaeid shrimp from the northern Bay of Bengal 238:124688

Ivar do Sul JA, Costa MF (2014) The present and future of microplastic pollution in the marine environment. Environ Pollut 185:352–364. https://doi.org/10.1016/j.envpol.2013.10.036

Kelsey L, Rogers Joan A, Carreres-Calabuig EG, Posth NR (2020) Micro-by-micro interactions: how microorganisms influence the fate of marine microplastics. Limnol Oceanogr Lett 5 (1):18–36. https://doi.org/10.1002/lol2.10136

Lattin GL, Moore CJ, Zellers AF, Moore SL, Weisberg SA (2004) A comparison of neustonic plastic and zooplankton at different depths near the southern California shore. Mar Pollut Bull 49:291–294. https://doi.org/10.1016/j.marpolbul.2004.01.020

Li J, Yang D, Li L, Jabeen K, Shi H (2015) Microplastics in commercial bivalves from China. Environ Pollut 207:190–195

Loulad S, Houssa R, Rhinane H, Boumaaz A, Benazzouz A (2017) Spatial distribution of marine debris on the seafloor of Moroccan waters. Mar Pollut Bull 124:303–313. https://doi.org/10.1016/j.marpolbul.2017.07.022

Mason SA, Garneau D, Sutton R, Chu Y, Ehmann K, Barnes J, Fink P, Papazissimos D, Rogers DL (2016) Microplastic pollution is widely detected in US municipal wastewater treatment plant effluent. Environ Pollut 218:1045–1054. https://doi.org/10.1016/j.envpol.2016.08.056

Mathalon A, Hill P (2014) Microplastic fibers in the intertidal ecosystem surrounding Halifax Harbor, Nova Scotia. Mar Pollut Bull 81(1):69–79

Moore CJ (2008) Synthetic polymers in the marine environment: a rapidly increasing, long-term threat. Environ Res 108:131–139. https://doi.org/10.1016/j.envres.2008.07.025

Murphy F, Ewins C, Carbonnier F, Quinn B (2016) Wastewater treatment works (WwTW) as a source of microplastics in the aquatic environment. Environ Sci Technol 50:5800–5808. https://doi.org/10.1021/acs.est.5b05416

Murray F, Cowie PR (2011) Plastic contamination in the decapod crustacean *Nephrops norvegicus* (Linnaeus, 1758). Mar Pollut Bull 62(6):1207–1217. https://www.scie ncedirect.com/science/article/pii/S0025326X11001755?via¼3Dihub

Neto JAB, Gaylarde C, Beech I, Bastos AC, Quaresma VS, Carvalho DG (2019) Microplastics and attached microorganisms in sediments of the Vitória bay estuarine system in SE Brazil. Ocean Coast Manag 169(1):247–253

Oberbeckmann S, Labrenz M (2020) Marine microbial assemblages on microplastics: diversity, adaptation, and role in degradation. Annu Rev Mar Sci 12:209–232. https://doi.org/10.1146/annurev-marine-010419-010633

Olomukoro JO, Anani OA (2019) Evaluation of aquatic macro-invertebrate populations: a model for emergent bio-monitoring guide for quantifying uncleanness of some Rivers in northern Central Nigeria. Nigerian J Tech Res 114(2):54–62

Papadopoulou A, Hecht K, Buller R (2019) Enzymatic PET degradation. Chimia (Aarau) 73:743–749. https://doi.org/10.2533/chimia.2019.743

Possatto FE, Barletta M, Costa MF, Ivar do Sul JA, Dantas DV (2011) Plastic debris ingestion by marine catfish: an unexpected fisheries impact. Mar Pollut Bull 62:1098–1102

Redondo-Hasselerharm PE, Falahudin D, Edwin THM, Peeters, Koelmans AA (2018) Microplastic effect thresholds for freshwater benthic macroinvertebrates. Environ Sci Technol 52 (4):2278–2286. https://doi.org/10.1021/acs.est.7b05367

Rochman CM, Browne MA, Underwood AJ, Van Franeker JA, Thompson RC, Amaral-Zettler LA (2016) The ecological impact of marine debris: unraveling the demonstrated evidence from what is perceived. Ecology 97:302–312. https://doi.org/10.1890/14-2070.1

Santana MFM, Ascer LG, Custódio MR, Moreira FT, Turra A (2016) Microplastic contamination in natural mussel beds from a Brazilian urbanized coastal region: rapid evaluation through bioassessment. Mar Pollut Bull 106(1):183–189

Scherer C, Brennholt N, Reifferscheid G, Wagner M (2017) Feeding type and development drive the ingestion of microplastics by freshwater invertebrates. Sci Rep 7:1–9. https://doi.org/10.1038/s4s1598-017-171911-7

Scientist in Mexico Creates Biodegradable Plastic from Prickly Pear Cactus (2020) Available online https://www.forbes.com/sites/scottsnowden/2019/07/14/scientist-in-mexico-creates-biodegradable-plastic-from-prickly-pear-cactus/#2dca0dc76c49. Accessed on 4 Feb 2020

Seeley ME, Song B, Passie R et al (2020) Microplastics affect sedimentary microbial communities and nitrogen cycling. Nat Commun 11:2372. https://doi.org/10.1038/s41467-020-16235-3

Sendra M, Carrasco-Braganza MI, Yeste PM et al (2020) Immunotoxicity of polystyrene nanoplastics in different hemocyte subpopulations of *Mytilus galloprovincialis*. Sci Rep 10: 8637. https://doi.org/10.1038/s41598-020-65596-8

Student Designs Sustainable Plastic Alternative Made of Fish Skin and Algae (2020). Available online http://www.ladbible.com/news/uk-student-designs-sustainable-plastic-alternative-made-using-fish-skin-20190605. Accessed 4 Feb 2020

Talsness CE, Andrade AJM, Kuriyama SN, Taylor JA, vom Saal FS (2009) Components of plastic: experimental studies in animals and relevance for human health. Philos Trans R Soc B Biol Sci 364:2079–2096. https://doi.org/10.1098/rstb.2008.0281

Thompson RC, Olson Y, Mitchell RP, Davis A, Rowland SJ, John AW et al (2004) Lost at sea: where is all the plastic? Science 304:838. https://doi.org/10.1126/science.1094559

Thushari GGN, Senevirathna JDM (2020) Plastic pollution in the marine environment. Heliyon 6 (8):e04709. https://doi.org/10.1016/j.heliyon.2020.e04709

Thushari GGN, Senevirathna JDM, Yakupitiyage A, Chavanich S (2017) Effects of microplastics on sessile invertebrates in the eastern coast of Thailand: an approach to coastal zone conservation. Mar Pollut Bull 124:349–355

Urbanek AK, Rymowicz W, Mironczuk AM (2018) Degradation of plastics and plastic-degrading bacteria in cold marine habitats. Appl Microbiol Biotechnol 102:7669–7678. https://doi.org/10.1007/s00253-018-9195-y

Van Cauwenberghe L, Claessens M, Vandegehuchte MB, Janssen CR (2015) Microplastics are taken up by mussels (*Mytilus edulis*) and lugworms (*Arenicola marina*) living in natural habitats. Environ Pollut 199:10–17

vom Saal FS, Parmigiani S, Palanza PL, Everett LG, Ragaini R (2008) The plastic world: Sources, amounts, ecological impacts and effects on development, reproduction, brain and behavior in aquatic and terrestrial animals and humans. Environ Res 108:127–130. https://doi.org/10.1016/j.envres.2008.03.008

Weber S, Traunspurger W (2015) The effects of predation by juvenile fish on the meiobenthic community structure in a natural pond. Freshwater Biol 60:2392–2409. https://doi.org/10.1111/fwb.12665

Wijethunga HNS, Athawuda AMGAD, Dias PCB, Abeygunawardana AP, Senevirathna JDM, Thushari GGN, Liyanage NPP, Jayamanne SC (2019) Screening the effects of microplastics on selected invertebrates along southern coastal belt in Sri Lanka: a preliminary approach to coastal pollution control. In: Proceedings of International Research Conference of Uva Wellassa University

Wright SL, Rowe D, Thompson RC, Galloway TS (2013) Microplastic ingestion decreases energy reserves in marine worms. Curr Biol 23:1031–1033. https://doi.org/10.1016/j.cub.2013.10.068

Xanthos D, Walker TR (2017) International policies to reduce plastic marine pollution from single-use plastics (plastic bags and microbeads): a review. Mar Pollut Bull 118:17–26. https://doi.org/10.1016/j.marpolbul.2017.02.048

Yang Y, Liu W, Zhang Z et al (2020) Microplastics provide new microbial niches in aquatic environments. Appl Microbiol Biotechnol 104:6501–6511. https://doi.org/10.1007/s00253-020-10704-x

Ziajahromi S, Neale PA, Rintoul L, Leusch FD (2017) Wastewater treatment plants as a pathway for microplastics: development of a new approach to sample wastewater-based microplastics. Water Res 112:93–99. https://doi.org/10.1016/j.watres.2017.01.042

Impact of Plastic Waste on the Coral Reefs: An Overview

13

Romana Akhtar, Mohd. Yaseen Sirwal, Khalid Hussain, Mudasir A. Dar, Mohd Shahnawaz, and Zhu Daochen

Abstract

Plastic is highly used polymer, with 330 million metric tons annual global production. The extensive use of plastic from kitchen to industry level leads to accumulation of around 50–90 million metric tons plastic litter per year. Plastic is nondegradable and needs various centuries to get mineralized. Most of the plastic waste gets transported via different water bodies, and wind and enters in to the oceans. After reaching the marine environment, plastic waste reported deaths of billions of marine flora and fauna annually. Coral reef is one the diverse ecosystem of the marine environment and represents around 25% of all the marine inhabitants. Plastic wastes in all forms are reported to impact the coral reefs significantly. So, in the current chapter, it was attempted to overview the trend and impact of marine plastic litter on ecology, economic value, geographical distribution, and health of the coral reefs. At the end a set of strategies were also enlisted to minimize the emission of inland plastic litter into the oceans.

All authors have contributed equally to this work and are all first authors.

R. Akhtar
Department of Zoology, Govt. Degree College Kishtwar, Kishtwar, Jammu and Kashmir, India

M. Y. Sirwal (✉)
Department of Chemistry, Govt. Degree College Kishtwar, Kishtwar, Jammu and Kashmir, India

K. Hussain
Center for Biodiversity and Taxonomy, Department of Botany, University of Kashmir, Hazratbal, Srinagar, Jammu and Kashmir, India

M. A. Dar · M. Shahnawaz · Z. Daochen
Biofuels Institute, School of the Environment and Safety Engineering, Jiangsu University, Zhenjiang, China

M. Shahnawaz et al. (eds.), *Impact of Plastic Waste on the Marine Biota*,
https://doi.org/10.1007/978-981-16-5403-9_13

Keywords

Plastic waste · Oceans · Plastic weathering · Marine inhabitants · Coral reefs

13.1 Introduction

Plastics have replaced all the traditional polymers used in our daily life to meet our needs from kitchen to industry level (Shah et al. 2008; Sangale et al. 2012). In 2016, the annual global production of plastic was 330 million metric tons (MMT) (Lebreton and Andrady 2019). The demands of plastic-based products are increasing exponentially and may get doubled in next 20 years. The extensive usage of plastics has resulted in significant increase in the generation of plastic waste. As per estimate, around 60 and 99 million metric tons of the plastic waste was accumulated in 2015 (Lebreton and Andrady 2019). Researchers have also predicted that by 2025 the world's non-rural population is estimated to produce 6 million metric tons of solid waste on daily basis (Hoornweg et al. 2013). Due to lack of proper efficient waste management system, most of the plastic wastes finally manages to enter into the ocean through different routes, viz., rivers (Lebreton et al. 2017), water canals (Tasseron et al. 2020), etc. As per report (Jambeck et al. 2015), around 4.8–12.7 million metric tons of plastic waste managed to enter in the marine environment annually. In the oceans, due to various environmental forces, the plastic waste started deteriorating by following varied pathways and leads to the fragmentation of plastic into different sizes >25 mm, 5–25 mm, <5 mm, and <0.1 μm and are referred as macroplastic, mesoplastic, microplastic, and nanoplastic, respectively (Arthur et al. 2009; Andrady 2011; Ng et al. 2018; Lamb et al. 2018; Boyle and Ormeci 2020). In the marine environment, these plastic particles, viz., macroplastic, mesoplastic, microplastic, and nanoplastic reported to threaten the marine life by different mechanisms, viz., ingestion, entanglement, and absorption of plastic particles over the body surfaces, and leads to death of around billions of marine flora and fauna annually.

In the marine environment, among the various ecosystems, there is one important underwater ecosystem with reef-building corals called as coral reefs. Although it represent less than 0.1% of worlds ocean area, still it is considered as one of the most diverse systems of ecosystem also known as rainforests of the oceans (Mulhall 2008) and represents around 25% of all marine species (Spalding and Grenfell 1997; Spalding et al. 2001). Coral reefs have tendered great services to ecosystem to benefit people which leads to the generation of good economic value ($375 billion) annually through tourism, coastal protection, and fisheries at global level (Costanza et al. 1997; Cesar et al. 2003; Burke et al. 2011). Coral reefs are delicate and are sensitive to the changing environmental conditions of the oceans, viz., excess nutrients, increasing temperature, acidification of the oceans, overfishing, and emission of varied pollutants from land to the oceans (Danovaro et al. 2008; Jangly 2013; Carlson et al. 2019).

Among the various pollutants, plastic waste has also been reported to affect the coral reefs ecosystem at an alarming rate (Lasut et al. 2018; Lamb et al. 2018; Tan et al. 2020; Huang et al. 2021), etc. In the current chapter, we have attempted to pool all the reports on the impact of various plastic particles on the coral reefs under the following directions: (i) to overview the generation of the plastic waste; (ii) to enlist the transportation routes and pathways of the inland plastic waste to the marine environment; (ii) to overview the fragmentation of plastic waste into various plastic particles, viz., macro-, meso-, micro-, and nanoplastic, and impact of these plastic particles on marine life; (iii) to discuss the ecology, economic value, geographical distribution, and eco-toxic effect of plastic waste on Coral reefs; and (iv) to suggest the strategies to be followed to minimize the entry of plastic waste into the marine environment.

13.2 Generation of Plastic Waste

At present, we all depend on plastic-based products for our basic needs from kitchen to industry at global level. Plastic has replaced all the traditionally used polymers, based on its various attributes and benefits. During the last decade, the demand of plastic-based products increased exponentially and resulted in the manufacturing of around 330 million metric tons (Lebreton and Andrady 2019) of the plastic-based products annually. Due to ultrahigh usage of the plastics (mostly single-use plastic) at global level, it has resulted in accumulation of a huge quantity of plastic litter. Plastic is highly inert polymer and can withstand in the environment for thousands of years (Sangale et al. 2012). Plastics litter pollution is reported to affect the flora and fauna in both terrestrial and marine environment due to their nondegradable nature (Gall and Thompson 2015). Huge increment (2–381 metric tons) in the production of plastic resin and fibers alone in just 65 years led to the generation of overall plastic waste (7800 metric tons) at an alarming rate (Europe 2008; Jambeck et al. 2015). Maximum plastic waste (50%) is reported to be generated by the Asian countries, whereas the European countries and North America share 18–19% each in the global plastic waste production.

1950s witnessed commercial production of plastics and showed exceptional growth, and as per the estimate in 2016, around 330 million metric tons plastic were produced (Jambeck et al. 2015). The present growth rate of plastic generation is estimated to get doubled within next 20 years. This extensive use of plastics will result in a concomitant increase in postconsumer plastic waste. For instance, by 2025 the global urban population is estimated to generate 6 Mt. of solid waste daily (Hoornweg et al. 2013). The broad estimation of global plastic production or municipal plastic waste (MPW) generation is not sufficient to assess the regional impacts of plastic waste on the ecosystem that would dictate the need for its mitigation. There are various strategies to tackle the plastic waste, i.e., landfill (65%) (Ling et al. 1998), incineration (25%), recycling (10%) (Sangale et al.

2012; Shahnawaz et al. 2019a), pyrolysis to produce petrochemical products (Sarker et al. 2011; Miandad et al. 2016), construction of roads and buildings (Panyakapo and Panyakapo 2008; Jalaluddin 2017; Preethi et al. 2019), degradation, and bioremediation (Shah et al. 2008; Sangale et al. 2012; Shahnawaz et al. 2019a). Till date no method reports 100% degradation of the plastic except biodegradation of polyethylene terephthalate (PET) by *Ideonella sakaiensis* (Yoshida et al. 2016).

13.3 Transportation Routes and Pathways of the Inland Plastic Waste to Marine Environment

Across the world, plastic pollution is considered as an anthropogenic issue in the coastal and marine ecosystems irrespective of developed or underdeveloped countries (Vince and Hardesty 2017; Thushari and Senevirathna 2020). Continuous discharge and dumping of plastic debris by humans into the world's oceans caused interruption to ecosystem structure functions, services, and values which ultimately lead to marine pollution (Joyner and Frew 1991). Due to a strong interconnectivity of aquatic and terrestrial ecosystems, changes in one ecosystem have drastic impacts on the other one. Anthropogenic activities including pollution, physical destruction of the environment, and plastic litter accumulation are considered the most dominant factors that have affected coastal and marine ecosystems (Richmond 1993; Adams 2005; Thushari and Senevirathna 2020). Plastic litter has long shelf life with high persistence in the ocean basins due to its unique potential of transportation by water currents and wind (Nualphan 2013; Rosevelt et al. 2013; Thushari and Senevirathna 2020). Nearly 50 years ago, plastic contamination was reported for the first time in the marine environment (Law 2017).

As per some recent studies (Eriksen et al. 2014; Law 2017; Thushari and Senevirathna 2020), a huge quantity of the plastic waste (equivalent to 260,000 tons) is floating on the surface of marine water due to lack of proper plastic waste management strategies. Different kinds of sources directly and indirectly result in the generation of plastic wastes in the aquatic ecosystems. The land- and ocean-based plastic wastes are the main sources of plastic litter in coastal and marine environment (Thushari et al. 2017; Law 2017). During the cargo transportation of plastic pellets (beads) from one country to other, sometimes the ships met with accidents and led to the spillage of plastic pellets directly in the marine water (Duxbury 1992). Among the marine plastic litter, plastic resin pellets were detected for the first time in the beaches and the marine water (Carpenter and Smith 1972; Hirai et al. 2011). Since 1980s, the abundance of plastic pellets have steeply declined, and the reason could be attributed to preventive measures to minimize the entry of the plastic pellets into oceans (Duxbury 1992; Van Franeker and Law 2015; Law 2017). During the transformation of resin to plastic, the resulted manufactured plastic resin products get released in the environment. Dumping of the plastic waste in the open dumping sites and landfills leads to the generation of unmanaged plastic waste. To minimize the emission of plastic waste to environment, it should be collected and managed properly. A study conducted to estimate per capita generation, composition, and disposal rate of plastic in 192 different coastal countries reported maximum emission of the plastic waste from the inland sources to the marine environment is the improper unmanaged plastic waste (Hoornweg and Bhada-Tata 2012; Jambeck et al. 2015). Plastic particles (microbeads) also enters (spills) into the oceans unintentionally due to accidents of cargo during the maritime transportation (Law 2017). Furthermore, huge amount of plastic along with other debris also gets transported to the oceans due to weather calamities, viz., tsunamis, hurricanes, or floods (Law 2017). More than 75% of plastic litter reached in the oceans is from land-based sources like freshwater input, domestic and residential activities, tourism, harbor operations, and other economic activities (Andrady 2011; Thushari and Senevirathna 2020). Single-use plastic, viz., packaging plastic, plastic bottles, wrappers, carry bags, and other toiletry items, directly reaches to the sea coasts through wind and water bodies and also finally enters to the oceans (Browne et al. 2007). As per reports around 2 billion plastic particles from fresh water of rivers and streams enter into the marine environment around California alone in just 3 days' time (Moore 2008). The tourism and recreational activities also contribute a significant portion of coastal and marine plastic accumulation. A study revealed that the recreational and tourism activities along the eastern coastal beaches of Thailand lead to the generation of more than 60% of the plastic litter (Thushari et al. 2017). Cole et al. (2011) reported that plastic litter at beaches is transported to marine environment as microplastic fragments and secondary microplastics. The weather disasters, viz., floods, tsunami, hurricanes, and violent tornadoes, result in enhanced accumulation of plastic waste in the marine environment (Thompson et al. 2005). Various offshore activities like commercial fishery, waste disposal, navigation actions, and discarded fishing gears also add significant amount of plastic waste in the oceans (Thushari and Senevirathna 2020).

13.4 Fragmentation of Plastic Waste and Generation of Macro-, Meso-, Micro-, and Nanoplastic and Their Impact on Marine Life

After reaching the marine environment, plastic waste gets exposed to various environmental and biogeochemical forces, which results in fragmentation of plastic litter into varied sizes of plastic particles of macro- (>25 mm), meso (5–25 mm), micro- (<5 mm), and nano (<0.1 μm)-scale level and is referred as macroplastic, mesoplastic, microplastic, and nanoplastic, respectively (Arthur et al. 2009; Andrady 2011; Ng et al. 2018; Lamb et al. 2018; Boyle and Ormeci 2020). These plastic particles are not generated only in the marine environment but manufactured either intentionally for value addition of the consumer products, e.g., cosmetics (used as abrasive agents, e.g., microbeads) or generated unintentionally due to usage of the products (Shahnawaz et al. 2019b).

All of these small-scale plastic particles posed serious threat to the inhabitants of the ocean (Issac and Kandasubramanian 2021). These plastic particles either manage to enter inside the body of the living organisms through food stuff or drinking water or get absorbed through the body by passing through the dermal tissues and pose serious threat to the inhabitants of ocean (Lusher et al. 2017; Anbumani and Kakkar 2018). So, the transport of these plastic particles through food chain across different trophic levels leads to cause threat to marine biota as well as human being (van Raamsdonk et al. 2020).

Marine animals, birds, and fishes that consume plastic waste as food mistakenly suffer from liver damage and fail to metabolize drugs, pesticides, and other pollutants (Rochman et al. 2013; Smith et al. 2018). This plastic waste blocks the digestive tracts, decreases the urge of food, alters feeding behaviors, reduces growth and reproductive output, and finally results in death of these marine inhabitants (Critchell and Hoogenboom 2018).

Additives are added to the plastic using two main pathways (Andrady and Neal 2009); in the first pathway, it is added during the manufacturing process to enhance certain features of the plastic (e.g., increase tensile strength, fire resistance, color) (Andrady and Neal 2009; Rani et al. 2015). However, the second pathway is when the plastic particles adsorb dissolved chemicals (additives) already available in the marine environment (Van Franeker and Law 2015), and the concentration of these chemicals gets escalated due to weathering of the plastic litter (Jahnke et al. 2017). So, when the marine animals, birds, reptiles, corals, etc. consume these plastic-based additives/plasticizers along with plastic, various diseases like cancer, mutations to DNA, hormone disruption, and damage to the liver, kidneys, heart, nervous system (including the brain), and reproductive system of the marine inhabitant are reported (Schubert 1972; Cingotti and Jensen 2019; Campanale et al. 2020). Smith et al. (2018) found microplastic in seafood and reported humans consuming these kinds of seafood regularly also develop cancer, congenital disabilities, immune system problems, and childhood development issues.

13.5 Ecology, Economic Value, Geographical Distribution, and Eco-Toxic Impact of Plastic Waste on the Coral Reefs

In the marine environment, all types of plastic litter are reported to affect the ecology, economic value, geographical distribution, and health of the coral reefs. Let's us have brief overview of *ecology*, economic value, and geographical distribution of coral reefs in the oceans, followed by the impact of plastic waste on Coral reefs.

13.5.1 Ecology and Economic Value of Coral Reefs

Corals are animals belonging to the phylum Cnidaria and class Anthozoa. Corals are armed with hard carbonated exoskeletons that support and protect them. Corals were reported for the first time during the early Ordovician period about 485 million years ago when they displaced sponge reefs (Lee et al. 2015). Majority of the coral reefs are less than 10,000 years old. Some of the coral reefs are found in the deep sea, around the oceanic islands which are mostly volcanic in origin. Coral reefs are among the world's most diversified ecosystems, even sometimes referred to as rainforests of the sea (NOAA 2020). Coral reefs are underwater habitats constructed by polyps of corals. Constructing reefs take diverse shapes and forms such as large sized reef-building colonies or even small, solitary forms and graceful flowing fans. To date, thousands of coral species have been discovered thriving in diverse habitats such as warm or shallow waters, some live under dark conditions deep down on the ocean surfaces. The topography of coral reefs is changing continuously. Each reef is characterized by irregularities having different bare rocks, sand, algae, and sessile animals. The size, shape, and relative abundance of these patches alter from year to year in response to environmental factors favoring one patch over another. Coral polyp colonies are bound together by calcium carbonate to build reefs. The majority of coral reefs are made up of stony corals, which have polyps that cluster together.

Coral reefs prevent the erosion of coastlines and protect them from storms. Coral reefs are a good source of new medications and foods. Although coral reefs occupy less than 0.1% of the ocean areas worldwide, they provide shelter to more than 25% of all marine species, like fishes, mollusks, sponges, and even some cnidarians (Spalding and Grenfell 1997; Hoover 1998; Spalding et al. 2001; Mulhall 2008). More than 7000 species of fishes, invertebrates, plants, sea turtles, birds, and marine mammals were reported in the coral reef ecosystem. Deep-sea reefs or mounds, though less famous, do support a diverse range of aquatic life in an otherwise barren environment. Coral reefs are source of food and money to about half a billion people (Annomyous 2020). The net economic value of the world's coral reefs is estimated to be US$9.9 trillion annually (Costanza et al. 2014). These habitats have a cultural importance for indigenous tribes. Corals that live in shallow water have developed a symbiotic association with photosynthetic alga, zooxanthellae, that dwells in their tissues. Corals provide shelter and raw molecules required by zooxanthellae for photosynthesis, while the algae exchange carbohydrates and oxygen to the corals apart from excretion of the wastes. However, corals living in deep seas or colder waters lack zooxanthellae. In contrast to shallow-water relatives, deep-sea corals obtain most of their energy from planktons and organic materials. Broadly, coral reef ecosystems are divided into three zones, viz., fore reef, reef crest, and rear reef that host diverse habitats. The rear reef is a completely enclosed area containing small reef patches and least affected by wave actions (Moyle and Cech 2004). These zones are interconnected physically as well as ecologically for the exchange of seawater, nutrients, sediments, and living forms. The reef surface is the shallowest portion of the coral reef ecosystems which is often subjected to tides. When tides pass over these shallow reefs, a process of water disturbance occurs, known as shoal which is ideal for the growth of corals. Moreover, these shallow reefs receive sufficient sunlight required for photosynthesis by zooxanthellae, while the water currents drift planktons as nutrition for corals.

13.5.2 Ecogeographical Distribution of the Coral Reefs in Marine Habitats

Coral reefs are distributed over 284,300 km^2 out of the 0.1% of total ocean surface area. It can be found in all oceans of the globe from tropics to the poles. Reef-building corals are distributed all over the Western Atlantic and Indo-Pacific waters, mostly between 30° N and 30° S latitudes. About 91.9% of the world's coral reefs are distributed in Indo-Pacific region, while Southeast Asia and Australia account for 32.3% and 40.8%, respectively. The Asia-Pacific area is home to 55.5% of the world's coral reefs (Burke et al. 2011). Together the coral reefs of the Atlantic and Caribbean comprise about 7.6% of the worldwide coral reefs (Spalding et al. 2001). Although corals may occur in both temperate and tropical environments, shallow-water reefs can only be found between 30° N and 30° S of the equator. At depths of more than 50 m, tropical corals do not develop. Most coral reefs thrive at temperatures between 26 °C and 27 °C, with only a few reefs surviving at temperatures below 18 °C (Achituv and Dubinsky 1990). However, some corals like those growing in Persian Gulf can sustain up to 13 °C in the winter and 38 °C in the summer. Most of the corals can thrive at water temperatures between 23 °C and 29 °C, and some species may endure higher temperatures like 40 °C but for brief periods of time. Majority of the species also require salinity of 32–42 parts per thousand, alongside clean waters that allow maximum penetration of light. More than 37 species of scleractinian corals are known to live in and around Larak Island (Samiei et al. 2013). The deep-water corals found at higher latitudes can grow under colder temperatures. Despite the fact that deep-sea corals may build reefs, little is known about them. Corals are rare along the coastline of South Asia, northern South America as well as west coast of America and Africa (Spalding et al. 2001). The incidences of low diversity of corals in these areas are based on upwelling and mighty cold coastal currents that results in decreasing the temperatures in these areas. The "Great Barrier Reef" is considered as the largest coral reef of the world and represents around 900 individual reefs and is spread across 900 islands covering an area of 2600 km in Queensland, Australia, while Raja Ampat in Indonesia is the healthiest coral reef of the world. On the basis of closeness to the land, structure, and shape, coral reefs are mainly categorized into three main types such as barrier reefs, fringing reefs, and coral atolls. Coral reefs usually grow faster in clear waters where light can easily reach to the symbiotic algae, zooxanthellae. This alga is believed to prompt calcification of the polyps. However, some corals can absorb light which helps them to survive under dim lights or in deeper seas. Reef building is optimal under warm temperatures of 20–28 °C with strong wave actions. However, warm waters flow across the eastern shores of large land masses. Tides supply nutrients including oxygen to the reefs, transfer coral larvae, and prevent sedimentation on the coral reefs. The formation of the skeleton of coral polyp's requires calcium precipitation which is possible under high salinity and temperatures with low CO_2 levels. These features are typical for warm, shallow, and tropical waters having hard substratum. Corals have developed an intricate relationship with many organisms (Fuchs 2013). At high tides, fringing reefs that usually live below low tides save mangrove forests and seagrass from strong waves and water currents. However, mangroves and sea grasses protect the corals from large influxes of silt, water, and other pollutants (Barnes and Mann 2009). Some of the largest coral reefs of the world are enlisted in Table 13.1.

13.5.3 Effect of Plastic Particles on Coral Reef

Coral reefs are delicate and sensitive to water conditions and later get deteriorated systematically (Carlson et al. 2019). Coral reefs are threatened by several factors including natural and man-made. Natural threats to coral reefs include diseases, storms, and predators, while pollutants like plastics, sedimentation, and fishing practices are some other examples of anthropogenic risks. Climate change is another factor which is increasing ocean temperature causing acidification.. During the past 30 years, 50% of the world's corals have been lost due to climate change, overfishing, and plastic pollution of the marine habitats. Similarly, during 2014–2017 about 70% of the coral reef ecosystems were damaged globally by warm waters (Lamb et al. 2018). Some areas such as Australia's Great Barrier Reef were damaged heavily bleaching hundreds of miles of corals. However, coral reefs show recovery potentials too although taking many years to recover fully from the bleaching events (Evans et al. 2020). Scientists are continuously striving with new experimentations using novel methods to enhance the recovery potential of corals. Soma coral species are grown in nurseries and then transplanted to damaged places/shores.

Plastic wastes show varying impact on the corals, depending upon their shape and size. Recently many researchers have speculated that plastic pollution in marine environments promotes diseases of corals. The chances of disease on a coral reef are enhanced by 22-fold by plastics. In case of branched and complex shaped corals, plastics are retained for longer periods, thereby making these organisms prone and susceptible to opportunistic infections and diseases. The common diseases of corals, viz., white syndromes, skeletal eroding band, blackband, growth abnormalities, atramentous necrosis, and brown band, are blamed to be prevalent due to contamination by plastics. Despite the fact that more research is needed to thoroughly understand these findings, some of the prepositions have already been put forth. Plastics are thought to have a direct impact on disease progression by inflicting physical damage to the coral tissues (Sweet et al. 2019). Moreover, plastics act as vectors to transport disease causing pathogens between corals thereby spreading infections (Lamb et al. 2018) .

Many types of pollutants are wreaking havoc on coral reefs. The pollutants such as pesticides, fertilizers, oil spills, sewage, and silt from degraded landscapes make it difficult for coral to survive, thus damaging the complex relationships of these beneficial creatures and other factors existing in coral reef ecosystems (Carlson et al. 2019). Roger Bradbury, an ecologist at the Australian National University, wrote that "acidification of oceans, overfishing, and plastic pollution are driving coral reefs into oblivion" (Hay and Rasher 2010). Each of those forces triggers

Table 13.1 Some of the largest coral reefs of the world (Annomyous 2018)

Sr. no.	Name of the coral reef	Length/area (km^2)	Geographical coordinates	Country/region
1	Great Barrier Reef	2600.00	018.1710° S, 147.4200° E	Australia
2	Mesoamerican barrier reef system	1000.00	018.3439° N, 087.4218° W	Yucatán peninsula to Honduras
3	Amazon reef	0970.00	001.0000° N, 49.0000° W	French Guiana and northern Brazil
4	Apo reef	26/26000	012.6617° N, 120.4144° E	Philippines
5	Bar reef	0306.70	008.2300° N, 079.4400° E	Sri Lanka
6	Belize barrier reef	0300.00	017.1856° N, 087.3240° W	Belize
7	Daintree reef	0110.00	016.0329° S, 145.4350° E	Australia
8	Filippo reef	0001.60	005.3000° S, 151.5000° W	Pacific Ocean
9	Flinders reef	0005.00	26°59′18″ S, 153°26′31.4″ E	Australia
10	Kingman reef	0017.00	006.2300° N, 162.2500° W	North Pacific Ocean
11	Lyra Reef	0459.00	001.4500° S, 153.2000° E	Brazil
12	Maro Reef	1935.00	025.4150° N, 170.5900° W	Northwestern Hawaiian Islands
13	Miami Terrace Reef	0065.00	025.4500° N, 079.5500° W	Florida, USA
14	Minerva Reefs	0010.40	023.3800° S, 178.5400° W	Fiji an Tonga
15	New Caledonia Barrier Reef	15743.00	020.2443° S, 164.3359° E	New Caledonia
16	North East Reef	344,400	028.2513° S, 113.4841° E	Australia
17	Osprey Reef	0195.00	13°54′29″ S, 146°36′55″ E	Australia
18	Palancar Reef	0005.00	020.2967° N, 087.0185° W	Mexico
19	Reunion Island's coral reef	0063.00	021.0851° S, 55.2214° E	Réunion
20	Røst Reef	0040.00	067.4840° N, 002.018849° E	Norway
21	Scorpion Reef	0100.00	022.2258° N, 089.40578° W	Mexico
22	Sula Reef	0013.00	64°10′30.92″ N, 8°17′6.17 W	Norway
23	Tubbataha Reef	0970.00	008.5500° N, 119.5500° E	Philippines

collapse of coral reefs throughput the world, when combined they assure it fast to happen. According to United Nations Environment Programme, plastics make up to 60–80% of the total waste in the oceans impacting over 800 marine species (Annomyous 2021). Every year more than 8 million tons of plastics end up in oceans, which is equivalent to emptying a truck of plastic wastes every minute. If the use of plastics continues at current rate, we may generate 33 billion tons of plastics by 2050 most of which may end up in seas and oceans, where it will persist for decades (Annomyous 2021). In 2018 a huge survey of the 159 coral reefs across Asia-Pacific region showed that over 11.1 billion plastic particles are entangling the corals (Lamb et al. 2018). However, this number is estimated to increase dramatically 40% by 2025 (Jambeck et al. 2015). The authors further stated that out of 124,000 individual reef-building corals, 89% were susceptible to diseases due to plastics, while only 4% were found free from plastic pollution. Plastic wastes deprive oxygen supply as well as light to the corals and releases toxins that help bacteria and viruses to invade. Lamb et al. (2018) further speculated that the plastic pollution was more intense in Asian seas showing diverse pollutant such as bottles, diapers, cotton swabs, and food wrappers. They also stated that corals having plastic particles in the body were sick. In another study, researchers observed a concerning alarming trend in the ingestion of plastic particles by wild marine animals (Reddy 2018). There are several examples of the toxicity and fatality of plastic waste, particularly microplastics against marine species which mistakenly take it as food. In one of the studies, authors have suggested that corals deliberately consume plastic due to its flavor or smell, which matches with their foods they sense through chemoreception mechanisms (Allen et al. 2017). However, the exact mechanism of the attraction of corals toward plastic particles needs further investigation.

Plastic pollution poses serious risks to reef-building corals potentially acting as a vector or transportation portal for white syndrome causing bacteria, which damages coral tissues. Infections due to these bacteria are further augmented by absence of the light and oxygen, when plastics deposition on the corals, and make it prone to black-band infection, which is another lethal disease. Plastics are generally destructive to branched and structurally complex corals, which provide crucial habitats to many species of fish in the Asia-Pacific region. Primarily, plastics block the light and oxygen supply to corals depriving their metabolism. Secondly, plastics provide substratum for pathogens thus encourage the establishment of diseases that damage coral reefs. Moreover, plastics damage the coral's skin by tearing it, which allows the infection to start. Structurally, complex and branched corals like staghorn corals are eight times more likely to trap plastics with body parts. This is worrisome when it comes to marine biodiversity because the branched corals serve as nursery habitats providing shelter to a variety of fish species. Some regions like Australian reefs have less plastic probably due to comprehensive waste management system. Other Pacific countries have little control over what ends up in the waste stream. Many marine biologists suggested that ocean waters containing a lot of plastic trash may also include other contaminants that contribute to increased incidences of coral illness (Lamb et al. 2018).

Plastic particles host pathogens that can implicate disease outbreaks on coral reefs (Nugues et al. 2004; Smith et al. 2006; Morrow et al. 2012; Sweet et al. 2013; Oberbeckmann et al. 2015; De Tender et al. 2015; Kirstein et al. 2016). For example, many species of the bacterial genus, *Vibrio* (Zettler et al. 2013), an opportunistic pathogen, is commonly associated with polypropylene in the marine environments. The bacterial group, *Vibrio*, is a devastating pathogen of corals known to cause white syndrome diseases globally (Bourne et al. 2015). In marine environments, microbes including pathogens that raft on plastic debris stay active for longer periods (Fazey and Ryan 2016) particularly in tropical regions which are hot spots for coral reefs, thus suggesting their high risks of diseases to coral reef ecosystems (Barnes 2002). Lamb et al. (2018) observed plastic debris on 17 genera and 8 families of reef-forming corals. Although the mechanism of disease spread by plastic debris in marine wastes is yet to be elucidated, it is likely that the impact of plastic debris on disease development may vary significantly throughout the world depending on the nature of pathogen as well as coral type. Plastic debris may cause physical damage to the corals by exhausting the resources for wound-healing process (Mydlarz et al. 2006) thereby making the corals weak and prone to opportunistic infections (Lamb et al. 2016) through facilitated invasion. Many experimental evidences indicate that wound and abrasions caused by plastics to corals are subsequently infected by the ciliated protozoan, *Halofolliculina corallasia*, causing skeletal eroding band disease (Page and Willis 2008). In some cases, plastic particles may directly introduce foreign pathogens or indirectly alter the composition of symbiotic microflora of corals. The polyvinylchloride (PVC) is commonly colonized by *Rhodobacterales* (Dang et al. 2008), which constitutes a group of potentially pathogenic bacteria which is famous for outbreak of various coral diseases (Soffer et al. 2015). Furthermore, recent studies have signposted that shade and low-light microenvironments created due to plastic waste in marine ecosystems can lead to anaerobic conditions favoring the emergence of polymicrobial mats, a typical symptom of the blackband disease (Glas et al. 2012). Reduction in the composition and abundance of coral reefs particularly branched and complex structured reef-building corals will have serious impacts for the overall marine biodiversity and human health (Worm et al. 2006). Coral reefs harbor a tremendous diversity of fishes and other organisms in marine environment; thus, the loss of structural habitats (coral reefs) will reduce the productivity of fisheries by threefold (Rogers et al. 2014). Over 275 million people depend directly or indirectly on coral reef ecosystems for coastal protection, food, tourism, and heritage/cultural significance (Burke et al. 2011). Thus, the protection and worldwide conservation of coral reefs are direly important; therefore efforts should be taken to reduce the incidences of disease outbreaks in oceans which might prove helpful for human and ecological well-being. Improving the waste management strategies and policies are critical for preventing the release of voluminous amounts of plastics into oceans which may support the recovery of coral reefs.

13.6 Strategies to Minimize the Entry of Plastic Waste to the Marine Environment

To meet the demands of the plastic-based products at global level, production and utility of the plastic products with significant pace result in generation of the plastic waste at an alarming rate. The land sources are the main cause of marine plastic litter (80%), and thus to mitigate the emission of plastic litter into the marine environment becomes a national and international interest across the globe (Gamaralalage and Onogawa 2019). The lack of proper regulation/guideline and efficient methods to tackle the plastic wastes especially in low- and middle-income countries has led to the generation of huge mismanaged plastic pollution (Shahnawaz et al. 2019c). Researchers and policy-makers have suggested seven key strategies/regulations/guidelines to minimize the emission of plastic waste into the oceans and have grouped these strategies into three main categories, viz., short-term, medium-term, and long-term strategies (Sherrington et al. 2016). The short-term strategies that mitigate the emission of plastic waste into the oceans are as follows: (1) prevent generation of plastic waste, (2) enhance the efficiency of the plastic waste collection/transportation methods, and (3) improve the plastic dumping sites. Whereas, the medium-term strategies enhance the plastic waste recycling recovery using three strategies, viz., (1) separate different types of plastic waste at the site of plastic waste generation, (2) enhance the efficiency rate of plastic waste recycling/recovery, and (3) use the suitable method for treatment of the plastic waste to recover energy. However, designing ecofriendly methods to degrade plastic waste and develop sustainable plastic production end-user society is only regarded as the long-term strategy. If the above key strategies will be followed, we will get success in reducing 80% of the plastic waste from entering into the oceans (Gamaralalage and Onogawa 2019) to save the marine life including coral reefs.

13.7 Conclusion

Every year around 330 million metric tons of plastic is manufactured, which results in the accumulation of around 99 million metric tons of the plastic waste, out of which 4.8–12.7 million metric tons is emitted into the world's oceans. In the marine environment, plastic litter gets fragmented into macroplastic, mesoplastic, microplastic, and nanoplastic and leads to death of around billions of marine flora and fauna annually. Coral reefs is one of the important marine ecosystems that represent less than 0.1% world's ocean surface area and contain around 25% of all marine animals. Coral reefs are delicate and sensitive and are badly affected by the marine plastic litter. Researchers and policy-makers have suggested seven key strategies/regulations/guidelines to minimize the emission of plastic waste into the oceans and have grouped these strategies into three main categories, viz., short-term (3), medium-term (3), and long-term (1) strategies. The short-term strategies can be employed to mitigate the emission of plastic waste into the oceans. Whereas, the medium-term strategies lead to enhance the plastic waste recycling recovery.

However, designing ecofriendly method to degrade plastic waste and develop sustainable plastic production end-user society is regarded as the long-term strategy. If these seven strategies will be followed, emission of around 80% of the plastic litter into the oceans will be stopped to save coral reefs.

Acknowledgments RA acknowledges Prof. MK Jyoti and Prof. Kadambari Gupta for their support and guidance. MYS is thankful to his research guide (Prof. KK Kapoor) and teachers (Dr. HN Sheikh, Prof. Satya Paul, Prof. Sunil Uppal, Prof. PK Sharma) for their support and guidance. RA and MYS also acknowledge the support, valuable time, and encouragement of their respective parents, spouse, and kids. KH is thankful to Department of Biotechnology (DBT), GoI for financial assistance, and his elder brother Dr. Tasadoq Hussain Jaweed for his kind support and encouragement. MAD acknowledges the grants received from Jiangsu University, China (5363000606), and the National Natural Science Foundation of China (31900367). MS is thankful to the authorities of the Jiangsu University for providing an opportunity to work as an Associate Professor under Invited Talent Programme. MS and ZD are highly indebted to the financial support of the National Key R&D Program of China (Grant No.2018YFE0107100) and the Primary Research and Development Plan of Jiangsu Province.

References

Achituv Y, Dubinsky Z (1990) Evolution and zoogeography of coral reefs. In: Dubinsky Z (ed) Ecosystems of the world, vol 25. Elsevier, Amsterdam, pp 1–9

Adams SM (2005) Assessing cause and effect of multiple stressors on marine systems. Mar Pollut Bull 51:649–657

Allen AS, Seymour AC, Rittschof D (2017) Chemoreception drives plastic consumption in a hard coral. Mar Pollut Bull 124:198–205

Anbumani S, Kakkar P (2018) Ecotoxicological effects of microplastics on biota: a review. Environ Sci Pollut Res 25:14,373–14,396

Andrady AL (2011) Microplastics in the marine environment. Mar Pollut Bull 62:1596–1605

Andrady AL, Neal MA (2009) Applications and societal benefits of plastics. Philos Trans R Soc B Biol Sci 364:1977–1984

Annomyous (2018) Global distribution of coral reefs, compiled from multiple sources including the Millennium Coral Reef Mapping Project. UNEP-WCMC, World Fish Centre, WRI, TNC

Annomyous (2020) Seven ways you're connected to coral reefs. UN Environment Programme. https://www.unep.org/news-and-stories/story/seven-ways-youre-connected-coral-reefs. Accessed 26 June 2021

Annomyous (2021) Our planet is drowning in plastic pollution—it's time for change. Accessed 26 June 2021 https://www.unep.org/interactive/beat-plastic-pollution/

Arthur C, Baker JE, Bamford HA (2009) Proceedings of the international research workshop on the occurrence, effects, and fate of microplastic marine debris, September 9–11, 2008. University of Washington Tacoma, Tacoma

Barnes DKA (2002) Invasions by marine life on plastic debris. Nature 416:808–809

Barnes RSK, Mann KH (2009) Fundamentals of aquatic ecology. John Wiley & Sons, Hoboken, NJ

Bourne DG, Ainsworth TD, Pollock FJ, Willis BL (2015) Towards a better understanding of white syndromes and their causes on Indo-Pacific coral reefs. Coral Reefs 34:233–242

Boyle K, Ormeci B (2020) Microplastics and nanoplastics in the freshwater and terrestrial environment: a review. Water 12:2633

Browne MA, Galloway T, Thompson R (2007) Microplastic-an emerging contaminant of potential concern? Integr Environ Assessment Manag Int J 3:559–561

Burke L, Reytar K, Spalding M, Perry A (2011) Reefs at risk revisited. World Resources Institute, Washington, DC

Campanale C, Massarelli C, Savino I, Locaputo V, Uricchio VF (2020) A detailed review study on potential effects of microplastics and additives of concern on human health. Int J Environ Res Public Health 17:1212

Carlson RR, Foo SA, Asner GP (2019) Land use impacts on coral reef health: a ridge-to-reef perspective. Front Mar Sci 6:562

Carpenter EJ, Smith KL (1972) Plastics on the sargasso sea surface. Science 175:1240–1241

Cesar H, Burke L, Pet-Soede L (2003) The economics of worldwide coral reef degradation. International Coral Reef Action Network, Cebu City, Philippines

Cingotti N, Jensen GK (2019) Health and Environment Alliance (HEAL). Food contact materials and chemical contamination. Health and Environment Alliance, Brussels, Belgium

Cole M, Lindeque P, Halsband C, Galloway TS (2011) Microplastics as contaminants in the marine environment: a review. Mar Pollut Bull 62:2588–2597

Costanza R et al (1997) The value of the world's ecosystem services and natural capital. Nature 387: 253–260

Costanza R et al (2014) Changes in the global value of ecosystem services. Glob Environ Chang 26: 152–158

Critchell K, Hoogenboom MO (2018) Effects of microplastic exposure on the body condition and behaviour of planktivorous reef fish (Acanthochromis polyacanthus). PLoS One 13:e0193308

Dang H, Li T, Chen M, Huang G (2008) Cross-ocean distribution of Rhodobacterales bacteria as primary surface colonizers in temperate coastal marine waters. Appl Environ Microbiol 74:52–60

Danovaro R et al (2008) Sunscreens cause coral bleaching by promoting viral infections. Environ Health Perspect 116:441–447

De Tender CA, Devriese LI, Haegeman A, Maes S, Ruttink T, Dawyndt P (2015) Bacterial community profiling of plastic litter in the Belgian part of the North Sea. Environ Sci Technol 49:9629–9638

Duxbury MA (1992) Plastic pellets in the aquatic environment: sources and recommendations. United States Environmental Protection Agency Office of Water (WH-556F),

Eriksen M et al (2014) Plastic pollution in the world's oceans: more than 5 trillion plastic pieces weighing over 250,000 tons afloat at sea. PLoS One 9:e111913

Europe (2008) The compelling facts about plastics: an analysis of plastics production, demand and recovery for 2006 in Europe

Evans RD et al (2020) Early recovery dynamics of turbid coral reefs after recurring bleaching events. J Environ Manage 268:110,666

Fazey FMC, Ryan PG (2016) Biofouling on buoyant marine plastics: an experimental study into the effect of size on surface longevity. Environ Pollut 210:354–360

Fuchs T (2013) Effects of habitat complexity on invertebrate biodiversity. Immed Sci Ecol 2:1–10. https://doi.org/10.7332/ise2013.2.1.dsc

Gall SC, Thompson RC (2015) The impact of debris on marine life. Mar Pollut Bull 92:170–179

Gamaralalage PJD, Onogawa K (2019) Strategies to reduce marine plastic pollution from land-based sources in low and middle-income countries. United Nations Environment Programme (UN Environment), Nairobi

Glas MS, Sato Y, Ulstrup KE, Bourne DG (2012) Biogeochemical conditions determine virulence of black band disease in corals. ISME J 6:1526–1534

Hay ME, Rasher DB (2010) Coral reefs in crisis: reversing the biotic death spiral. F1000 Biol Rep 2

Hirai H et al (2011) Organic micropollutants in marine plastics debris from the open ocean and remote and urban beaches. Mar Pollut Bull 62:1683–1692

Hoornweg D, Bhada-Tata P (2012) What a waste: a global review of solid waste management. Urban Development & Local Government Unit World Bank, Washington, DC

Hoornweg D, Bhada-Tata P, Kennedy C (2013) Environment: waste production must peak this century. Nature News 502:615

Hoover JP (1998) Hawai'i's sea creatures. A guide to Hawai'i's marine invertebrates. Mutual Publishing, Honolulu

Huang W et al (2021) Microplastics in the coral reefs and their potential impacts on corals: a mini-review. Sci Total Environ 762:143,112. https://doi.org/10.1016/j.scitotenv.2020.143112
Issac MN, Kandasubramanian B (2021) Effect of microplastics in water and aquatic systems. Environ Sci Pollut Res:1–19
Jahnke A et al (2017) Reducing uncertainty and confronting ignorance about the possible impacts of weathering plastic in the marine environment. Environ Sci Technol Lett 4:85–90
Jalaluddin M (2017) Use of plastic waste in civil constructions and innovative decorative material (eco-friendly) MOJ. Civ Eng 3:01–10
Jambeck JR et al (2015) Plastic waste inputs from land into the ocean. Science 347:768–771
Jangly GP (2013) How stress affects secondary metabolite production and the associated bacterial community of the octocoral Plexaura homomalla. University of Houston-Clear Lake, Houston
Joyner CC, Frew S (1991) Plastic pollution in the marine environment. Ocean Dev Int Law 22:33–69
Kirstein IV, Kirmizi S, Wichels A, Garin-Fernandez A, Erler R, Loder M, Gerdts G (2016) Dangerous hitchhikers? Evidence for potentially pathogenic Vibrio spp on microplastic particles. Mar Environ Res 120:1–8
Lamb JB, Wenger AS, Devlin MJ, Ceccarelli DM, Williamson DH, Willis BL (2016) Reserves as tools for alleviating impacts of marine disease. Philos Trans R Soc B Biol Sci 371:20150210
Lamb JB et al (2018) Plastic waste associated with disease on coral reefs. Science 359:460–462
Lasut MT et al. (2018) From coral triangle to trash triangle–how the hot spot of global marine biodiversity is threatened by plastic waste. In: Proceedings of the International Conference on Microplastic Pollution in the Mediterranean Sea, Springer, pp 107–113
Law KL (2017) Plastics in the marine environment annual review of marine. Science 9:205–229
Lebreton L, Andrady A (2019) Future scenarios of global plastic waste generation and disposal. Palgrave Commun 5:1–11
Lebreton LCM, Van Der Zwet J, Damsteeg J-W, Slat B, Andrady A, Reisser J (2017) River plastic emissions to the world's oceans. Nat Commun 8:1–10
Lee J-H, Chen J, Chough SK (2015) The middle-late Cambrian reef transition and related geological events: a review and new view Earth Sci Rev 145:66–84 doi:doi:https://doi.org/10.1016/j.earscirev.2015.03.002
Ling HI, Leshchinsky D, Mohri Y, Kawabata T (1998) Estimation of municipal solid waste landfill settlement. J Geotech Geoenviron Eng 124:21–28
Lusher A, Hollman P, Mendoza-Hill J (2017) Microplastics in fisheries and aquaculture: status of knowledge on their occurrence and implications for aquatic organisms and food safety vol technical paper no. 615. FAO fisheries and aquaculture, Rome, Italy
Miandad R, Barakat MA, Aburiazaiza AS, Rehan M, Nizami AS (2016) Catalytic pyrolysis of plastic waste: A review. Process Safety Environ Protection 102:822–838
Moore CJ (2008) Synthetic polymers in the marine environment: a rapidly increasing, long-term threat. Environ Res 108:131–139
Morrow KM, Ritson-Williams R, Ross C, Liles MR, Paul VJ (2012) Macroalgal extracts induce bacterial assemblage shifts and sublethal tissue stress in Caribbean corals. PLoS One 7:e44859
Moyle PB, Cech JJ (2004) Fishes: an introduction to ichthyology, 5th edn. Pearson, London
Mulhall M (2008) Saving the rainforest of the sea: An analysis of international efforts to conserve coral reefs. Duke Envtl L & Pol'y F 19:321
Mydlarz LD, Jones LE, Harvell CD (2006) Innate immunity, environmental drivers, and disease ecology of marine and freshwater invertebrates. Annu Rev Ecol Evol Syst 37:251–288
Ng E-L, Lwanga EH, Eldridge SM, Johnston P, Hu H-W, Geissen V, Chen D (2018) An overview of microplastic and nanoplastic pollution in agroecosystems. Sci Total Environ 627:1377–1388
NOAA (2020) Coral reefs. NOAA National Ocean Service. https://coralreef.noaa.gov/. Accessed 20 June 2021
Nualphan (2013) Types and sources of marine debris in BangSaen Beach. Chulalongkorn University, Chonburi Province

Nugues MM, Smith GW, Van Hooidonk RJ, Seabra MI, RPM B (2004) Algal contact as a trigger for coral disease. Ecol Lett 7:919–923
Oberbeckmann S, Loder MGJ, Labrenz M (2015) Marine microplastic-associated biofilms-a review. Environ Chem 12:551–562
Page CA, Willis BL (2008) Epidemiology of skeletal eroding band on the Great Barrier Reef and the role of injury in the initiation of this widespread coral disease. Coral Reefs 27:257–272
Panyakapo P, Panyakapo M (2008) Reuse of thermosetting plastic waste for lightweight concrete. Waste Manag 28:1581–1588
Preethi S, Prasannadevi R, Preethi K, Prasad SR, Banuchander TR (2019) Reuse of plastic bottles in construction of buildings. Ratio 1:10
Rani M, Shim WJ, Han GM, Jang M, Al-Odaini NA, Song YK, Hong SH (2015) Qualitative analysis of additives in plastic marine debris and its new products. Arch Environ Contam Toxicol 69:352–366
Reddy S (2018) Plastic pollution affects sea life throughout the ocean The Pew Charitable Trusts, p 14
Richmond RH (1993) Coral reefs: present problems and future concerns resulting from anthropogenic disturbance. Am Zool 33:524–536
Rochman CM, Hoh E, Kurobe T, Teh SJ (2013) Ingested plastic transfers hazardous chemicals to fish and induces hepatic stress. Sci Rep 3:1–7
Rogers A, Blanchard JL, Mumby PJ (2014) Vulnerability of coral reef fisheries to a loss of structural complexity. Curr Biol 24:1000–1005
Rosevelt C, Los Huertos M, Garza C, Nevins HM (2013) Marine debris in Central California: quantifying type and abundance of beach litter in Monterey Bay, CA. Mar Pollut Bull 71:299–306
Samiei JV, Dab K, Ghezellou P, Shirvani A (2013) Some scleractinian corals (Scleractinia: Anthozoa) of Larak Island. Persian Gulf Zootaxa 3636:101–143
Sangale MK, Shahnawaz M, Ade AB (2012) A review on biodegradation of polythene: the microbial approach. J Bioremed Biodegr 3:1–9
Sarker M, Kabir A, Rashid MM, Molla M, Mohammad ASMD (2011) Waste polyethylene terephthalate (PETE-1) conversion into liquid fuel. J Fundam Renew Energy Appl 1
Schubert J (1972) A program to abolish harmful chemicals. Ambio:79–89
Shah AA, Hasan F, Hameed A, Ahmed S (2008) Biological degradation of plastics: a comprehensive review. Biotechnol Adv 26:246–265
Shahnawaz M, Sangale MK, Ade AB (2019a) Bioremediation technology for plastic waste, vol 1. Springer, Singapore
Shahnawaz M, Sangale MK, Ade AB (2019b) Microplastics. In: Bioremediation technology for plastic waste. Springer, Singapore, pp 11–19
Shahnawaz M, Sangale MK, Ade AB (2019c) Policy and legislation/regulations of plastic waste around the globe. In: Bioremediation technology for plastic waste. Springer, Singapore, pp 113–126
Sherrington C, Darrah C, Hann S, Cole G, Corbin M (2016) Study to support the development of measures to combat a range of marine litter sources. European Commission DG Environment
Smith JE et al (2006) Indirect effects of algae on coral: algae-mediated, microbe-induced coral mortality. Ecol Lett 9:835–845
Smith M, Love DC, Rochman CM, Neff RA (2018) Microplastics in seafood and the implications for human health. Curr Environ Health Rep 5:375–386
Soffer N, Zaneveld J, Vega Thurber R (2015) Phage-bacteria network analysis and its implication for the understanding of coral disease. Environ Microbiol 17:1203–1218
Spalding M, Spalding MD, Ravilious C, Green EP (2001) World atlas of coral reefs. Univ of California Press
Spalding MD, Grenfell AM (1997) New estimates of global and regional coral reef areas. Coral Reefs 16:225–230

Sweet M, Stelfox M, Lamb J (2019) Plastics and shallow water coral reefs: synthesis of the science for policy-makers. University of Derby. University of California, Irvine. doi: https://doi.org/10.13140/RG.2.2.29699.14880

Sweet MJ, Bythell JC, Nugues MM (2013) Algae as reservoirs for coral pathogens. PLoS One 8: e69717

Tan F et al (2020) Microplastic pollution around remote uninhabited coral reefs of Nansha Islands, South China Sea. Sci Total Environ 725:138,383. https://doi.org/10.1016/j.scitotenv.2020.138383

Tasseron P, Zinsmeister H, Rambonnet L, Hiemstra A-F, Siepman D, van Emmerik T (2020) Plastic hotspot mapping in urban water systems. Geosciences 10:342

Thompson R, Moore C, Andrady A, Gregory M, Takada H, Weisberg S (2005) New directions in plastic debris. Science 310:1117–1118

Thushari GGN, Chavanich S, Yakupitiyage A (2017) Coastal debris analysis in beaches of Chonburi Province, eastern of Thailand as implications for coastal conservation. Mar Pollut Bull 116:121–129

Thushari GGN, Senevirathna JDM (2020) Plastic pollution in the marine environment. Heliyon 6: e04709

Van Franeker JA, Law KL (2015) Seabirds, gyres and global trends in plastic pollution. Environ Pollut 203:89–96

van Raamsdonk LWD et al (2020) Current insights into monitoring, bioaccumulation, and potential health effects of microplastics present in the food chain. Foods 9:72

Vince J, Hardesty BD (2017) Plastic pollution challenges in marine and coastal environments: from local to global governance. Restoration Ecol 25:123–128

Worm B et al (2006) Impacts of biodiversity loss on ocean ecosystem services. Science 314:787–790

Yoshida S et al (2016) A bacterium that degrades and assimilates poly (ethylene terephthalate). Science 351:1196–1199

Zettler ER, Mincer TJ, Amaral-Zettler LA (2013) Life in the “plastisphere”: microbial communities on plastic marine debris. Environ Sci Technol 47:7137–7146

Ecotoxicological Impact of Plastic Waste on Marine Flora

14

Bhushan P. Bhusare, Vasudeo P. Zambare, Tasadoq Hussain Jaweed, and Mohd Shahnawaz

Abstract

Worldwide, plastic pollution is a prioritized, pressing and environmental concern. Due to the multisectoral application of plastic, globally plastic production is 359 metric tons (MT) in 2018. Out of these 12% plastic are incinerated, only 9% are recycled and the rest are discarded or landfilled, with illegal and mismatched practices. Biodegradation, hydrolysis, mechanical abrasion, photodegradation or UV radiations affected on larger plastic debris converted into microplastic (1 μm to 5 mm) and nanoplastic (1–1000 nm). These are considered as major contaminants of emerging concern, and since a decade an exceptional research has been done to identify it and for its strategic solution. Different macrophytes can adsorb or chanellized plastic particles which is the function of particle size, charge as well as plant surfaces. In the food web, these macrophytes are at the lowest point but very important item of the diet of different

The original version of this chapter was revised: has affiliation of the author Vasudeo P. Zambare updated. The correction to this chapter can be found at https://doi.org/10.1007/978-981-16-5403-9_17.

B. P. Bhusare (✉)
R&D Department, Excellence Teaching and Research Academy, Ahmednagar, Maharashtra, India

V. P. Zambare
R&D Department, Om Biotechnologies, Nashik, Maharashtra, India

Aesthetika Eco Research Pvt Ltd, Nashik, Maharashtra, India

T. H. Jaweed
Govt. Higher Secondary School, Kharpora Kokernag, Jammu and Kashmir UT, India

M. Shahnawaz
Biofuels Institute, School of Environment and Safety Engineering, Jiangsu University, Zhenjiang, China

M. Shahnawaz et al. (eds.), *Impact of Plastic Waste on the Marine Biota*,
https://doi.org/10.1007/978-981-16-5403-9_14

organism which uptakes the micro-, meso- and nanoplastic through adsorption and the contaminated macrophytes consequences thereof for food security and stress conditions. Present chapter discussed the current knowledge of plastic and its different forms of contaminant such as macro-, micro- and nanoplastic and its ecotoxicological impact on marine macrophytes.

Keywords

Aquatic macrophytes · Ecotoxic · Macro · Meso · Nano

14.1 Plastic

Plastic is a synthetic or semi-synthetic organic compounds of long-chain polymeric molecules of high molecular mass (Shah et al. 2008). Generally, plastic is a complex polymer of long-chain linked molecular group and blend of some organic and inorganic additives, plasticizers and fillers (Zheng et al. 2005). Plastic is composed of elements like carbon, oxygen, hydrogen, chlorine, nitrogen and bromine. Plastics are inert chemicals that showed resistance to decompose. It is durable because of complex polymeric nature, lightweight and malleability. It is very flexible and has good moulding property to convert it into different sizes and shapes (Zheng et al. 2005; Plastics Europe 2019). In the daily routine, plastic has gained important roles in many products and technologies which are related to plastic and plastic originated products because of its versatile nature, high stability and easy way of manufacturing. The 'Plastic Era' was beginning in the middle of the nineteenth century and a rapid advancement in twentieth century for manufacture of items in each and every section of science and technology (Law 2017). Fifty different types of polymers have been discovered and utilized for more than 60,000 plastic blends (Shashoua 2008), and the great amounts of used are low-density polyethylene (LDPE), high-density polyethylene (HDPE), polyvinyl chloride (PVC), polypropylene (PP), polystyrene (PS) and polyethylene terephthalate (PET) (Mohee and Unmar 2007; Gutow et al. 2015; Li et al. 2016; Plastics Europe 2016; Mateos-Cardenas et al. 2019). Particles of plastic are detected in all three life support systems like air, water and soils worldwide. Respective plastic have specific properties and are used for manufacturing of bottles, bags, container, pipes, packing materials, dishes, furnitures, filers, mats, textile, lenses, construction material, ornamental materials, electronics, food and cosmetic packing, etc. Being a foreign material, it has some serious health effects too which are summarized in Table 14.1.

In different industries such as healthcare, constructions, households, environmental, biotechnology, food, pharmaceutical, chemical, shipping and related service industries, plastic is widely used (Sudesh and Iwata 2008). Globally, total 8100 MT of total plastic was manufactured from 1950 to 2019. Current plastic manufacturing exceeds 320 million tons per year and is likely to be exceed two times in next 20 years. Of this, up to 94% ends up in landfills but due to the administrative and other related mismanagement leads to release of 21–42% of total mass from landfills into the environment (Chae and An 2020; Li et al. 2020; Lian et al. 2020). As per the properties, the plastic will have some adverse effect on

Table 14.1 Different types of plastic blends and their uses

Sr. no.	Plastic formulations	Abbreviation	Specific gravity (g cm−3)	Uses	Health effects	References
1	Polyethylene	PE	0.91–0.96	Plastic bottle, supermarket bags	No information	Andrady (2011), Ghosh et al. (2013), Li et al. (2016)
2	High-density polyethylene	HDPE	0.94–0.96	Bottles: chemicals, laundry detergents, shampoo, milk, juice	Alteration of female oestrogen	Andrady (2011), Ghosh et al. (2013), Li et al. (2016)
3	Low-density polyethylene	LPDE	0.91–0.93	Bags: refrigerated bags	No information	Andrady (2011), Ghosh et al. (2013), Li et al. (2016)
4	Polyethylene terephthalate	PET	1.37	Pipe: agricultural pipe, pumping pipe	Potential human carcinogen	Andrady (2011), Ghosh et al. (2013), Li et al. (2016)
5	Polypropylene	PP	0.83–0.85	Squeeze bottles	No information	Li et al. (2016)
6	Polyamides (Nylons)	PA	1.13–1.35	Clack irrigation tube, wraps, black cover films, garbage bins and bags	Headache, skin allergy, dizziness, Lead to cancer, spinal pain	Li et al. (2016)
7	Polycarbonate	PC	1.20–1.22	Wrappers and biscuit trays, water and beverages bottles, salad dressing and food containers	Dysfunctioning of the liver and brain, insulin resistance, alteration in reproductive system due to bisphenol	Li et al. (2016)
8	Polyester	PES	1.38–1.40	Food wrappers, restaurant food boxes and microwavable dishes, furnitures, ice cream containers, dip bottles packing tapes	Ophthalmic, respiratory and dermatological irritations	Li et al. (2016)
9	Polystyrene	PS	1.04	High-strength fishing fibres, toothbrush, fishing line, bag, luggaging items, car interior, bristles, food packaging films	Eyes, nose and throat irritations, dizziness, unconsciousness, accumulation in body fat, cancers	Li et al. (2016)

(continued)

Table 14.1 (continued)

Sr. no.	Plastic formulations	Abbreviation	Specific gravity (g cm−3)	Uses	Health effects	References
10	High-impact polystyrene	HIPS	1.04–1.07	Electronic CD's, construction utilities, traffic signal lights, eyeglasses, lenses, compact discs, etc.	No information	Li et al. (2016)
11	Polyvinyl chloride	PVC	1.37–1.39	Textile fibres	Cancer, bronchitis, ophthalmic, ulcers, liver dysfunction, dermal diseases, reproduction birth defects, genetic changes and indigestion	Li et al. (2016)
12	Acrylonitrile butadiene styrene	ABS	1.06–1.08	Electronic CD and video cases, foot cutleries, cheap toys, imitation crystal glassware	Oxidative stress, heart, respiratory, and lung diseases due to ultrafine particles	Li et al. (2016)

health and environment, where the produced plastic needs to be converted or utilized by recycling process. Globally a trend of plastic recycling is increasing, but still proper handling of plastic disposal is challenging (Plastics Europe 2019; Barría et al. 2020). Plastic-making monomers are propylene and ethylene, derived from fossil-based hydrocarbons (Geyer et al. 2017). Hydrocarbon-based plastics are non-biodegradable and gets accumulated in the landfills followed by their percolation into groundwater. Thus, it contaminates the water and ultimately the aquatic ecosystem. Plastic pollution in the ocean is a global concern and found everywhere and throughout the ocean with concentration about 580,000 plastic pieces per square kilometre (Willis et al. 2018). To overcome this problem, many procedures and methods are being tried and implemented by many researchers, and one of the effective and environment friendly research output is the manufacturing of bio-based and biodegradable plastic as an alternative to the plastics. Nowadays many local industries are manufacturing different types of biodegradable plastic bags, containers with carbohydrates or partially degradable items with some additives with carbohydrates. Biodegradable plastics are degraded by either ultraviolet (UV) radiation or by microorganisms (Mohee and Unmar 2007). Bio-based plastic will be an alternative to these petroleum-based plastics, however its current and worldwide production is only 4 MT and needs to be explored more. Mitigation of the plastic waste is very important, and its strategic management to control it is through legislative, educational, engineering and biotechnological remediation (Fig. 14.1).

14.2 Macro- to Nanoplastic

Ubiquitously, plastic and plastic materials are consumed in each and every sector such as electronics, transportation, textile, energy, homecare, healthcare and related packagings which showed a 359 MT of plastic production globally, and this is reported on the basis of a study conducted in 2018 (Plastics Europe 2019). Among the total plastic generated 12% gets incinerated, 9% recycled, and the rest gets discarded or landfilled, with illegal and mismatched practices (Geyer et al. 2017). In the marine enviornment and at the plastic dumping sites, plastic waste gets exposed to various enviornmental forces viz. weathering, UV-radiations, biological metabolism and gets fragmented into plastic particles of the varied sizes viz, megaplastic, macroplastic, meso plastic, microplastic and nanoplastic (Schwaferts et al. 2019). Barnes et al. (2009), Blettler et al. (2017), and Boyle and Örmeci (2020) classified plastic as megaplastic (>100 mm), macroplastic (>20 mm), mesoplastic (5-20 mm), microplastic ($\leq$5 mm) and nanoplastic (<1 μm) (Table 14.2). Classification-wise, microplastic blogs to particles size $\leq$5 mm, but other sizes <10 mm, 2–6 mm, <2 mm and <1 mm are also considered as microplastic (da Costa et al. 2019). While, Schwaferts and his co-researcher stated 1 μm to 5 mm particle size belongs to microplastic (Schwaferts et al. 2019).

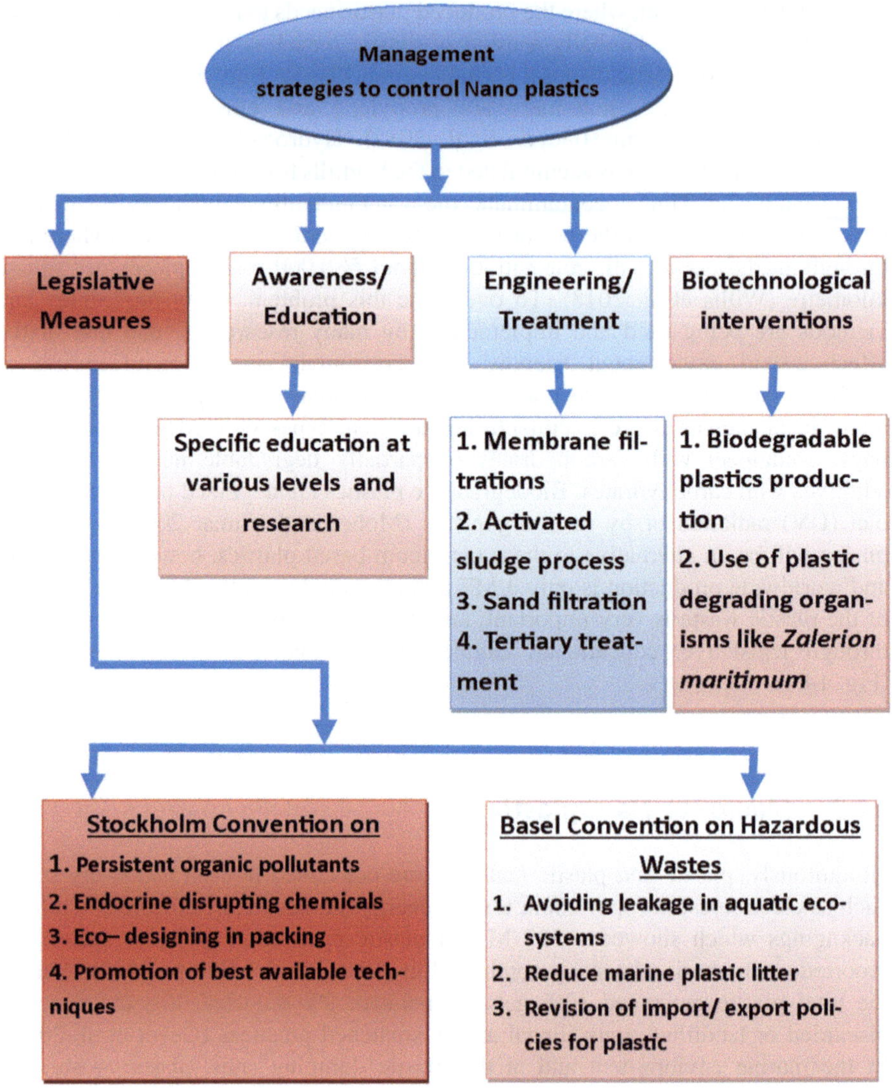

Fig. 14.1 An outline of plastic waste management strategies for mitigation of plastic waste

14.2.1 Macroplastic

Mishandling, mismanagement (e.g., spills or spreads from waste handling sites) and direct littering by individuals are main sources introducing the macroplastic in the environment. According to Ryberg et al. (2018), mismanagement is the dominant source of macroplastic. The macroplastic fibres belongs to virgin and recycled PP, HDPE or PET fibres, and the physico-chemical properties of microplastic depend upon the manufacturing techniques (Fraternali et al. 2011). A melt spinning plastic

Table 14.2 Different types of plastic (size basis) of marine plastic litter

Sr. no.	Types of plastic	Size	Source	Examples	Reference
1	Mega	>100 mm	Abandoned gear, catastrophic activities	Net and traps used for fishery	Barnes et al. (2009), GESAMP (2016)
2	Macro	>20 mm	Lost articles in rivers and ocean	Food wrapper, bags, balloons, fishing accessories	Barnes et al. (2009), GESAMP (2016)
3	Meso	5–20 mm	Large plastic fragmentation	Bottle caps	Barnes et al. (2009), GESAMP (2016)
4	Micro	≤5 mm	Primary and secondary microplastic	Textile, fibres, tyre dust	Blettler et al. (2017)
5	Nano	< 1 μm	Breakdown of macro- to nanometre-sized particles through microplastic	Cosmetic and healthcare products and plastic powder	Blettler et al. (2017), Boyle and Örmeci (2020)

granules to plastic filament followed by hot treatment, the monofilaments converted to fibres are the most popular and common technique (Fraternali et al. 2011). Ochi and his colleagues conducted a study of melting and extrusion of PET granules into monofilaments (60,000 dtex) followed by hot drawing of monofilaments. By using orientation unit, it converts into 5000 dtex (Ochi et al. 2007). Finally, 30–40-mm-long fibres were formed by cutting the microfilament. Thus, process like melt spinning and hot drawing resulted into high crystallinity of PET which results into improved Young's modulus and tensile strength. These methods showed highest 450 MaP tensile strength for PET and PP (Fraternali et al. 2011; Yin et al. 2015).

14.2.2 Microplastic

It has been categorized into primary and secondary microplastic.

14.2.2.1 Primary Microplastic

Cole et al. (2011) defined that primary microplastic is a plastic of microscopic size. Mostly, industrial and domestic packing products create primary microplastic (Betts 2008; Moore 2008) and some of them air blasting media, cosmetics and pharmaceutical drug container (Zitko and Hanlon 1991; Gregory 1996). Rust- and paint-removing chemical such as acrylic and melamine scrubbers generate 0.25–1.7 mm particle size polyester microplastic (Browne et al. 2007). A waste water treatment technology with new techniques removed up to 99% of microplastic, even though, large amount of microplastic was observed in regular effluents (Rochman 2015).

14.2.2.2 Secondary Microplastic

Through the conversion of large plastic (land and ocean) into smaller particle formed secondary microplastic (Ryan et al. 2009). Approx. thousand years is durability of this plastic; we can't guess the exact figure because from 60 years mass production of conventional plastic has been done (Zheng et al. 2005; Barnes et al. 2009). Browne et al. (2007) reported that, physical and biological processes mainly responsible for conversion of fragmentation of microplastic in the environment, which ultimately reduce the plastic structural integrity. However, Arthur et al. (2009) and Corcoran et al. (2009), reported weathering and chemomechanical weathering process for same. Barnes et al. (2009) reported oxidative degradation of through radiations generated by UV. Comparatively, the plastic degradation is quick and more on cold temperatures marine environment and beaches because availability of maximum O_2 and direct contact of sunlight, which results into loss and damage of plastic structural integrity (Browne et al. 2007). Use of mechanical forces for fragmentation of plastic is a risky process (Barnes et al. 2009). The source of microplastic could be a biodegradable plastic, and these are the blends of natural and synthetic polymer and polymers of plant and animal origin which are helping to reduce the time of degradation (Derraik 2002; Thompson et al. 2004; O'Brine and Thompson 2010). However, if such hybrid plastic are disposed of without a proper care, then the non-degrading synthetic polymer starts accumulating in the environment, especially in water stream.

14.2.3 Nanoplastic

Nanoplastic, is a debatable subject and defined as particle size of 1–100 nm (Ivleva et al. 2017), and some researcher considered a broad range of 1–1000 nm (Gigault et al. 2018; Schwaferts et al. 2019). Particles of <100 nm were considered as engineered nanomaterials (ENM), nanoplastic (EU Commission 2011) as well as particles smaller than 1 mm. However, as per European Union classification of nanoplastic, many researchers consider the former definition of <100 nm (EU Commission 2011). Even though it is debatable part of discussion, these definitions have a very great impact in both research and its implementation. Some researchers observed and suggested <20 nm particles as nanoplastic, and this categorization is mainly used to clarify plankton as nanoplankton (Wagner et al. 2014). In fact, these sizes of nanoplastic and ENM showed potential hazard to those organisms, which are the parts of various food webs. Hence practically, the <100 nm particles size is considered for nanoplastic classification. Many researchers published their observation on microplastics in terms of availability of sources, distribution of these, storage or absorption, movement, biological accumulation, toxicity, risk of environment and remedial measurements (Bradney et al. 2019; Chen et al. 2020; Fu et al. 2020; Guo et al. 2020; Ma et al. 2020; Xu et al. 2020; Zhang et al. 2020). However, the nanoplastic are less explored. It has been observed that the nanoscaling of microplastic is the function of physical and chemical properties. Apart from these, the downscaling is the function of some environmental factors such as migration and aggregation, bioaccumulation and toxicity. Further investigations on nanoplastic investigation is necessary (Wang et al. 2021) and has a scope for novel research of nanoplastic for ecotoxicological research.

14.3 Plastic in the Marine Environment

Worldwide, the annual plastic demand has consistently increased over the few years with current production of 245 million tonnes. It is used in many product applications due to its lightweight, flexibility, versatility, strength and transparent properties. Tin addition to this plastic is cheap; moisture/oxygen gas barrier properties and bio-inertness make it as excellent packaging materials. Conventional materials such as metal, paper and glass are being replaced by plastic low-cost plastic material with equivalent or superior-quality material. The most commonly used consumer disposable items are stumble across the beach debris (Andrady 2003). The amount of actual plastic waste generated per annum is uncleared that accumulates 75–80 million tonnes of plastic. Plastic transported from land to river/ ocean because of its unique properties of lightweight, hydrophobic nature and physico-chemical resistance. Plastic debris is found in the environmental media, including the water surfaces, sediments, soil and groundwater and in atmosphere too (Wang et al. 2021). It is estimated that over 80% of marine plastic are from land-based sources, through coastal landfill operations, nanoplastic percolation in rivers, compost, bio-solids and inappropriate discarding of sewage material (Ganesh Kumar et al. 2020). Carpenter and his colleague in 1972 reported presence of plastic in the ocean, and this is the first report (Carpenter et al. 1972). After this, many researchers worked on the same (Colton and Knapp 1974; Coe and Rogers 1996). In the following decades, environment impact of these plastic particles became an interesting topic for many researchers (Andrady 2011). Several broad plastic classes used in packaging as PE, PP, PS, PET and PVC. The high-volume usage of these plastic is reflected in their production, and consequently these in particular have high likelihood of ending up in the ocean environment. Among the all five subtropical gyres, about one third of total floating plastic is accumulated only in the North Pacific (Cózar et al. 2014; van Sebille et al. 2015). In the North Atlantic gyre was found to have maximum plastic, while in the North Atlantic Ocean, the reduced number of plastic was recorded (van Sebille et al. 2012). Through thermohaline circulation the plastic is migrated from North Atlantic to the Greenland and Barents Seas of the Arctic Ocean (Cózar et al. 2017). A density-driven global circulation is monitored by the formation of deep water by cooling in the Greenland and Barents Seas which acts as a suction pump of surface water to create motion for the TC (Kuhlbrodt et al. 2007). Hence, Arctic Ocean was considered as final point of plastic in the North Atlantic branch of the TC (Cózar et al. 2017). Also, surface water transported through TC to Ross and Weddell Seas, in the Pacific and Atlantic sectors of the Southern Ocean, respectively (Kuhlbrodt et al. 2007).

The floating plastic dispersed back to subtropical gyres of Southern Hemisphere by northward woman flow generated from the Antarctic Circumpolar Current of Arctic Ocean and observed from drifting buoys which one is tracked with the help of satellite (van Sebille et al. 2012). From the world's oceans, a numerical model of floating debris was developed by Lebreton et al. (2012), and Mediterranean and East Asian Seas around Japan (Cózar et al. 2015; Isobe et al. 2015) regions have observed significant accumulation of plastic. But these are high-pressure and hydrodynamics regions where local traps are required for floating pollutions. There is a very high concentration of plastic accumulation in these regions than other subtropical gyres (van der Hal et al. 2017). Hence to remove such floating plastic debris, ballasting and ingestion processes are used with a special focus on seafloor (Van Cauwenberghe et al. 2013). Furthermore, only half of all produced plastic in sea water is floating (APM 2011). Hence, still the worldwide marine plastic debris distribution needs to be explored more.

The entry of plastic into the oceans is through man-made recreation and fishing activities (Ribic et al. 2010). Fishing and aquaculture industry is one of the booming industries and contributes 18% of rubbished plastic in marine environment (Hinojosa and Thiel 2009). The fishing nets are made up of polyolefin's plastic (PP, PE) and nylons which are lost slowly during their every fishing activity (Klust 1982; Timmers et al. 2005; Watson et al. 2006). Apart from fishing some other contributors are land-based sources of beach wastes and loss of items during ocean transportation (Gregory 1996; Ogata et al. 2009). Sea water contains more salt, and hence it has high specific gravity (1.025), and some plastic material specific gravity is important to float it or sediment it in sea water. Pure plastic materials have specific gravity, but it was altered by some additives and use of recycle plastic. PP and PE are lightweight plastic and can float on surfaces but like nylon (denser) can be sedimented. It was observed that every year around 8MT of rubbished plastic entered in ocean belongs to various categories such as macroplastic (>25 mm), mesoplastic (5–25 mm), microplastic (1–5 mm), mini-microplastic (1 mm) and nanoplastic (<1 mm) (Jambeck et al. 2015; Crawford and Quinn 2017). To check the migration of plastic wastes, several hypothetical and practical numerical models were developed, and the major observation of redistribution of plastic particles is due to ocean currents which accumulates the major rubbish plastic in Atlantic, Pacific and Indian Oceans (Sherman and Van Sebille 2016; Mountford and Morales Maqueda 2019). This rubbished plastic has a lot of impact on ecology especially a toxic impact on aquatic life including microbes, algae, animals and fish (Rios Mendoza et al. 2018; Peng et al. 2020). The toxicological impact is either engulf of plastic berries or metabolic percolation of micro-/nanoplastic through food chain. So far the marine contamination of microplastic is widely discussed by several researchers, but nanoplastic is still need to explore more (Wang et al. 2021).

14.4 Marine Macrophytes

Marine macrophytes (seaweeds and seagrasses) are foundation species, vital to the structure and function of benthic marine ecosystems, and are among the most productive coastal communities in the world (Mann 1973; Christie et al. 2009). Marine benthic macrophytes (macroalgae and seagrasses) are a good bioindicators to assess for the ecological quality status of the coastal and transitional waters. Marine benthic macrophytes are split into two groups by several biotic macrophytes indices, generally (Orfanidis et al. 2001; Ballesteros et al. 2007; Sfriso et al. 2007; Sfriso et al. 2009; Orfanidis et al. 2011; Taşkın et al. 2018). The first group the sensitive, late-successional macroalgae (*Cystoseira* spp., *Sargassum* spp., *Padina* spp., calcareous red algae, etc.) and seagrasses (*Posidonia oceanica*, *Cymodocea nodosa*, *Zostera* spp.), are found common in pristine waters (Boudouresque 1969; Ballesteros 1990; Pergent 1991), while the second group the tolerant and opportunistic algae (filamentous *Cladophora*, *Ceramium*, *Ectocarpus*, *Polysiphonia*, sheet-like *Ulva*, blue green algal layer, etc.) are commonly distributed in the human-impacted coastal waters (Orfanidis et al. 2003). Furthermore, macrophytes provide important ecosystem services (Costanza et al. 2014; Bennett et al. 2016) including coastal protection, nutrient cycling, carbon sequestration and food production upon which human society closely depends (Christie et al. 2009; Mineur et al. 2015; Bennett et al. 2016). Macrophytes are very important in marine system due to several reasons such as dissolution of organic carbon (Findlay et al. 1986), predator protection, heterogeneity, complex habitat and nutritional supply (Carpenter and Lodge 1986). Thus, macrophytes are consumed by herbivores, and putting the habitat of other organisms in danger such as green turtles lowered the fish community in grass meadows grazed (Arthur et al. 2013). The primary aquatic ecosystem production especially in small lakes was majorly contributed by submerged and emergent macrophytes (Downing et al. 2006; Brothers et al. 2013; Verpoorter et al. 2014).

Marine macrophytes such as macroalgae and seagrasses are contributing for country's economy as they are the most productive and diversified marine ecosystems on the planet (Larkum et al. 2006; Klinger 2015; Teagle et al. 2017). Annually US$ 28.9 per hectare revenue is developed by such ecosystem (Costanza et al. 2014). In Asian regions, the most popularly cultivated macrophyte is macroalgae with an economy contribution of US$ 13.3 billion (Hu et al. 2021; FAO 2020). Thus, macrophytes are now has major boost for bio-economy development because of its profitability especially in European and American countries (Grebe et al. 2019; Araújo et al. 2021).

14.5 Sources of Plastic in Aquatic Environment

14.5.1 Sources of Macroplastic in Aquatic Environment

Since 1990s, macroplastic is one of the major issues in all around the world and currently covered by different national and international policies and regulations. Macroplastic pollution was reported from the aquatic environment such as freshwater and marine water (Shomura and Godfrey 1990). Clean-up campaigns were focused mostly on the macroplastic which are found in the coastal regions and beaches. Due to the regular usage of macroplastic in routine life, it has been regularly ingested in the marine animals such as fish and seabirds which results into the

Table 14.3 Occurrence and levels of macroplastic debris found in surface water

Location	Region	Debris load	Item unit	Plastic (%)	Reference
Atlantic Ocean	Bay of Seine	1.7 ± 0.1	Per hectare	89%	Galgani et al. (2000); Li et al. (2016)
	Celtic Sea	5.3 ± 2.5	Per hectare	29.5%	Galgani et al. (2000); Li et al. (2016)
	Channel East	1.2 ± 0.1	Per hectare	84.6%	Galgani et al. (2000); Li et al. (2016)
	North Sea	1.6 ± 0.4	Per hectare	48.3%	Galgani et al. (2000); Li et al. (2016)
	Offshore, Ireland	2.5	Per m^3	94%	Lusher et al. (2014); Li et al. (2016)
Baltic Sea	Baltic Sea	1.3 ± 0.8	Per hectare	35.7%	Galgani et al. (2000); Li et al. (2016)
Mediterranean Sea	Adriatic Sea	3.8 ± 2.5	Per hectare	69.5%	Galgani et al. (2000); Li et al. (2016)
	Gulf of Lion	1.4 ± 0.2	Per hectare	70.5%	Galgani et al. (2000); Li et al. (2016)
Pacific Ocean	North Pacific Central Gyre	334,271	Per m^3	98%	Moore et al. (2001)
	North Pacific Central Gyre	0.02	Per m^3	-	Carson et al. (2013)
	Waters around Australia	4256.4	Per km^2	80%	Reisser et al. (2013)
	The South Pacific subtropical gyre	26,898	Per km^2	88.8%	Eriksen et al. (2013b)
	Waters around Australia	4256.4	Per km^2	80%	Reisser et al. (2013)

retention of this plastic into their bodies (Romeo et al. 2015; Derraik 2002; Teuten et al. 2007). Ocean plastic pollution was majorly contributed by rivers and annually 0.4–2.75 million tons of plastic is exported to ocean by rivers (Lebreton et al. 2017; Schmidt et al. 2018). Nets, surface booms, manta trawls and sampling of coastal residues in rivers used to quantify the macro- and mesoplastic (Browne et al. 2010; Moore et al. 2011; Gasperi et al. 2014; Lechner et al. 2014; Morritt et al. 2014; Cózar et al. 2015; Dris et al. 2015, 2018; Faure et al. 2015).

Being in large surface area, most of the floating rubbished plastic belongs to macroplastic, and it is the most serious concern. Table 14.3 showed a briefly compile data of location- and region-wise percentage of contaminating plastic ad its load.

Winds, coastal region, current and anthropogenic factor were mainly responsible for entry of plastic in sea or marine environments (Barnes et al. 2009). Moore et al. (2001) reported that, North Pacific subtropical gyre showed plastic pollution measurement of N33 particles per km^2 which were in between the 0.3 and 4.7 mm. These plastic particles include plastic films, PP and some of the pellets and pieces of Styrofoam. The factors affecting plastic accumulation and distribution of rubbish plastic are sedimentation due to low turbulence, human activities such as fishing and tourism places (Jegou and Salomon 1991; Kanehiro et al. 1995; Galgani et al. 2000).

14.5.2 Sources of Microplastic in Aquatic Environment

Recently various researchers are focussing on microplastic and its environmental pollution in aquatic biosphere (Reisser et al. 2013; Eriksen et al. 2013a; Desforges et al. 2015). Plastic have floating property called buoyancy which is the function of specific gravity. Due to the anthropogenic activities, the occurrence of microplastic is abundant in coastal areas and is the major source of marine water contamination (Ribic et al. 2010; Doyle et al. 2011; Collignon et al. 2012; Dubaish and Liebezeit 2013). Table 14.4 showed a briefly compile data of location and region-wise percentage of contaminating plastic and its load.

Thus, such plastic are generally developed in high population area with heavy industrial activities carried out such as coal-based power plant sand blasting area (Eriksen et al. 2013a). Also some of the degradation product of plastic such as PP and PE were considered as microplastic which reported to contaminate the surface water in highly populated cities -Sydney and Brisbane of Australia (Reisser et al. 2013).

Table 14.4 Occurrence and levels of microplastic debris found in surface water

Location	Region	Debris Load	Items Unit	Plastic (%)	Reference
Atlantic Ocean	Offshore, Ireland	2.5	Per m^3	94%	Lusher et al. (2014), Li et al. (2016)
	Portuguese coast	0.02–0.04	Per m^3	43.8–91.7%	Frias et al. (2014), Li et al. (2016)
Pacific Ocean	North Pacific Central Gyre	334,271	Per km^2	98%	Moore et al. (2001)
	North Pacific Central Gyre	0.02	Per m^3	–	Carson et al. (2013)
	Waters around Australia	4256.4	Per km^2	80%	Reisser et al. (2013)
	The South Pacific subtropical gyre	26,898	Per km^2	88.8%	Eriksen et al. (2013b)
	Waters around Australia	4256.4	Per km^2	80%	Reisser et al. (2013)
	NE Pacific Ocean	8–9180	Per m^3	75%	Desforges et al. (2015)
	Geoje Island, South Korea	16,000	Per m^3	–	Song et al. (2014)

The plastic debris are altered in marine environment by formation of algal or bacterial biofilm on plastic surface and thus change its densities which allowed some plastic to float of surfaces or suspend in sediment (Lobelle and Cunliffe 2011). The microplastic entered into aquatic system directly or indirectly by consumer products, UV radiation, mechanical shear and biological degradation of larger plastic material and is a serious concern from environmental risk assessment point of view (Mattsson et al. 2015a, 2015b). Research studies have most dominantly reported for the presence of microplastic and its trees in various marine habitat (Cole et al. 2011), freshwater samples (Baldwin et al. 2016) and even in some of the mountain lakes (Imhof et al. 2013). Apart from this some reports showed presence of microplastic in freshwater lakes (Sighicelli et al. 2018), oceans (Zhang et al. 2017a, 2017b) and rivers (Anderson et al. 2017). Thus, the microplastic is most commonly accumulated in freshwater or marine organism which is moved from terrestrial area to aquatic area (Horton et al. 2017; Sharma and Chatterjee 2017; Alimi et al. 2018). Another important source of microplastic which directly entered into water stream is the personal care products (Baldwin et al. 2016; Bhattacharya et al. 2016; Carr et al. 2016). These microplastic were analysed in the form of different shapes, size (0.5–1 mm) and irregular fragments of microbeads containing PE monomers (Fendall and Sewell 2009). Most of these microbeads are retained in waste water treatment plants (WWTPs) but still considered as microplastic pollution (Carr et al. 2016; Mintenig et al. 2019). Moreover, waste water entered into freshwater stream by leakages, percolation or use of sewage-based fertilizer (Bhattacharya et al. 2016). Microplastic analysis profile of WWTP effluent showed blue PE particles that are generally used in toothpaste (Carr et al. 2016). Bhattacharya et al. (2016) reported that the sewage overflow and run-off of agricultural fertilizers are main route of microplastic in the freshwater. It was also reported that the polyethylene was predominant material Rhine, Dalålven, Danube and Po Rivers of European country (SFRA0025 2015).

14.5.3 Sources of Nanoplastic in Aquatic Environment

Klaine et al. (2012) define that if the size of plastic particle if below than 100 nm, then it is nanosized plastic. There are some reasons for the formation of nanoplastic (a) breakdown of microplastic, (b) degradation of plastic material or objects and (c) degradation of industrial plastic material in manufacturing process. Like microplastic, it may have the primary or secondary origin. Some of the beauty or cosmetics products, medicines and creams are a source of Nps into the soils and water, and it was also reported that the most of these products were removed from waste water in treatment plants of waste water (Mateos-Cárdenas et al. 2019). Biodegradation, hydrolysis, mechanical abrasion, photodegradation and UV radiations are coming under the environmental weathering processes. Due to the different environmental processes, larger plastic particle degrades in the environment, and this is the secondary origin of nanoplastic. It was also reported that other sources such agricultural wastes (plastic mulch), textile industry (microfibre), and rubber industries (car tyre) contributed as biggest plastic source in the ocean (Fig. 14.2). Nanosized plastic easily transported through soil, air and water medium, so it is very difficult to detect (Wieczorek et al. 2018).

14.6 Ecotoxicological Impact of Macro-, Micro- and Nanoplastic on Aquatic Macrophytes

Plastic debris ranging from macroplastic to smallest nanoplastic particles can pose a threat to marine fauna through both direct and indirect way of absorption (Ward and Shumway 2004). Bouwmeester et al. (2015) reported that various researcher reported the impact of polystyrene microplastic on vertebrates and invertebrates of freshwater and marine water. Mazurais et al. (2015) reported that a very rare data of polythene Mps. The effect of these plastic particles on the plants is under study; only reports on the adverse impact of these plastic on phytoplanktons were available. A negative impact of nanoplastic (polystyrene) on the photosynthesis and increase in the ROS (reactive oxygen species) formation were observed in the *Chlorella* and *Scenedesmus* sp. (Bhattacharya et al. 2010) (Table 14.5).

A reduction in the growth of *Dunaliella tertiolecta*, *Thalassiosira pseudonana* and *Chlorella vulgaris* were observed due to the particle of polystyrene nanoplastic (Sjollema et al. 2016). Less reports were available currently on the impact of microplastic on higher aquatic plants. Eerkes-Medrano et al. (2015) reported that the microplastic (<1 g per cm^3) abundantly presents on the water surface. This density of microplastic majorly affects on the floating plants because their body is permanently in contact with this plastic because of movement of water surface (Prokin et al. 2015).

Different members of Lemnaceae family are very important for food for different marine animals (fishes), and these microplastic are found habiting places on these floating duckweeds (Zhang et al. 2015; Van Hoeck et al. 2015). Few reports were available on the microplastic interaction with floating plants, but some more investigation is necessary because of the importance of these pants in water ecosystem. Fendall and Sewell (2009) reported that there are no any effects of microbeads found on the *Lemna minor* (Duckweed). 0.02 mm concentration of nanoplastic (polystyrene) reduces the photosynthesis of different algae (Bhattacharya et al. 2010). It is anticipated that microplastic block the access of light by algae and microbeads affect the root growth. Decreases in the root length were observed as root surface absorbed the microbeads because of the electrostatic forces the charged particle (positive) attracted to the cellulose, and it also increases the roughness on plant surfaces. With this it also negatively effects on the permeability of the root cell membrane. The shape and surface roughness are key factors of mechanical damage to organisms. The PE microplastic which are present in the water showed negative effect on the macrophyte *Lemna minor* roots. Change in the colour (slightly yellowish) when comparing the control treatment was observed when treating these shoots with 3% nanoplastic solution (van Weert et al. 2019). Microplastic showed negative impact on the DW (dry reight) of *Elodea* sp. *M. spicatum*. In freshwater ecosystems,

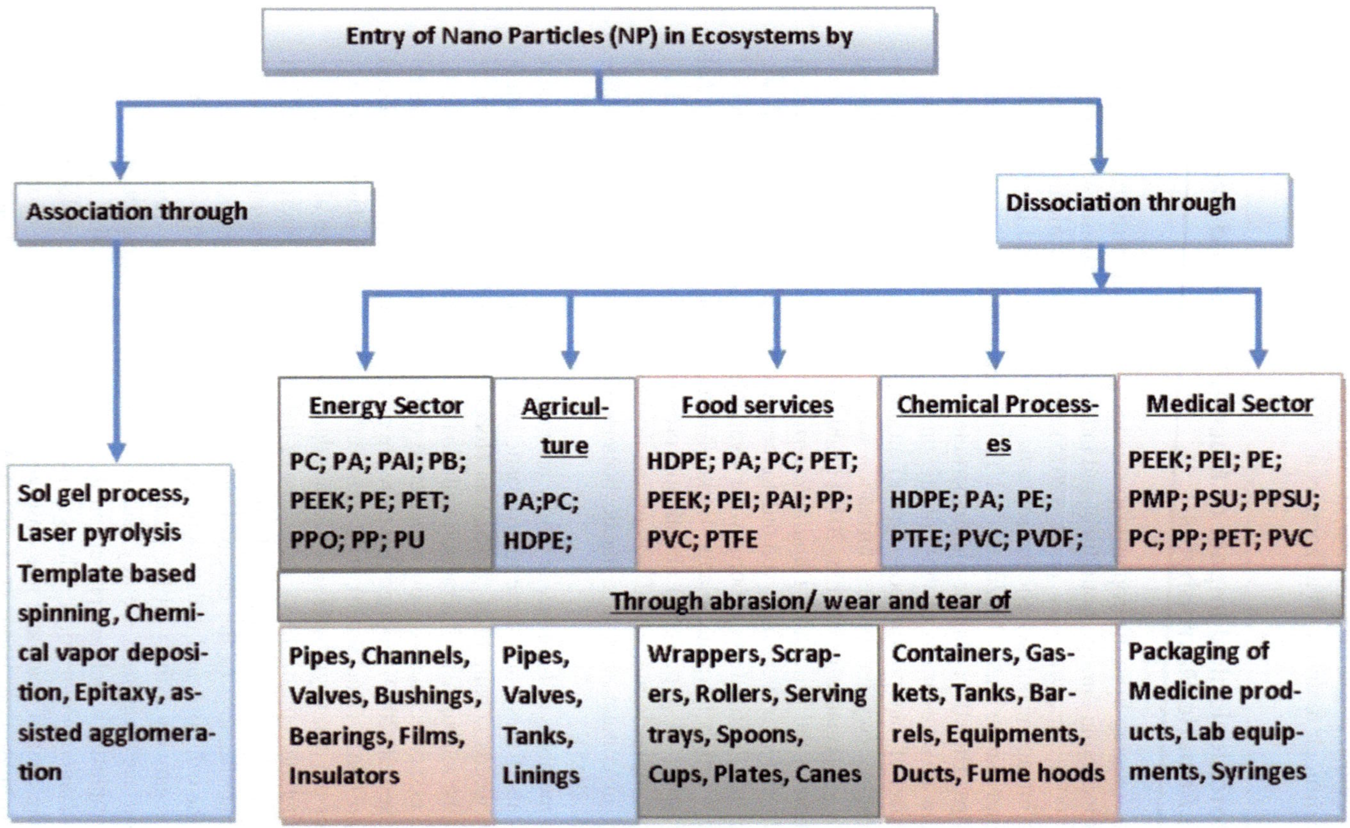

Fig. 14.2 Applications of plastic polymers and productions of NPs from various fields. *Abbreviations: PA* Polyamide, *PB* Polybutylene, *PC* Polycarbonate, *PE* Polyethylene, *PI* Polyimide, *PP* Polypropylene, *PS* Polystyrene, *PU* Polyurethane, *PAI* Polyamide imide, *PBT* Polybutylene terephthalate, *PEI* Polyester imide, *PET* Polyethylene terephthalate, *PMP* Polymethylpentene, *PPO* Polyphenylene oxide, *PPS* Polyphenylene sulphide, *PSU* Polysulfone, *PVC* Polyvinyl chloride, *CPVC* Chlorinated polyvinyl chloride, *HDPE* High-density polyethylene, *LDPE* Low-density polyethylene, *PEEK* Polyether ether ketone, *PPSU* Polyphenylsulfone, *PTFE* Polytetrafluoroethylene, *PVDF* Polyvinylidene fluoride

Table 14.5 Impact of plastic (micro and nano) on algal members

Microalgae	Polymer type	Particle size	Used conc.	EC_{50}/ IC50 value	Tested parameters	Effects on algae	References
Chlorella	PS	20 nm	0.08–0.8 mg/mL	Not applicable	(i) Cellular respiration (ii) Production of ROS	(i) Photosynthesis of algae is found to be inhibited (ii) Adsorption of plastic shows ROS production	Sjollema et al. (2016), Zhang et al. (2017a, 2017b)
Dunaliella tertiolecta	PS	0.05 μm, 0.5 μm and 6 μm	25, 250 mg/L	Not applicable	(i) Effect on photosynthesis (ii) Growth	(i) At high conc. (250 mg/L) resulted negatively on growth (ii)Slight effect found on photosynthesis	Sjollema et al. (2016)
Microcystis aeruginosa and *Dolichospermum flos-aquae*	N/A	<200 μm	66.7 mg/L	Not applicable	(i)Size of the colony (ii) Abundance	(i) No effect found on the growth. (ii) No effect on biomass (iii) It was seen that the size of algal particle is decreased at specific point of life cycle	Yokota et al. (2017)
Pseudokirchneriella subcapitata	PS	110 nm	0–100 mg/L	>100 mg/L (72 h)	(i) Only effect on the growth was measures	(I) The particle size of PS-COOH particles (110 nm) are negatively affected on growth at 10 mg/L.	Casado et al. (2013)
	PS-PEI	55 nm and 110 nm	0–100 mg/L	0.54 and 0.58 μg/L (72 h)	(i) Effect on rate of growth	(i) 110 nm particle size was found more toxic as compared to 55 nm (ii) Decrease in the growth was noticed	Casado et al. (2013)
Scenedesmus obliquus	PS	~70 nm	44–1100 mg/L	N/A	(i) Total biomass content (ii) Photosynthesis	(i) Decrease in total content of chlorophyll was seen	Zhang et al. (2017a, 2017b)
Tetraselmis chui	PS	1–5 μm	0.02–0.64 m/L	0.145 mg/L (72 h)	(i) Rate of growth	(i) Significant effect was not found on microplastic (ii) Significant difference was not observed on the toxicity	Yokota et al. (2017)

reduction in the length of shoots of *M. spicatum* was reported because of microplastic (Yu et al. 2020; van Weert et al. 2019). Nutrition adsorbed through the bladders in *Utricularia vulgaris* as this plant is rootless. Kibriya and Iwan Jones (2007) reported the highest concentration of microplastic debris in bladders of *Utricularia vulgaris*. Because of this microplastic accumulation, *U. vulgaris* difficulty in nutrition absorption was seen. It was concluded that the growth rate of *U. Vulgaris* depends on the dose (concentration) of microplastic (Sirova et al. 2009). Toxicity of plastic in the plant mainly resulted in the total number of green leaves into yellow leaves and decreasing chlorophyll content (Du et al. 2015; Miralles et al. 2012). If the dose of microplastic increases, it changes the total chlorophyll and protein which resulted the reduce chlorophyll content (Franzellitti et al. 2019; Kim et al. 2019). Because of shading, adsorption of plastic to the body of algae and aggregation of this plastic in the root showed negative impact on the photosynthesis of algae (Zhang et al. 2016). The injury to different parts of different plant was observed because of excess amount of ROS species (Jiang et al. 2019). By increasing the concentration of microplastic, decrease in the content of soluble protein was reported. The reason behind this is microplastic alters the expression of gene which is required for synthesis of protein during the different stages of growth (Singh et al. 2006; Mushtaq et al. 2008). Yu et al. (2020) reported the increase in the content of malondialdehyde (MDA) when treating with 70–140 mg L^{-1} of 1 and 2 mm microplastic. Higher stress condition resulted damage and altered lipid peroxidation in *U. vulgaris*.

Nanoplastic can affect the aquatic environment through different ways starting from phytoplanktons their growth and photosynthesis, growth and zooplankton reproduction and finally the ocean carbon stocks (Wieczorek et al. 2018). In energy flow and transformation of genetic information by the food chain, marine zooplankton is important which eventually affect particulate organic carbon (POC) and its extent of relation with other biogeochemical cycles. Furthermore, nanoplastic influence the CO_2 cycling in marine water (Shen et al. 2019a) and eventually perturb marine productivity and potential for carbon sequestration. Disruption in the sequestration of carbon dioxide (CO_2) and marine productivity changes the global climate in various magnitudes (Shen et al. 2019b; Fig. 14.3).

Plastic can affect plants externally or internally. Macrophytes have the ability to adsorb and/or internalize NPs particularly in the form of aggregates. The impact of NPs on aquatic ecosystem and various organisms was studied previously; presently adsorption of NPs in marine environments is largely overlooked (Bosker et al. 2019; Yuan et al. 2019). Though, some research work has found that cellulose present in the cell walls of green plants and algae acts as an initial point for particle introduction into the food web in marine environments. Nanoplastic also aggregates with itself or with other biological organic matter as illustrated in Fig. 14.3 in the surrounding environment (Mattsson et al. 2015a, 2015b). The available information recommends that size-dependent properties remain same after the particles have entered the organisms. Plastic nanoparticles are the most common to be adsorbed by different marine macrophytes (Li et al. 2020). Different rates of adsorption and/or localization of tissue have found to be reported for a variety of species like *Cladophora* sp.,

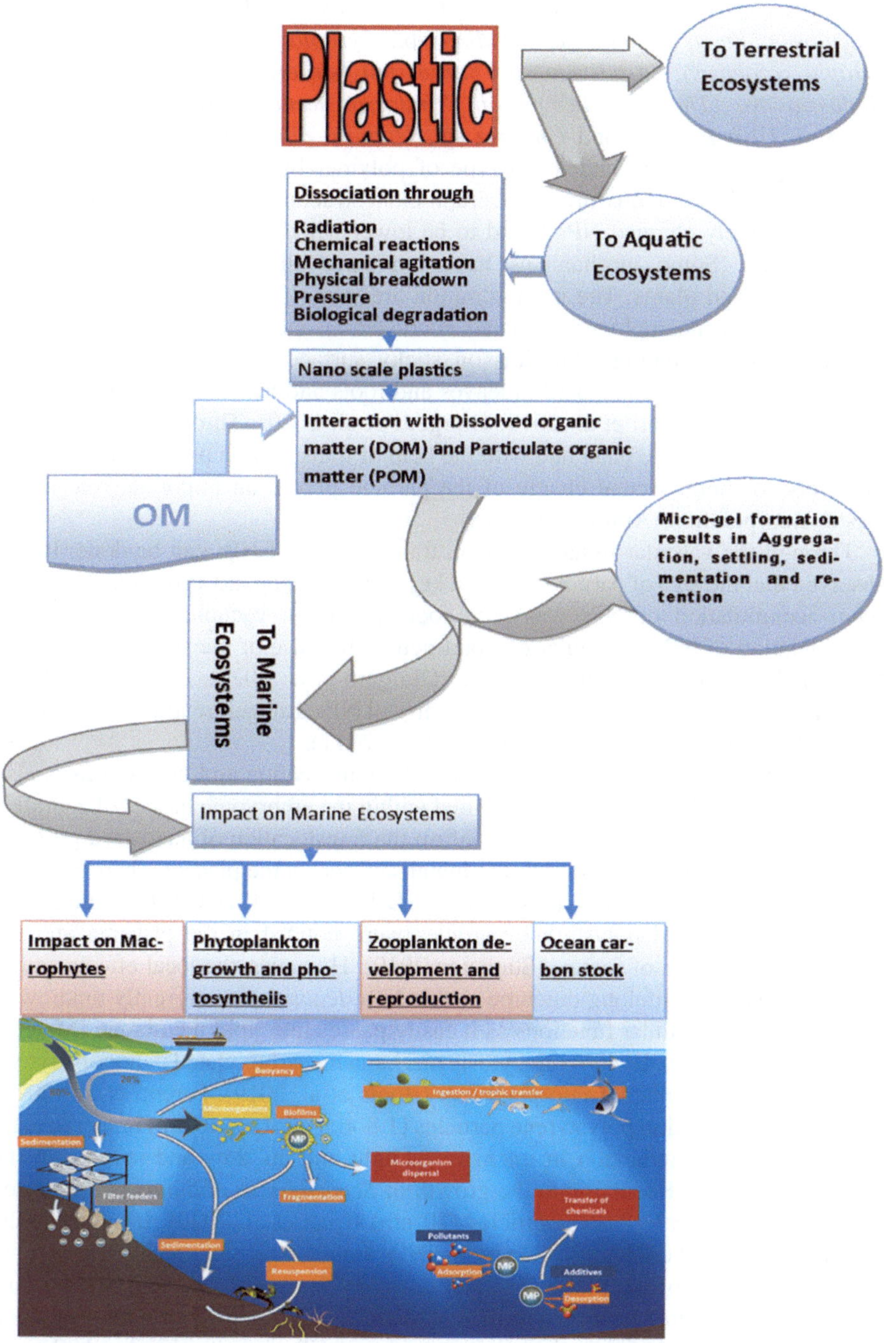

Fig. 14.3 Degradation of plastic from bulk through macro- and micro- to nanosize and entering food chains through various routes and mechanisms

Cymodocea sp. (*Cymodocea rotundata, Cymodocea serrulata*), *Fucus* sp. (*Fucus vesiculosus, Fucus vesiculosus*), *Padina* sp., *Pyropia* spp., (*Pyropia yezoensis, Pyropia yezoensis*), *Sargassum* sp. (*Sargassum horneri, Sargassum ilicifolium*), *Thalassia* sp. (*Thalassia hemprichii, Thalassia testudinum*), *Ulva* sp. (*Ulva prolifera, Ulva pertusa*) and *Undaria* sp. (*Undaria pinnatifida*). Adsorption processes might not be dependent on type of polymer but also depends on further properties of plastic in mixture with certain plant structures. Uptake and internalization of positively charged NPs found to be lower when compared with negatively charged NPs. Some mechanisms have explained the attachment of NPs to macrophytes and plants. The most common are (a) surface adsorption, (b) surface entrapment, (c) macrophyte-associated encrustation and entrapment by epibionts and (d) bacterial coatings (biofilms on sticky surface). Engineered NPs can be taken up by the plant and stored in leaves and roots and translocate to other part of plant or tissues (Parenti et al. 2019; Giorgetti et al. 2020). Through the root stele, these beads enter in the plant and with the help of transpiration pooling it to transfer from root to shoot. Actual charge of the particle played an important role in the plastic incorporation (Alimi et al. 2018).

Through the vascular tissues of plant, translocation of NPs can be done. It was observed that the size of particle played a key role in plastic incorporation. In *Allium cepa*, accumulation of NPs was seen root epidermis: vacuoles and cytoplasm. Accumulation of NPs in cellular compartment is because of the small size of these NPs which can pass the cell membrane (Giorgetti et al. 2020). In root cell of rice (*Oryza sativa*) plant, aquaporins increase entry of NPs. Longer treatment resulted the partial shutdown of these channels (Zhou et al. 2011).

Secondary toxicity associated with particle translocation and depositions is also documented, and NPs have been identified within the embryonic tissues of zebrafish (Parenti et al. 2019). Recent studies confirm the translocation of nanosized polystyrene particles from the maternal to the foetal sac, across the placental barrier within 70 min of injection into the uterine artery (D'Errico et al. 2019). Further, 20 nm fluorescently labelled polystyrene particles were noticed in the placenta and the heart, liver and brain of foetus, (Stapleton 2019). However, the local effects within foetal tissues or the lifelong outcome of this NPs deposition is currently unknown.

NPs affect on cellular functions as it easily passes into lipid membrane (Mateos-Cárdenas et al. 2019). By the endocytosis mechanism, carboxylated nanopolystyrene (40–50 nm) easily entered into plant cell (Boots et al. 2019). Parenti et al. (2019) showed that adsorption polystyrene particles (1.8–6.5 mg/L of 20 nm) hindered algal photosynthesis by reduction in intensity of light and air flow and stimulated the formation of ROS. Reduction in the growth of *Scenedesmus obliquus* was reported when exposed to *polystyrene* particles (70 nm) (Lian et al. 2020). It indicates the inhibition chemical communication by the NPs among different species. To find out the exact location of the plastic, we can use plant and algae as bioindicator. This implies that macrophytes act as potential vectors which allow NPs into the food webs. Some studies have confirmed the transfer of NPs from freshwater duckweed *L. minor* to amphipods (Mateos-Cárdenas et al. 2019), from *Fucus vesiculosus* to the periwinkle *Littorina littorea* (Gutow et al. 2015). The NPs showed toxicity on biomass production, elongation growth, germination and photosynthesis, and other effects are reported in different magnitudes like eustress, distress, no effects or mixed effects. After exposure of NPs, eustress refers to a positive response, while distress for damage and mixed effects cited for eustress and distress responses. It has been observed that highest (ethylene propylene diene) doses had negative effects on growth elongation of *P. lanceolata,* whereas the lowest doses had positive effect. Thus, it has been suggested that the different concentration showed different responses and effects on plants. An example of NPs showed disturbance in *L. sativum* germination by blocking the seed capsule pores (Yuan et al. 2019). Variation in NPs exposure not only showed positive effect of growth but also have negative effects especially in plant roots. Total 21 days exposure of polystyrene nanoparticles (0.01–10 mg L^{-1} to 100 nm) on *Triticum aestivum* showed decreased shoot to root ratio (Lian et al. 2020). Polyethylene NPs showed a eustress response in *Lolium perenne* by increasing the root biomass after the exposure period of 30 days (Boots et al. 2019). Plant biomass growth is the function of nanoplastic polymer type, size or shape and variable concentration and exposure time. The plant root growth may increase because of competition for nutrient, water and the root growth may be decreased by contamination of NPs.

Some studies (Dovidat et al. 2019; Mateos-Cárdenas et al. 2019) reported external adsorption and aggregation of NPs has no effect of plant photosynthetic efficiency, but some reports showed negative photosynthetic efficacy. Similarly, decreased chlorophyll content in carnivorous plant *Utricularia vulgaris* was found by exposure of 1–5 μm sized polystyrene nanoplastic (Yu et al. 2019). Hence, based on a limited number of studies, it appears that impacts by NPs on photosynthesis are variable, with little concrete information on the reasons. In summary, the limited literature provides some evidence of effects of NPs to marine.

14.7 Conclusions and Future Recommendation

Plastic is the most damaging polymer which has macro-, micro- and nanoplastic classified based on its size. These are non-degrading plastic, but it can be fragmented by environmental factor, and the fragmented polymeric matter percolates in water stream and ended in ocean contamination. Macroplastic floats on surfaces; however, micro and nano entered into the marine ecosystem and macrophytes. Resultant contamination spoils the marine environment and toxicological impact on macrophytes.

To control plastic pollution, a combination of complete research attention with feasible lifestyles is necessary. Using the degradable plastic as a daily habit valued a large contribution to face the plastic issue. In sustainable lifestyles; through encouraging the people (use of reusable plastic, counselling to avoid or reduce the maximum plastic use and giving the proper environmental education to the people) can be improved to mitigate plastic waste. Government policies should be framed for the use of biodegradable and single-use plastic. Additionally, specific strategies should be made for the production of single-use plastic. Individual, institutional and government/semi-government or private sector effort will be definitely helpful to reduce this environmental issue.

Acknowledgement We acknowledge the Dr. Avinash B. Ade (Professor and Head, Department of Botany, Savitribai Phule Pune University) and Dr. T. D. Nikam (Emeritus Scientist, Department of Botany, Savitribai Phule Pune University) for their constant support and valuable guidance for writing present book chapter. Mohd Shahnawaz is thankful to the authorities of the Jiangsu University for providing an opportunity to work as an Associate Professor under invited Talent Programme.

References

Alimi OS, Farner Budarz J, Hernandez LM, Tufenkji N (2018) Microplastics and nanoplastics in aquatic environments: aggregation, deposition, and enhanced contaminant transport. Environ Sci Technol 52:1704–1724

Anderson PJ, Warrack S, Langen V, Challis JK, Hanson ML, Rennie MD (2017) Microplastic contamination in Lake Winnipeg, Canada. Environ Pollut 225:223–231

Andrady AL (2003) Plastics and the environment. John Wiley and Sons, Research Triangle Park, NC. ISBN: 0-471-09520-6

Andrady AL (2011) Microplastics in the marine environment. Mar Pollut Bull 62(8):1596–1605

APM (2011) Association of plastic manufacturers: plastics—the facts 2011: an analysis of european plastic production, demand and recovery for 2010. Plastic Europe, Brussels

Araújo R, Vázquez Calderón F, Sánchez López J, Azevedo IC, Bruhn A, Fluch S et al (2021) Current status of the algae production industry in Europe: an emerging sector of the blue bioeconomy. Front Mar Sci 7:626,389

Arthur CA, Baker J, Bamford H (2009) Proceedings of the international research workshop on the occurrence, effects, and fate of microplastic marine debris, September 9–11. NOAA Technical Memorandum NOS-OR&R-30, 2009, Washington

Arthur R, Kelkar N, Alcoverro T, Madhusudan MD (2013) Complex ecological pathways underlie perceptions of conflict between green turtles and fishers in the Lakshadweep Islands. Biol Conserv 167:25–34

Baldwin AK, Corsi SR, Mason SA (2016) Plastic debris in 29 Great Lakes tributaries: relations to watershed attributes and hydrology. Environ Sci Technol 50(19):10,377–10,385

Ballesteros E (1990) Structure and dynamics of the *Cystoseira caespitosa* (Fucales, Phaeophyceae) community in the North-Western Mediterranean. Sci Mar 54:155–168

Ballesteros E, Torras X, Pinedo S, García M, Mangialajo L, De Torres M (2007) A new methodology based on littoral community cartography dominated by macroalgae for the implementation of the European Water Framework Directive. Mar Pollut Bull 55(1-6):172–180

Barnes DK, Galgani F, Thompson RC, Barlaz M (2009) Accumulation and fragmentation of plastic debris in global environments. Philos Trans R Soc B Biol Sci 364(1526):1985–1998

Barría C, Brandts I, Tort L, Oliveira M, Teles M (2020) Effect of nanoplastics on fish health and performance: a review. Mar Pollut Bull 151:110,791

Bennett S, Wernberg T, Connell SD, Hobday AJ, Johnson CR, Poloczanska ES (2016) The 'Great Southern Reef': social, ecological and economic value of Australia's neglected kelp forests. Mar Freshw Res 67(1):47–56

Betts K (2008) Why small plastic particles may pose a big problem in the oceans. Environ Sci Technol 42:89–95

Bhattacharya M, Paramati SR, Ozturk I, Bhattacharya S (2016) The effect of renewable energy consumption on economic growth: evidence from top 38 countries. Appl Energy 162:733–741

Bhattacharya P, Lin S, Turner JP, Ke PC (2010) Physical adsorption of charged plastic nanoparticles affects algal photosynthesis. J Phys Chem C 114(39):16,556–16,561

Blettler MC, Ulla MA, Rabuffetti AP, Garello N (2017) Plastic pollution in freshwater ecosystems: macro-, meso-, and microplastic debris in a floodplain lake. Environ Monit Assess 189(11):1–3

Boots B, Russell CW, Green DS (2019) Effects of microplastics in soil ecosystems: above and below ground. Environ Sci Technol 53:11,496–11,506

Bosker T, Bouwman LJ, Brun NR, Behrens P, Vijver MG (2019) Microplastics accumulate on pores in seed capsule and delay germination and root growth of the terrestrial vascular plant Lepidium sativum. Chemosphere 226:774–781

Boudouresque CF (1969) Etude qualitative et quantitative d'un peuplement algal à *Cystoseira mediterranea* dans la région de Banyuls sur Mer. Vie et Milieu 20:437–452

Bouwmeester H, Hollman PC, Peters RJ (2015) Potential health impact of environmentally released micro-and nanoplastics in the human food production chain: experiences from nanotoxicology. Environ Sci Technol 49(15):8932–8947

Boyle K, Örmeci B (2020) Microplastics and nanoplastics in the freshwater and terrestrial environment: a review. Water 12(9):2633

Bradney L, Wijesekara H, Palansooriya KN, Obadamudalige N, Bolan NS, Ok YS, Rinklebe J, Kim KH, Kirkham MB (2019) Particulate plastics as a vector for toxic trace-element uptake by aquatic and terrestrial organisms and human health risk. Environ Int 131:104,937

Brothers SM, Hilt S, Meyer S, Köhler J (2013) Plant community structure determines primary productivity in shallow, eutrophic lakes. Freshw Biol 58(11):2264–2276

Browne M, Galloway T, Thompson R (2007) Microplastic—an emerging contaminant of potential concern? Integrated Environ Assess Manag Int J 3:559–561

Browne MA, Galloway TS, Thompson RC (2010) Spatial patterns of plastic debris along estuarine shorelines. Environ Sci Technol 44:3404–3409

Carpenter EJ, Anderson SJ, Harvey GR, Miklas HP, Peck BB (1972) Polystyrene spherules in coastal waters. Science 178(4062):749–750

Carpenter SR, Lodge DM (1986) Effects of submersed macrophytes on ecosystem processes. Aquat Bot 26:341–370

Carr SA, Liu J, Tesoro AG (2016) Transport and fate of microplastic particles in wastewater treatment plants. Water Res 91:174–182

Carson HS, Nerheim MS, Carroll KA, Eriksen M (2013) The plastic-associated microorganisms of the North Pacific Gyre. Mar Pollut Bull 75(1–2):126–132

Casado MP, Macken A, Byrne HJ (2013) Ecotoxicological assessment of silica and polystyrene nanoparticles assessed by a multitrophic test battery. Environ Int 51:97–105

Chae Y, An YJ (2020) Nanoplastic ingestion induces behavioral disorders in terrestrial snails: trophic transfer effects via vascular plants. Environ Sci Nano 7(3):975–983

Chen G, Feng Q, Wang J (2020) Mini-review of microplastics in the atmosphere and their risks to humans. Sci Total Environ 703:135,504

Christie H, Norderhaug KM, Fredriksen S (2009) Macrophytes as habitat for fauna. Mar Ecol Prog Ser 396:221–233

Coe J, Rogers D (1996) Marine debris sources, impacts, and solutions. Springer, New York

Cole M, Lindeque P, Halsband C, Galloway TS (2011) Microplastics as contaminants in the marine environment: a review. Mar Pollut Bull 62(12):2588–2597

Collignon A, Hecq JH, Glagani F, Voisin P, Collard F, Goffart A (2012) Neustonic microplastic and zooplankton in the North Western Mediterranean Sea. Mar Pollut Bull 64(4):861–864

Colton JB, Knapp FD (1974) Plastic particles in surface waters of the northwestern Atlantic. Science 185:491–497

Corcoran PL, Beisinger MC, Grifi M (2009) Plastic and beaches: a degrading relationship. Mar Pollut Bull 58:80–84

Costanza R, De Groot R, Sutton P, Van der Ploeg S, Anderson SJ, Kubiszewski I, Farber S, Turner RK (2014) Changes in the global value of ecosystem services. Glob Environ Chang 26:152–158

Cózar A, Echevarria F, Gonzalez-Gordillo JI, Irigoien X, Ubeda B, Hernandez-Leon S et al (2014) Plastic debris in the open ocean. Proc Natl Acad Sci 111:10,239–10,244

Cózar A, Martí E, Duarte CM, García-de-Lomas J, Van Sebille E, Ballatore TJ, Eguíluz VM, González-Gordillo JI, Pedrotti ML, Echevarría F, Troublè R (2017) The Arctic Ocean as a dead end for floating plastics in the North Atlantic branch of the thermohaline circulation. Sci Adv 3(4):e1600582

Cózar A, Sanz-Martín M, Martí E, González-Gordillo JI, Ubeda B, Gálvez J et al (2015) Plastic accumulation in the mediterranean sea. PLoS One 10:339–347

Crawford B, Quinn B (2017) Microplastics pollutants. Elsevier, Amsterdam, 330

D'Errico JN, Doherty C, Fournier SB, Renkel N, Kallontzi S, Goedken M, Fabris L, Buckley B, Stapleton PA (2019) Identification and quantification of gold engineered nanomaterials and impaired fluid transfer across the rat placenta via ex vivo perfusion. Biomed Pharmacother 117: 109,148

da Costa JP, Reis V, Paço A, Costa M, Duarte AC, Rocha-Santos T (2019) Micro (nano) plastics—analytical challenges towards risk evaluation. TrAC Trends Anal Chem 111:173–184

Derraik JGB (2002) The pollution of the marine environment by plastic debris: a review. Mar Pollut Bull 44:842–852

Desforges JP, Galbraith M, Ross PS (2015) Ingestion of microplastics by zooplankton in the Northeast Pacific Ocean. Arch Environ Contam Toxicol 69(3):320–330

Dovidat LC, Brinkmann BW, Vijver MG, Bosker T (2019) Plastic particles adsorb to the roots of freshwater vascular plant Spirodela polyrhiza but do not impair growth. Limnol Oceanogr Lett

Downing JA, Prairie YT, Cole JJ, Duarte CM, Tranvik LJ, Striegl RG, McDowell WH, Kortelainen P, Caraco NF, Melack JM, Middelburg JJ (2006) The global abundance and size distribution of lakes ponds, and impoundments. Limnol Oceanogr 51:2388–2397

Doyle MJ, Watson W, Bowlin NM, Sheavly SB (2011) Plastic particles in coastal pelagic ecosystems of the Northeast Pacific ocean. Mar Environ Res 71(1):41–52

Dris R, Gasperi J, Rocher V, Tassin B (2018) Synthetic and non-synthetic anthropogenic fibers in a river under the impact of Paris Megacity: sampling methodological aspects and flux estimations. Sci Total Environ 618:157–164

Dris R, Imhof H, Sanchez W, Gasperi J, Galgani F, Tassin B, Laforsch C (2015) Beyond the ocean: contamination of freshwater ecosystems with (micro-) plastic particles. Environ Chem 12:539–550

Du W, Gardea-Torresdey JL, Ji R, Yin Y, Zhu J, Peralta-Videa JR, Guo H (2015) Physiological and biochemical changes imposed by CeO2 nanoparticles on wheat: a life cycle field study. Environ Sci Technol 49(19):11,884

Dubaish F, Liebezeit G (2013) Suspended microplastics and black carbon particles in the jade system, Southern North Sea. Water Air Soil Pollut 224:1352

Eerkes-Medrano D, Thompson RC, Aldridge DC (2015) Microplastics in freshwater systems: a review of the emerging threats, identification of knowledge gaps and prioritisation of research needs. Water Res 75:63–82

Eriksen M, Mason S, Wilson S, Box C, Zellers A, Edwards W, Farley H, Amato S (2013a) Microplastic pollution in the surface waters of the Laurentian Great Lakes. Mar Pollut Bull 77:177–182

Eriksen M, Maximenko N, Thiel M, Cummins A, Lattin G, Wilson S, Hafner J, Zellers A, Rifman S (2013b) Plastic pollution in the South Pacific subtropical gyre. Mar Pollut Bull 68:71–76

EU Commission (2011) Commission recommendation of 18 October 2011 on the Definition of Nanomaterial (2011/696/EU), Off J Eur Union L275, 2011. https://ec.europa.eu/research/industrial_technologies/pdf/policy/commission-recommendation-on-the-definition-of-nanomater-18102011_en.pdf.

FAO (2020) The state of world fisheries and aquaculture 2020. Food and Agriculture Organization of the United Nations, Rome

Faure F, Demars C, Wieser O, Kunz M, De Alencastro LF (2015) Plastic pollution in Swiss surface waters: nature and concentrations, interaction with pollutants. Environ Chem 12:582–591

Fendall LS, Sewell MA (2009) Contributing to marine pollution by washing your face: microplastics in facial cleansers. Mar Pollut Bull 58:1225–1228

Findlay S, Carlough L, Crocker MT, Gill HK, Meyer JL, Smith PJ (1986) Bacterial growth on macrophyte leachate and fate of bacterial production. Limnol Oceanogr 31:1335–1341
Franzellitti S, Canesi L, Auguste M, Wathsala RH, Fabbri E (2019) Microplastic exposure and effects in aquatic organisms: a physiological perspective. Environ Toxicol Pharmacol 68:37–51
Fraternali F, Ciancia V, Chechile R, Rizzano G, Feo L, Incarnato L (2011) Experimental study of the thermo-mechanical properties of recycled PET fiber-reinforced concrete. Compos Struct 93(9):2368–2374
Frias JP, Otero V, Sobral P (2014) Evidence of microplastics in samples of zooplankton from Portuguese coastal waters. Mar Environ Res 95:89–95
Fu D, Chen CM, Qi H, Fan Z, Wang Z, Peng L, Li B (2020) Occurrences and distribution of microplastic pollution and the control measures in China. Mar Pollut Bull 153:110,963
Galgani F, Leaute JP, Moguedet P, Souplet A, Verin Y, Carpentier A, Goraguer H, Latrouite D, Andral B, Cadiou Y, Mahe JC (2000) Litter on the sea floor along European coasts. Mar Pollut Bull 40:516–527
Ganesh Kumar A, Anjana K, Hinduja M, Sujitha K, Dharani G (2020) Review on plastic wastes in marine environment—biodegradation and biotechnological solutions. Mar Pollut Bull 150: 110,733
Gasperi J, Dris R, Bonin T, Rocher V, Tassin B (2014) Assessment of floating plastic debris in surface water along the Seine River. Environ Pollut 195:163–166
GESAMP (2016) Sources, fate and effects of microplastics in the marine environment: part two of a global assessment. In: Kershaw PJ, Rochmann CM (eds). IMO/FAO/UNESCO-IOC/UNIDO/ WMO/IAEA/UN/UNEP/UNDP joint group of experts on the scientific aspects of marine environmental protection. Rep Stud No. 93, GESAMP
Geyer R, Jambeck J, Law KL (2017) Production, use, and fate of all plastics ever made. Sci Adv 3: 25–29
Ghosh SK, Pal S, Ray S (2013) Study of microbes having potentiality for biodegradation of plastics. Environ Sci Pollut Res 20:4339–4355
Gigault J, Ter Halle A, Baudrimont M, Pascal PY, Gauffre F, Phi TL, El Hadri H, Grassl B, Reynaud S (2018) Current opinion: what is a nanoplastic? Environ Pollut 235:1030–1034
Giorgetti L, Spanò C, Muccifora S, Bottega S, Barbieri F, Bellani L, Castiglione MR (2020) Exploring the interaction between polystyrene nanoplastics and *Allium cepa* during germination: internalization in root cells, induction of toxicity and oxidative stress. Plant Physiol Biochem 149:170–177
Grebe GS, Byron CJ, Gelais AS, Kotowicz DM, Olson TK (2019) An ecosystem approach to kelp aquaculture in the Americas and Europe. Aquac Rep 15:100,215. https://doi.org/10.1016/j.aqrep.2019.100215
Gregory M (1996) Plastic 'Scrubbers' in Hand Cleansers: a further (and minor) source for marine pollution identified. Mar Pollut Bull 32(12):867–871
Guo JJ, Huang XP, Xiang L, Wang YZ, Li YW, Li H, Cai QY, Mo CH, Wong MH (2020) Source, migration and toxicology of microplastics in soil. Environ Int 137:105,263
Gutow L, Eckerlebe A, Giménez L, Saborowski R (2015) Experimental evaluation of seaweeds as a vector for microplastics into marine food webs. Environ Sci Technol 50:915–923
Hinojosa IA, Thiel M (2009) Floating marine debris in fjords, gulfs and channels of southern Chile. Mar Pollut Bull 58:341–350
Horton AA, Walton A, Spurgeon DJ, Lahive E, Svendsen C (2017) Microplastics in freshwater and terrestrial environments: evaluating the current understanding to identify the knowledge gaps and future research priorities. Sci Total Environ 586:127–141
Hu ZM, Shan TF, Zhang J, Zhang QS, Critchley AT, Choi HG, Yotsukura N, Liu FL, Duan DL (2021) Kelp aquaculture in China: a retrospective and future prospects. Reviews in Aquaculture.
Imhof HK, Ivleva NP, Schmid J, Niessner R, Laforsch C (2013) Contamination of beach sediments of a subalpine lake with microplastic particles. Curr Biol 23:R867–R868
Isobe A, Uchida K, Tokai T, Iwasaki S (2015) East Asian seas: a hot spot of pelagic microplastics. Mar Pollut Bull 101(2):618–623

Ivleva NP, Wiesheu AC, Niessner R (2017) Microplastic in aquatic ecosystems. Angew Chem Int Ed 56(7):1720–1739

Jambeck JR, Geyer R, Wilcox C, Siegler TR, Perryman M, Andrady A, Naray R (2015) Plastic waste inputs from land into the ocean. Science 347:768–771

Jegou AM, Salomon JC (1991) Couplage imagerie thermique satellitaire-modeles numeriques. application a la manche. Oceanol Acta 91(11):55–61

Jiang X, Chen H, Liao Y, Ye Z, Li M, Klobučar G (2019) Ecotoxicity and genotoxicity of polystyrene microplastics on higher plant Vicia faba. Environ Pollut 250:831–838

Kanehiro H, Tokai T, Matuda K (1995) Marine litter composition and distribution on the sea-bed of Tokyo Bay. Fish Eng 3:195–199

Kibriya S, Iwan Jones J (2007) Nutrient availability and the carnivorous habit in *Utricularia vulgaris*. Freshw Biol 52(3):500–509

Kim SW, Chae Y, Kim D, An YJ (2019) Zebrafish can recognize microplastics as inedible materials: quantitative evidence of ingestion behavior. Sci Total Environ 649:156–162

Klaine SJ, Koelmans AA, Horne N, Carley S, Handy RD, Kapustka L, Nowack B, von der Kammer F (2012) Paradigms to assess the environmental impact of manufactured nanomaterials. Environ Toxicol Chem 31:3–14

Klinger T (2015) The role of seaweeds in the modern ocean. Perspect Phycol 2:31–39. https://doi.org/10.1127/pip/2015/0024

Klust G (1982) Netting materials for fishing gear, 2nd ed. Fraham, Surrey, Fishing News Books Ltd., UK

Kuhlbrodt T, Griesel A, Montoya M, Levermann A, Hofmann M, Rahmstorf S (2007) On the driving processes of the Atlantic meridional overturning circulation. Rev Geophys 45:RG2001/2004

Larkum AWDD, Orth RJ, Duarte CM (2006) Seagrasses: biology, ecology, and conservation. Mar Ecol 27:431–432

Law KL (2017) Plastics in the marine environment. Ann Rev Mar Sci 9:205–229

Lebreton LC, van der Zwet J, Damsteeg J-W, Slat B, Andrady A, Reisser J (2017) River plastic emissions to the world's oceans. Nat Commun 8:15,611

Lebreton LM, Greer SD, Borrero JC (2012) Numerical modelling of floating debris in the world's oceans. Mar Pollut Bull 64:653–661

Lechner A, Keckeis H, Lumesberger-Loisl F, Zens B, Krusch R, Tritthart M, Glas M, Schludermann E (2014) The Danube so colourful: a potpourri of plastic litter outnumbers fish larvae in Europe's second largest river. Environ Pollut 188:177–181

Li Q, Feng Z, Zhang T, Ma C, Shi H (2020) Microplastics in the commercial seaweed nori. J Hazard Mater 388:122,060

Li WC, Tse HF, Fok L (2016) Plastic waste in the marine environment: a review of sources, occurrence and effects. Sci Total Environ 566–567:333–349

Lian J, Wu J, Xiong H, Zeb A, Yang T, Su X, Su L, Liu W (2020) Impact of polystyrene nanoplastics (PSNPs) on seed germination and seedling growth of wheat (*Triticum aestivum* L.). J Hazard Mater 385:121,620

Lobelle D, Cunliffe M (2011) Early microbial biofilm formation on marine plastic debris. Mar Pollut Bull 62(1):197–200

Lusher AL, Burke A, O'Connor I, Officer R (2014) Microplastic pollution in the Northeast Atlantic Ocean: validated and opportunistic sampling. Mar Pollut Bull 88(1–2):325–333

Ma H, Pu S, Liu S, Bai Y, Mandal S, Xing B (2020) Microplastics in aquatic environments: toxicity to trigger ecological consequences. Environ Pollut 261:114,089

Mann KH (1973) Seaweeds—their productivity and strategy for growth. Science 182:975–981

Mateos-Cárdenas A, Scott DT, Seitmaganbetova G, van Pelt Frank NA et al (2019) Polyethylene microplastics adhere to *Lemna minor* (L.), yet have no effects on plant growth or feeding by *Gammarus duebeni* (Lillj.). Sci Total Environ 689:413–421

Mattsson K, Ekvall MT, Hansson LA, Linse S, Malmendal A, Cedervall T (2015a) Altered behavior, physiology, and metabolism in fish exposed to polystyrene nanoparticles. Environ Sci Technol 49:553–561

Mattsson K, Jocic S, Doverbratt I, Hansson LA (2015b) Nano-plastics in the aquatic environment. Environ Sci: Processes Impacts 17:1712–1721

Mazurais D, Ernande B, Quazuguel P, Severe A, Huelvan C, Madec L, Mouchel O, Soudant P, Robbens J, Huvet A, Zambonino-Infante J (2015) Evaluation of the impact of polyethylene microbeads ingestion in European sea bass (*Dicentrarchus labrax*) larvae. Mar Environ Res 112:78–85

Mineur F, Arenas F, Assis J, Davies AJ, Engelen AH, Fernandes F, Malta EJ, Thibaut T, Van Nguyen T, Vaz-Pinto F, Vranken S (2015) European seaweeds under pressure: consequences for communities and ecosystem functioning. J Sea Res 98:91–108

Mintenig SM, Löder MG, Primpke S, Gerdts G (2019) Low numbers of microplastics detected in drinking water from ground water sources. Sci Total Environ 648:631–635

Miralles P, Church TL, Harris AT (2012) Toxicity, uptake, and translocation of engineered nanomaterials in vascular plants. Environ Sci Technol 46(17):9224–9239

Mohee R, Unmar G (2007) Determining biodegradability of plastic materials under controlled and natural composting environments. Waste Manag 27:1486–1493

Moore CJ (2008) Synthetic polymers in the marine environment: a rapidly increasing, long-term threat. Environ Res 108:131–139

Moore CJ, Lattin GL, Zellers AF (2011) Quantity and type of plastic debris flowing from two urban rivers to coastal waters and beaches of Southern California. Revista de Gestão Costeira Integrada-J Integr Coast Zone Manag 11:65–73

Moore CJ, Moore SL, Leecaster MK, Weisberg SB (2001) A comparison of plastic and plankton in the North Pacific central gyre. Mar Pollut Bull 42:1297–1300

Morritt D, Stefanoudis PV, Pearce D, Crimmen OA, Clark PF (2014) Plastic in the Thames: a river runs through it. Mar Pollut Bull 78:196–200

Mountford AS, Morales Maqueda MA (2019) Eulerian modeling of the three-dimensional distribution of seven popular microplastic types in the Global Ocean. J Geophys Res Oceans 124: 8558–8573

Mushtaq R, Katiyar S, Bennett J (2008) Proteomic analysis of drought stress responsive proteins in rice endosperm affecting grain quality. J Crop Sci Biotechnol 11(4):227–232

O'Brine T, Thompson RC (2010) Degradation of plastic carrier bags in the marine environment. Mar Pollut Bull 60:2279–2283

Ochi T, Okubo S, Fukui K (2007) Development of recycled PET fiber and its application as concrete-reinforcing fiber. Cement Concrete Comp 29(6):448–455

Ogata Y, Takada H, Mizukawa K, Hirai H, Iwasa S, Endo S, Mato Y, Saha M, Okuda K, Nakashima A, Murakami M (2009) International pellet watch: global monitoring of persistent organic pollutants (POPs) in coastal waters. Initial phase data on PCBs, DDTs, and HCHs1. Mar Pollut Bull 58(10):1437–1446

Orfanidis S, Panayotidis P, Stamatis N (2001) Ecological evaluation of transitional and coastal waters: a marine benthic macrophytes-based model. Mediterr Mar Sci 2:45–65

Orfanidis S, Panayotidis P, Stamatis N (2003) An insight to the ecological evaluation index (EEI). Ecol Indic 3:27–33

Orfanidis S, Panayotidis P, Ugland K (2011) Ecological Evaluation Index continuous formula (EEI-c) application: a step forward for functional groups, the formula and reference condition values. Mediterr Mar Sci 12:199–231

Parenti CC, Ghilardi A, Della Torre C, Magni S, Del Giacco L, Binelli A (2019) Evaluation of the infiltration of polystyrene nanobeads in zebrafish embryo tissues after short-term exposure and the related biochemical and behavioural effects. Environ Pollut 254:112,947

Peng L, Fu D, Qi H, Lan CQ, Yu H, Ge C (2020) Micro- and nano-plastics in marine environment: source, distribution and threats—a review. Sci Total Environ 698:134,254

Pergent G (1991) Les indicateurs écologiques de la qualité du milieu marin en Méditerranée. Oceanis 17(4):341–350
Plastics Europe (2016) The facts—an analysis of European latest plastics production, demand and waste data, Plastics Europe, Association of Plastic Manufacturers, Brussels, 2016. Available at http://www.plasticseurope.org.
Plastics Europe (2019) Plastics-the facts 2019. An analysis of European production, demand and waste data. https://www.plasticseurope.org/application/files/
Prokin AA, Dubov PG, Bolotov SE (2015) Formation of macroinvertebrates communities in duckweed (Lemnaceae) and artificial surface-floating substrate: results of the experiment under natural conditions. Inland Water Biol 8:373–383
Reisser J, Shaw J, Wilcox C, Hardesty BD, Proietti M, Thums M, Pattiaratchi C (2013) Marine plastic pollution in waters around Australia: characteristics, concentrations, and pathways. PLoS One 11:e80466
Ribic CA, Sheavly SB, Rugg DJ, Erdmann ES (2010) Trends and drivers of marine debris on the Atlantic coast of the United States 1997–2007. Mar Pollut Bull 60:1231–1242
Rios Mendoza LM, Karapanagioti H, Alvarez NR (2018) Micro(nanoplastics) in the marine environment: current knowledge and gaps. Curr Opin Environ Sci Health 1:47–51
Rochman CM (2015) The complex mixture, fate and toxicity of chemicals associated with plastic debris in the marine environment. In: Bergmann M, Gutow L, Klages M (eds) Marine anthropogenic litter. Springer, Berlin, pp 117–140
Romeo T, Pietro B, Pedà C, Consoli P, Andaloro F, Fossi MC (2015) First evidence of presence of plastic debris in stomach of large pelagic fish in the Mediterranean Sea. Mar Pollut Bull 95(1): 358–361
Ryan PG, Moore CJ, Van Franeker JA, Moloney CL (2009) Monitoring the abundance of plastic debris in the marine environment. Philos Trans R Soc B Biol Sci 364(1526):1999–2012
Ryberg MW, Laurent A, Hauschild M (2018) Mapping of global plastics value chain and plastics losses to the environment: with a particular focus on marine environment
Schmidt C, Krauth T, Wagner S (2018) Correction to export of plastic debris by rivers into the sea. Environ Sci Technol 52:927
Schwaferts C, Niessner R, Elsner M, Ivleva NP (2019) Methods for the analysis of submicrometer- and nanoplastic particles in the environment. TrAC Trends Anal Chem 112:52–65
SFRA0025 (2015) Identification and assessment of riverine input of (Marine) litter. Final report for the European Commission DG Environment under Framework Contract No. ENV.D.2/FRA/ 2012/0025
Sfriso A, Facca C, Ghetti PF (2007) Rapid quality index, based mainly on macrophyte associations (R-MaQI), to assess the ecological status of the transitional environments. Chem Ecol 23:1–11
Sfriso A, Facca C, Ghetti PF (2009) Validation of the Macrophyte Quality Index (MaQI) set up to assess the ecological status of Italian marine transitional environments. Hydrobiologia 617:117–141
Shah AA, Hasan F, Hameed A, Ahmed S (2008) Biological degradation of plastics: a comprehensive review. Biotechnol Adv 26:246–265
Sharma S, Chatterjee S (2017) Microplastic pollution, a threat to marine ecosystem and human health: a short review. Environ Sci Pollut Res Int 24:21,530–21,547
Shashoua Y (2008) Conservation of plastics—materials science, degradation and preservation. Elsevier/Butterworth Heinemann, Amsterdam. (286 pp.)
Shen M, Zhang Y, Zhu Y, Song B, Zeng G, Hu D, Wen X, Ren X (2019a) Recent advances in toxicological research of nanoplastics in the environment: a review. Environ Pollut 252:511–521
Shen M, Zhu Y, Zhang Y, Zeng G, Wen X, Yi H, Ye S, Ren X, Song B (2019b) Micro (nano) plastics: Unignorable vectors for organisms. Mar Pollut Bull 139:328–331
Sherman P, Van Sebille E (2016) Modeling marine surface microplastic transport to assess optimal removal locations. Environ Res Lett 11:1–6

Shomura RS, Godfrey ML (1990) Proceedings of the Second International Conference on Marine Debris: 2–7 April, 1989, Honolulu, Hawaii. US Department of Commerce, National Oceanic and Atmospheric Administration, National Marine Fisheries Service

Sighicelli M, Pietrelli L, Lecce F, Iannilli V, Falconieri M, Coscia L, Di Vito S, Nuglio S, Zampetti G (2018) Microplastic pollution in the surface waters of Italian Subalpine Lakes. Environ Pollut 236:645–651

Singh S, Eapen S, D'Souza SF (2006) Cadmium accumulation and its influence on lipid peroxidation and antioxidative system in an aquatic plant, Bacopa monnieri L. Chemosphere 62(2): 233–246

Sirova D, Borovec J, Erna B, Rejmankova EK, Adamec L, Vrba J (2009) Microbial community development in the traps of aquatic Utricularia species. Aquat Bot 90(2):129–136

Sjollema SB, Redondo-Hasselerharm P, Leslie HA, Kraak MH, Vethaak AD (2016) Do plastic particles affect microalgal photosynthesis and growth? Aquat Toxicol 170:259–261

Song YK, Hong SH, Jang M, Kang JH, Kwon OY, Han GM, Shim WJ (2014) Large accumulation of micro-sized synthetic polymer particles in the sea surface microlayer. Environ Sci Technol 48(16):9014–9021

Stapleton PA (2019) Nanoplastic translocation between the maternal-fetal environment. Rutgers University, New Brunswick, NJ

Sudesh K, Iwata T (2008) Sustainability of biobased and biodegradable plastics. Clean Soil Air Water 36:433–442

Taşkın E, Tsiamis K, Orfanidis S (2018) Ecological quality of the Sea of Marmara (Turkey) assessed by the Marine Floristic Ecological Index (MARFEI). J Black Sea/Mediterranean Environ 24:97–114

Teagle H, Hawkins SJ, Moore PJ, Smale DA (2017) The role of kelp species as biogenic habitat formers in coastal marine ecosystems. J Exp Mar Biol Ecol 492:81–98

Teuten EL, Rowland SJ, Galloway TS, Thompson RC (2007) Potential for plastics to transport hydrophobic contaminants. Environ Sci Technol 41:7759–7764

Thompson RC, Olsen Y, Mitchell RP, Davis A, Rowland SJ, John AWG, McGonigle D, Russell AE (2004) Lost at sea: where is all the plastic? Science 304(5672):838

Timmers MA, Kistner CA, Donohue MJ (2005) Marine debris of the Northwestern Hawaiian Islands: ghost net identification. Hawaii Sea Grant Publication

Van Cauwenberghe L, Vanreusel A, Mees J, Janssen CR (2013) Microplastic pollution in deep-sea sediments. Environ Pollut 182:495–499

van der Hal N, Ariel A, Angel DL (2017) Exceptionally high abundances of microplastics in the oligotrophic Israeli Mediterranean coastal waters. Mar Pollut Bull 116(1–2):151–155

Van Hoeck A, Horemans N, Monsieurs P, Cao HX, Vandenhove H, Blust R (2015) The first draft genome of the aquatic model plant *Lemna minor* opens the route for future stress physiology research and biotechnological applications. Biotechnol Biofuels 8:188

van Sebille E, England MH, Froyland G (2012) Origin, dynamics and evolution of ocean garbage patches from observed surface drifters. Environ Res Lett 7(4):1–6

Van Sebille E, Wilcox C, Lebreton L, Maximenko N, Hardesty BD, Van Franeker JA, Eriksen M, Siegel D, Galgani F, Law KL (2015) A global inventory of small floating plastic debris. Environ Res Lett 10(12):124,006

van Weert S, Redondo-Hasselerharm PE, Diepens NJ, Koelmans AA (2019) Effects of nanoplastics and microplastics on the growth of sediment-rooted macrophytes. Sci Total Environ 654:1040–1047

Verpoorter C, Kutser T, Seekell DA, Tranvik LJ (2014) A global inventory of lakes based on high-resolution satellite imagery. Geophys Res Lett 41:6396–6402

Wagner M, Scherer C, Alvarez-Muñoz D, Brennholt N, Bourrain X, Buchinger S, Fries E, Grosbois C, Klasmeier J, Marti T, Rodriguez-Mozaz S (2014) Microplastics in freshwater ecosystems: what we know and what we need to know. Environ Sci Eur 26(1):1–9

Wang L, Wu WM, Bolan NS, Tsang DC, Li Y, Qin M, Hou D (2021) Environmental fate, toxicity and risk management strategies of nanoplastics in the environment: current status and future perspectives. J Hazard Mater 401:123,415

Ward JE, Shumway SE (2004) Separating the grain from the chaff: particle selection in suspension- and deposit-feeding bivalves. J Exp Mar Biol Ecol 300(1–2):83–130

Watson R, Revenga C, Kura Y (2006) Fishing gear associated with global marine catches I. Database development. Fish Res 79(1–2):97–102

Wieczorek AM, Morrison L, Croot PL, Allcock AL, MacLoughlin E, Savard O, Brownlow H, Doyle TK (2018) Frequency of microplastics in mesopelagic fishes from the Northwest Atlantic. Front Mar Sci 5:127

Willis K, Maureaud C, Wilcox C, Hardesty BD (2018) How successful are waste abatement campaigns and government policies at reducing plastic waste into the marine environment? Mar Policy 96:249

Xu S, Ma J, Ji R, Pan K, Miao AJ (2020) Microplastics in aquatic environments: occurrence, accumulation, and biological effects. Sci Total Environ 703:134,699

Yin S, Tuladhar R, Shi F, Combe M, Collister T, Sivakugan N (2015) Use of macro plastic fibres in concrete: a review. Construct Build Mater 93:180–188

Yokota K, Waterfield H, Hastings C, Davidson E, Kwietniewski E, Wells B (2017) Finding the missing piece of the aquatic plastic pollution puzzle: interaction between primary producers and microplastics. Limnol Oceanogr Lett 2(4):91–104

Yu H, Zhang X, Hu J, Peng J, Qu J (2020) Ecotoxicity of polystyrene microplastics to submerged carnivorous *Utricularia vulgaris* plants in freshwater ecosystems. Environ Pollut 265:114,830

Yu S, Shen M, Li S, Fu Y, Zhang D, Liu H, Liu J (2019) Aggregation kinetics of different surface-modified polystyrene nanoparticles in monovalent and divalentelectrolytes. Environ Pollut 255: 113,302

Yuan W, Zhou Y, Liu X, Wang J (2019) New perspective on the nanoplastics disrupting the reproduction of an endangered fern in artificial freshwater. Environ Sci Technol 53: 12,715–12,724

Zhang K, Gong W, Lv J, Xiong X, Wu C (2015) Accumulation of floating microplastics behind the three Gorges dam. Environ Pollut 204:117–123

Zhang C, Chen X, Wang J, Tan L (2016) Toxic effects of microplastic on marine microalgae Skeletonema costatum: interactions between microplastic and algae. Environ Pollut 220: 1282–1288

Zhang C, Chen X, Wang J, Tan L (2017a) Toxic effects of microplastic on marine microalgae Skeletonema costatum: interactions between microplastic and algae. Environ Pollut 220:1282–1288

Zhang W, Zhang S, Wang J, Wang Y, Mu J, Wang P, Lin X, Ma D (2017b) Microplastic pollution in the surface waters of the Bohai Sea, China. Environ Pollut 231:541–548

Zhang X, Chen J, Li J (2020) The removal of microplastics in the wastewater treatment process and their potential impact on anaerobic digestion due to pollutants association. Chemosphere 251: 126,360

Zheng Y, Yanful EK, Bassi AS (2005) A review of plasticwaste biodegradation. Crit Rev Biotechnol 25:243–250

Zhou P, Huang C, Fang H, Cai W, Li D, Li X, Yu H (2011) The abundance, composition and sources of marine debris in coastal seawaters or beaches around the northern South China Sea (China). Mar Pollut Bull 62(9):1998–2007

Zitko V, Hanlon M (1991) Another source of pollution by plastics: skin cleansers with plastic scrubbers. Mar Pollut Bull 22:41–42

Ecotoxic Effects of the Plastic Waste on Marine Fauna: An Overview

15

Mudasir A. Dar, Neeraja P. Dhole, Kiran D. Pawar, Rongrong Xie, Mohd. Shahnawaz, Radhakrishna S. Pandit, and Jianzhong Sun

Abstract

Although plastics are widely used polymers in our daily life as a blessing, however due to the lack of proper waste management system, every year million tons of plastic wastes get generated in the environment. Besides posing serious threats to the terrestrial life, most of the plastic waste manages to enter marine environments and threatens the life as well as the ecological balance of the oceans. Every year the plastic litter enters the marine environments at an alarming rate. Marine animals ingest plastic waste as a food and leads to chocking of their stomach followed by death. In some cases, marine animals get entangled with the plastic waste, and due to scarcity of food, these animals also get killed. As per reports plastic wastes lead to the death of millions of marine animals annually. In the present chapter, an attempt was made to overview the impact of plastic waste

Mudasir A. Dar and Neeraja P. Dhole have contributed equally to this work and are both first authors.

M. A. Dar (✉)
Biofuels Institute, School of the Environment and Safety Engineering, Jiangsu University, Zhenjiang, China

Department of Zoology, Savitribai Phule Pune University, Ganeshkhind, Pune, Maharashtra, India
e-mail: mudasir.dar@unipune.ac.in

N. P. Dhole · R. S. Pandit
Department of Zoology, Savitribai Phule Pune University, Ganeshkhind, Pune, Maharashtra, India

K. D. Pawar
School of Nanoscience and Biotechnology, Shivaji University, Vidyanagar, Kolhapur, Maharashtra, India

R. Xie · M. Shahnawaz · J. Sun
Biofuels Institute, School of the Environment and Safety Engineering, Jiangsu University, Zhenjiang, China

M. Shahnawaz et al. (eds.), *Impact of Plastic Waste on the Marine Biota*,
https://doi.org/10.1007/978-981-16-5403-9_15

on the marine fauna under the following lines: (1) to discuss the demand, usage and generation of plastic waste at global level, (2) to enlist the routes followed by the plastic waste to reach oceans, (3) to overview the deterioration of the plastic waste and generation of plastic particles in the marine environment due to different environmental forces and (4) to report the interaction of the plastic particles and marine fauna, viz. ingestion and entanglement. Lastly the probable mechanism of death of the marine fauna due to ingestion and entanglement of plastics is also elaborated.

Keywords

Plastic pollution · Impact · Marine animals · Entanglement · Ingestion · Disease risks · Biodegradation

15.1 Introduction

The planet Earth is excessively clustered with a gregarious substance called, plastics. Plastics are fossil oil-derived synthetic polymers widely used in the food packaging, electronics, construction, health and chemical industries. Due to all the advantages offered by plastics (Table 15.1) over other materials such as glass, wood and metals, the demand for plastic has tripled in the past 25 years (Feil and Pretz 2020). During the last 50 years, plastic production has risen at an exponential rate (Bombelli et al. 2017) producing more than 830 million tons (Boucher et al. 2019). In 2018, over 359 million tons of plastics were manufactured globally, and a substantial growth is anticipated in near future. The global plastic market size was valued at USD 568.9 billion in 2019. The annual compound growth rate (CAGR) of the plastic market is projected to increase at 3.2% between 2020 and 2027 (Grand Review Research 2020). The overall production of plastic materials has exceeded at an alarming rate comprising most of the materials as single-use convenience products. These single-use products accumulate in enormous quantities and eventually result in environmental pollution (Jambeck et al. 2015; Plastics Europe 2018; Cassone et al. 2020). The four dominant plastic products, polyethylene (PE), polystyrene (PS), polyurethane (PUR) and polypropylene (PP), represent over 92% of plastic productions. Polyethylene type of plastic is largely utilized for packaging, representing ~40% of total demand for plastic products (www.plasticseurope.org) with over a trillion plastic bags used every year. Polystyrene, a synthetic polymer [poly(1-phenylethene)] being the fifth most widely used plastics, is made up of styrene monomers (Kim et al. 2020). It is used in the packaging industry, apart from making a variety of products like CD cases, plastic cutlery, petri dishes, etc. (Maul et al. 2007). Polystyrene is a polymeric product used in packaging, disposable cups and insulating products, accounting for around 6.6 percent of overall plastic use (Plastic Europe 2018). Approximately 1 trillion plastic bags are used worldwide,

Table 15.1 Some of the common types of plastics and their day-to-day applications

Sr. no.	Type of the plastic	Composition	Applications	References
1.	Acrylonitrile butadiene styrene (ABS)	Acetonitrile, butadiene, styrene	Pipes, electronics, vehicle spares, protective head gears	Moore (1973)
2.	Polyamide	Amides, α, ω-amino acids, diamine, diacid	Carpets, garments, ropes, gears, seatbelts	https://omnexus.specialchem.com/selection-guide/
3.	Polyethylene	Ethylene	Wire insulations, bottles, toys, bags	
4.	Polyimide	Dianhydride, diamine, diisocyanate	Medical tubing, high temperature adhesive	
5.	Polymethyl methacrylate	Propylene, benzene	Construction, automotive industry, medical	
6.	Polypropylene	Propylene	Automotive industry, packaging for consumer products, textiles	
7.	Polystyrene	Styrene	Food packaging, laboratory ware, electronics, automotive parts	https://www.chemicalsafetyfacts.org/polystyrene/
8.	Polytetrafluoroethylene	Tetrafluoroethylene	Cookware, buildings, dental clinics, pipes, machinery, gaskets	https://www.smartechonline.com/uses-ptfe/
9.	Polyvinyl chloride	Vinyl chloride	Construction, health care, electronics, automobile	https://www.bpf.co.uk/plastipedia/polymers/pvc
10.	Acrylic resins	Acrylic acid, methacrylic acid, other related compounds	Decorative panels, adhesives, coatings, tiles	
11.	Epoxy resins	Epoxy monomers, acidic hydroxy groups, epichlorohydrin	Adhesives, plastics, paints, coatings, sealers	Jin et al. (2015)
12.	Polyester resin	Dibasic organic acids, polyhydric alcohols	Boats, flat roofing, pond building, arts, crafts, surfboards, custom mould,	https://www.elkem.com/silicones/technologies/fluids/resin/
13.	Polyurethane resin	Polyol, isocyanate	Furniture, automotive interiors, packaging	
14.	Silicone resin	Oligosiloxanes	Automotive, electronics, packaging, toys	
15.	Vinyl resin	Styrene, methacrylic acid, epoxy	Water pipe, window frames, gutters and downspouts, tiles, electronics	https://polymerdatabase.com/Polymer%20Brands/PVE.html

demanding production of over 45 million tons annually (Kluser et al. 2010). Despite the availability of advanced technologies, only 9% of all plastics are recycled, while 12% are directly incinerated. Even though 95% of used PET is collected in some European countries, only 30% is recycled under ideal conditions. Similarly, in 2017, out of the 8.3 billion tons of plastic waste, less than 9% was recycled globally. Moreover, PS is known to be difficult to decompose due to its unique and complex molecular structure. This low level of recycling is due to several reasons such as cost, consumer acceptance, safety regulations surrounding recycled material, etc. (Wierckx et al. 2015). Over the years, more than half of the plastics are carelessly discarded into the environment, while majority of the plastics end up in landfills or oceans causing negative impacts on the environment as well as animal life.

15.2 Impact of Plastics on the Marine Animals

With increasing and continuous production, approximately 5–13 million metric tons of plastics end up in the ocean every year (Jambeck et al. 2015). Hitherto, 51 trillion bits of plastics equivalent to 51 trillion dangerous chemicals discharged in the water bodies including oceans that pose threat to the survival of marine life. More than 1.8 trillion bits of plastic, weighing an estimated 80,000 tons, have accumulated in the Great Pacific garbage patch alone, with no end in sight (Jambeck et al. 2015; Derraik 2002; Cózar et al. 2014; Lebreton et al. 2018). Many scientists have envisaged that by 2050, the total volume of plastic in the oceans will exceed that of fish (https://www.theguardian.com/lifeandstyle/2017/feb/14/sea-to-plate-plastic-got-into-fish). Plastic tempts organisms of all sizes, small or large, which mistake it for food. The taste, shape, odour and even size of some plastics are deceptively similar to the foods of many animals. Plastic pollution harms all organisms that come into its contact including tiny corals to majestic whales (Fig. 15.1). Plastic has been found in abundance within the guts of diverse organisms ranging from seabirds (Wilcox et al. 2015) and fish (Lusher et al. 2013) to marine and freshwater zooplanktons (Wright et al. 2013; Hurley et al. 2017; Cole et al. 2013). To date more than 700 marine species have been killed by ingestion of plastic or entanglement (Laist 1997), resulting in millions of animal deaths each year (Fig. 15.2). However, the dire consequences of the plastic consumption and its effects on the metabolism of animals are still elusive. Recently, a study reported that (https://thinkprogress.org/zooplankton-are-eatingplastic-and-thats-bad-news-for-ocean-life-dbe31be8ae2b/) almost one-third of the copepods were found to have consumed microplastics. These small creatures invisible to the naked eyes are consumed by shrimp, krill and small fishes, which are in turn preyed upon by larger predators. It is intriguing to the scientists, how these plastic pollutants harm the organisms at different strata of the food chains. The devastating quantity of plastics accumulating in the oceans poses serious and negative consequences for the animal life and ecosystem.

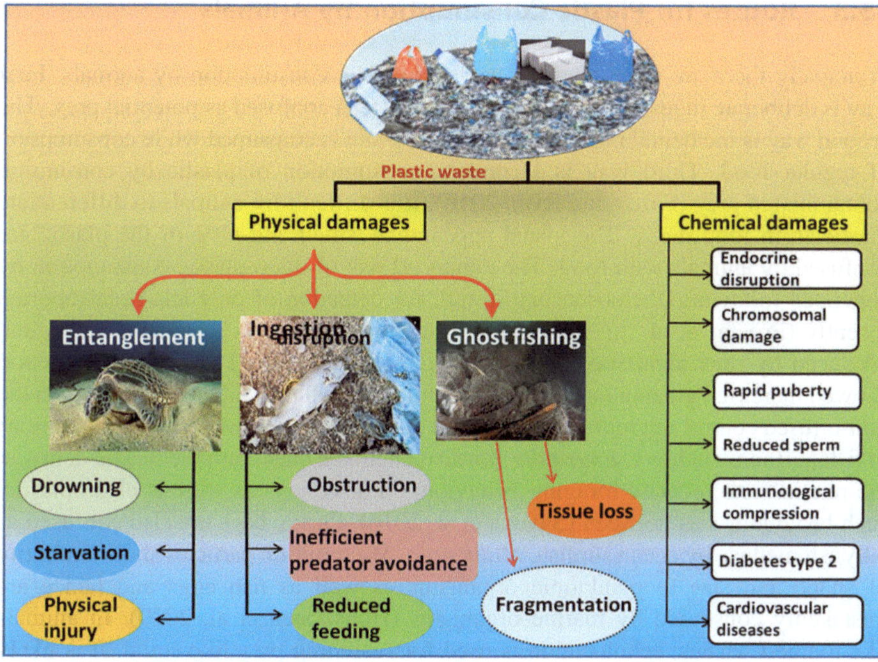

Fig. 15.1 Physical, chemical and physiological impacts of plastic pollution on animals

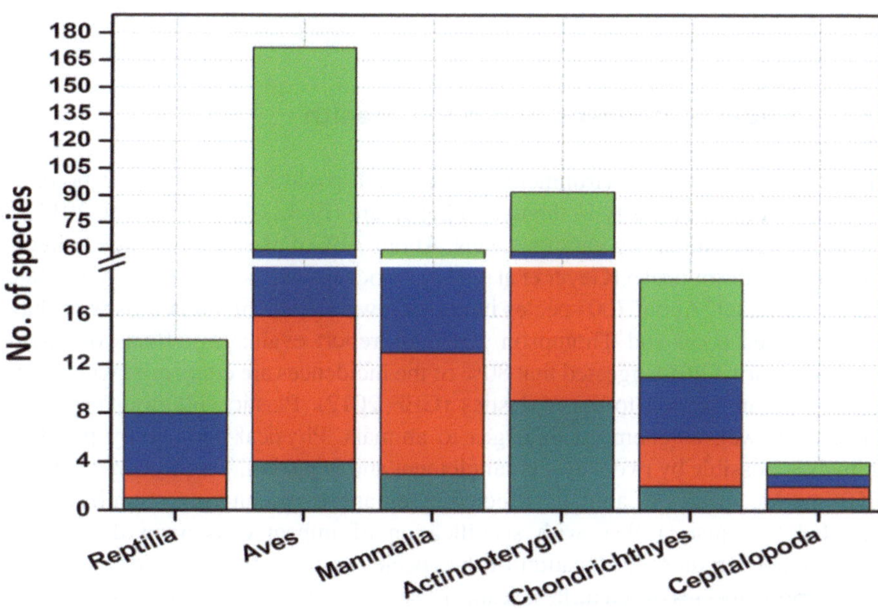

Fig. 15.2 The number of the species of marine animals affected by plastic pollution (adopted from Laist 1997 and López-Martínez et al. 2021)

15.3 Routes for Plastic Consumption by Animals

Principally there are three possible ways of plastic consumption by animals. First way is deliberate ingestion, in which plastic waste is confused as potential prey. The second way is incidental ingestion, in which plastic is consumed while consumption of regular food. Third way is secondary consumption of plastic by consuming contaminated prey (Fazey and Ryan 2016). It is difficult for animals to differentiate plastic from food items. The visual appearance, smell and taste of the plastic are confused by animals with food. The enhanced risk of microplastic consumption by copepods following chemosensory stimuli for detection of prey has been reported recently (Procter et al. 2019). Dimethyl sulphide (DMS) is a volatile compound produced by prey algal species of *Calanus helgolandicus*. The calonoid copepod showed increased grazing rates by 0.7 to 3-folds to the DMS infused plastics than the non-infused control microplastic (Procter et al. 2019). In the case of wild populations of fish northern anchovy, *Engraulis mordax*, chemical cues from the plastic debris in the photic zone induced foraging behaviours. For control as well as clean plastic, such behaviour was not seen (Savoca et al. 2016). Plastic bags are also confused as jellyfish medusa by ocean sunfish, *Mola mola*. Microplastic particles show similarity of shape and size to planktonic organisms as well as fish eggs and hence are mistakenly consumed by marine organisms (Huuskonen et al. 2020). In animals like toothed whales, echolocation is used for searching prey like squid. It is likely that plastic garbage sounds like their food to them. Lobsters, *Nephrops norvegicus*, are seen to consume plastic while burrowing and foraging (Murray and Cowie 2011).

15.4 Impacts of Plastic on Animal Health

Plastic pollution is top environmental issue which could influence human potential to save biological diversity in the upcoming future (Eerkes-Medrano et al. 2015). It can affect land animals by various means. About 79% of plastic is discarded directly into the landfills or ocean (Geyer et al. 2017). In oceans plastics have become a major marine pollutant. About 700 species interacting with plastic in numerous ways have been reported (Gall and Thompson 2015). A report evaluating effects of marine debris on marine life suggested that 80% of the incidences are concerned with plastic while 11% are related to microplastics (GEF 2012). Plastic pollution may cause physical as well as chemical damages to animals. Physical damage by plastic in animals is possible by two ways: entanglement and ingestion. Two hyndred and fifty (250) marine species of animals affected by entanglement and ingestion have been recorded by Laist (1997), with specification of impact caused on the species, material-causing damage, location of the species and source. The affected animals were various species of sea turtles, seabirds, sea mammals, fish, crab and squid (Laist 1997).

15.4.1 Entanglement

Before 1950s ropes and cordages used for activities such as fishing were made from natural materials such as Indian or Manila hemp and cotton. Due to various disadvantages of these natural materials, these materials were replaced by plastic, which shows desired properties. These very advantageous properties of plastic along with its persistence, disposable nature and improper disposal are responsible for causing damage to marine life. Small as well as large marine animals such as turtles, fishes, crabs and mammals are drawn or entangled in the plastic. Entanglement is entrapping of animals in plastic waste. Entanglement affects mobility of the animals. It may be responsible for threat to life of affected animals, and many animals die due to drowning, starvation or physical injuries (Gregory 2009). It may cause overheating and suffocation. All the species of turtle, two third of seal species, one third of whales and many sea bird species are affected by entanglement (Welden 2020). Animals like seals, sea lions are attracted to cut and uncut plastic loops. To date a number of studies have stated the entanglement caused by plastics to animals in the marine environments. California sea lions, northern elephant seals, Steller sea lions, Pacific Harbour seals and Northern fur seal are the species that have been entangled or constricted by plastic materials (Hanni and Pyle 2000). Similarly, entanglement of Hawaiian monk seals was reported by Henderson (2001). Entanglement of New Zealand fur seals in marine debris was reported by Boren et al. (2006). Reports of entanglement of juvenile Brazilian sharp nose sharks in southwest Atlantic were made by Sazima et al. (2002). Plastic rings were found around mouth and gill area of the affected individuals. Serious abrasion with growth of animal, effects on feeding and ventilation were the effects observed on the sharks. Accidental slipping of heads into plastic loops or holes leads to formation of lethal necklaces around their heads. They are responsible for constriction of animal with subsequent growth. Many dugongs such as *Dugong dugon* are also reported to be affected from entanglement in marine debris (Poonian and Lopez 2016). Reduction in performance of swimming, reduced feeding and antipredatory behaviour were observed in several tropical reef fishes in response to plastic debris collar wrapping affecting the overall survival fitness of the individuals (Nunes et al. 2018). In the case of four species of baleen whales, it was observed that entanglement was responsible for inability to come to surface ultimately causing death and reduced foraging leading to starvation, infections due to open entanglement wounds and haemorrhages in many cases (Cassoff et al. 2011). About 69 species of freshwater birds are affected by entanglement (Ryan 2018). In the case of terrestrial environment, debris like decorations and footballs may cause entanglement (Ryan 2018). Entanglement, laceration and mortality were caused by plastic netting in snake species *Masticophis flagellum* and *Pituophis catenifer* (Stuart et al. 2001).

Abandoned or lost comparatively unimpaired fish nets or monofilaments may lead to ghost fishing by still continuing to capture same fish species or by capturing nontarget species for a long duration, which leads to mortality lacking human control. Tissue loss and fragmentation have been recorded in response to abandoned fishing gears in coral reefs in Koh Tao gulf in Thailand. Most common coral species found to colonize such gears were *Tubastraea micranthus* and fungiids (Ballesteros et al. 2018). In a study conducted in Louisiana, 1073 ghost blue crab traps were found to contain 19 species. The most common animals found entrapped were oysters, marine fishes, stone crabs and diamondback terrapins (Anderson and Alford 2014). In a ghost fishing experiment done in deception bay part of larger Moreton Bay, Queensland, pot fishing was found to cause huge ghost fishing for the target species as well as many nontarget species (Campbell and Sumpton 2009).

15.4.2 Ingestion

Ingestion of plastics is one of the major ways by which animals are affected by plastic. There are different ways by which plastic affects animals which are internal and external injuries, obstruction of alimentary canal leading to the feeling of satiation, starvation and overall weakness ultimately affecting survival quality and fertility and may be fatal. The other effects might be drowning, inefficiency of predator avoidance and reduced feeding capacity. The plastic beads may also absorb toxic compounds and cause threats to animals (Gregory 2009). Plastic ingestion also results in inefficiency to absorb volatile fatty acids and reduced fattening (Sheferaw et al. 2014). Plastic ingestion can also negative impact on fore coming generations (Lusher et al. 2013). In seabirds, blockage of gastric enzyme secretion, lowering of feeding stimuli, reduced secretion of sex hormones and reproduction failure were reported (Azzarello and Van Vleet 1987). About 100 species of seabirds are affected by ingestion and entanglement of plastic (Laist 1997). Microplastic consumption by zooplankton, *Centropages typicus* was shown by Cole et al., in 2013, which was reported to affect functioning and health of the individuals. Economically important species of zooplankton, *N. cristatus,* and *E. pacifica* were also shown to ingest plastic (Desforges et al. 2015). In the case of tropical fiddler crab, *Uca rapax* accumulation and probable effects on the hepatopancreas, gills and stomach were reported (Brennecke et al. 2015). Exposure to polystyrene microplastic in oyster, *Crassostrea gigas*, reduced egg production and decreased larvae production and sperm velocity (Sussarellu et al. 2016). Incidence of plastic ingestion by mesopelagic fishes in North Pacific subtropical gyre was reported in 2011 (Davison and Asch 2011). Frequency of occurrence of plastic in fishes among 15 species of the North Sea were found to be 1.8% (Kühn and Van Franeker 2020). In the case of albatross chicks, they are unable to regurgitate swallowed plastic (Ryan 2018). Weight of albatrosses was found to be inversely proportional to plastic load. In Laysan albatrosses from Midway Island proventricular obstruction and ulcerations were seen (Fry et al. 1987). Six out of seven marine turtle species of world are reported to eat plastic (Ceccarelli 2009). They are vulnerable or threatened species (Schuyler et al. 2014). Turtles which had greater amounts of plastic in gastrointestinal tracts had greater probability of mortality (Wilcox et al. 2015). Numerous records of seals and whales consuming plastic are also there. Zhao and colleagues reported presence of plastic in the alimentary canals of 94% birds, having varying foraging behaviours in China (Zhao et al. 2016). Effects of plastic ingestion on ruminants such as in appetence, simple indigestion, ruminal impaction, ruminal tympany, polybezoars, traumatic reticulopericarditis, local or diffused peritonitis, poor production, chemical leaching, immunosuppression, heavy metal toxicity, oestrogenic activity and reproductive problems, carcinogenicity and teratogenicity have been discussed recently (Priyanka and Dey 2018). In the case of humans, oesophageal obstruction and laceration from accidental consumption of plastic and blockage and perforation of small bowel from consumption of bag clips were reported (Newell et al. 2000). Uptake of microplastics by mice may lead disorders related to amino acids, bile acid and liver lipids (Jin et al. 2019; Lu et al. 2018).

15.5 Disease Risks Due to Plastic Pollution

The chemical effects of plastic include several serious effects. Plasticizers are released from leaching of plastic. These compounds such as phthalates and bisphenol A are highly oestrogenic in function, which lead to endocrine disruption. These endocrine disruptors may cause broad spectrum of long-term effect. Bisphenol A polymer released from plastic can mimic oestrogen, and in humans effects such as ovarian chromosomal damage, rapid puberty reduced sperm production, immunological alterations, diabetes type 2, cardiovascular diseases and obesity are seen. It has also shown to increased risk of breast cancer, prostate gland, pain, metabolic disorders, etc. In women it causes obesity, endometrial hyperplasia, recurrent miscarriages, sterility and polycystic ovary syndrome (Warner et al. 2002; Rayner et al. 2004; Eskenazi et al. 2007). Poorly soluble biopersistent microplastics have ability to interact with cell membrane, organelles and other molecules which may result in inflammation, alterations in membrane permeability and oxidative stress. Nanoplastics adsorbed on cell surface may disrupt the homeostasis of the cell internally (De Souza Machado et al. 2018). Upon internalization of nanoplastics inside the cell, the interaction of nanoplastics with genetic material and other cellular machinery is possible. Alterations in gene expressions, inflammation, biochemical responses and carcinogenesis are reported in various model systems (Forte et al. 2016; Kato et al. 2003). Human intestinal cells as well as isolated loops of chicken intestine when exposed to plastic nanoparticles showed reduced iron absorption (Mahler et al. 2012). In the case of crucian carp, *Carassius carassius*, when exposed to nanoplastics, weight loss, triglycerides to cholesterol ratio in blood serum and distribution of cholesterol in the muscles and liver were observed (Mahler et al. 2012). Chemicals present in plastic are also found to cause gut dysbiosis (Rosenfeld 2017).

15.6 Conclusion

Plastic waste after entering into the marine environment gets fragmented into different plastic particles of varied length due to various environmental forces. In the marine environment, the marine fauna either ingest the plastic particles as food or gets entangled in the plastic waste. Every year billions of marine fauna gets killed due to this plastic litter. To save the marine fauna from the menace of plastic pollution, the emission of plastic particles from land to marine environment needs to be reduced either by employing efficient eco-friendly waste-managing bioremediation technology or by using alternatives of the plastic, e.g. bioplastic.

Acknowledgements The authors acknowledge the grants received from Jiangsu University, China (5363000606), and the National Natural Science Foundation of China (31900367. NPD thanks the Council of scientific and industrial research (CSIR), Govt. of India, for fellowship. RSP acknowledges S. P. Pune University, Pune, for the grants received under DRDP, UGC-CAS III and DST-PURSE schemes. KDP acknowledges the DBT-BUILDER-SUK programme sanctioned to Shivaji University, Kolhapur, India. MS is thankful to the authorities of the Jiangsu University for providing an opportunity to work as an Associate Professor under invited the talent programme. The funders had no role in study design, data collection, analyses, decision to publish or preparation of the manuscript.

Competing Interests: The authors declare no conflict of interest.

Author Contributions:

MAD and NPD designed the study, surveyed literature and prepared original draft.

KDP and XRR, MS prepared original draft, writing, review and editing.

RSP and JZS funding acquisition, validation, writing, review and editing.

All authors approved the final version of the manuscript.

Data Availability: All the data sets generated for this research are included in the manuscript or supplement files.

References

Anderson JA, Alford AB (2014) Ghost fishing activity in derelict blue crab traps in Louisiana. Mar Pollut Bull 79:261–267. https://doi.org/10.1016/j.marpolbul.2013.12.002

Azzarello MY, Van Vleet ES (1987) Marine birds and plastic pollution. Mar Ecol Prog Ser 37:295–303

Bombelli P, Howe CJ, Bertocchini F (2017) Polyethylene bio-degradation by caterpillars of the wax moth *Galleria mellonella*. Curr Biol 27:R283–R293. https://doi.org/10.1016/j.cub.2017.02.060

Boren LJ, Morrissey M, Muller CG, Gemmell NJ (2006) Entanglement of New Zealand fur seals in man-made debris at Kaikoura, New Zealand. Mar Pollut Bull 52:442–446. https://doi.org/10.1016/j.marpolbul.2005.12.003

Boucher J, Dubois C, Kounina A, Puydarrieux P (2019) Review of plastic footprint methodologies: laying the foundation for the development of a standardized plastic footprint measurement tool. IUCN Publication, Gland, Switzerland

Brennecke D, Ferreira EC, Costa TM, Appel D, da Gama BA, Lenz M (2015) Ingested microplastics (>100 μm) are translocated to organs of the tropical fiddler crab *Uca rapax*. Mar Pollut Bull 96:491–495. https://doi.org/10.1016/j.marpolbul.2015.05.001

Campbell MJ, Sumpton WD (2009) Ghost fishing in the pot fishery for blue swimmer crabs *Portunus pelagicus* in Queensland, Australia. Fish Res 95:246–253. https://doi.org/10.1016/j.fishres.2008.09.026

Cassoff RM, Moore KM, McLellan WA, Barco SG, Rotstein DS, Moore MJ (2011) Lethal entanglement in baleen whales. Dis Aquat Org 96:175–185

Cassone BJ, Grove HC, Elebute O, Villanueva SMP, LeMoine CMR (2020) Role of the intestinal microbiome in low-density polyethylene degradation by caterpillar larvae of the greater wax moth, *Galleria mellonella*. Proc R Soc B Biol Sci 287(1922):9. https://doi.org/10.1098/rspb.2020.0112

Ceccarelli DM (2009) Impacts of plastic debris on Australian marine wildlife. In: C&R Consulting (ed), Report by C&R consulting for the department of the environment, water, heritage and the arts, p 83
Cole M, Lindeque P, Fileman E, Halsband C, Goodhead R, Moger J (2013) Microplastic ingestion by zooplankton. Environ Sci Technol 47(12):6646–6655. https://doi.org/10.1021/es400663f
Cózar A, Echevarría F, González-Gordillo JI, Irigoien X, Úbeda B, Hernández-León S, Palma ÁT, Navarro S, García-de-Lomas J, Ruiz A, Fernández-de-Puelles ML, Duarte CM (2014) Plastic debris in the open ocean. Proc Natl Acad Sci U S A 111:10239–10244. https://doi.org/10.1073/pnas.1314705111
Davison P, Asch RG (2011) Plastic ingestion by mesopelagic fishes in the North Pacific subtropical gyre. Mar Ecol Prog Ser 432:173–180. https://doi.org/10.3354/meps09142
de Souza Machado AA, Kloas W, Zarfl C, Hempel S, Rillig MC (2018) Microplastics as an emerging threat to terrestrial ecosystems. Glob Chang Biol 24:1405–1416. https://doi.org/10.1111/gcb.14020
Derraik J (2002) The pollution of the marine environment by plastic debris: a review. Mar Pollut Bull 44:842–852. https://doi.org/10.1016/S0025-326X(02)00220-5
Desforges JPW, Galbraith M, Ross PS (2015) Ingestion of microplastics by zooplankton in the Northeast Pacific Ocean. Arch Environ Contam Toxicol 69:320–330. https://doi.org/10.1007/s00244-015-0172-5
Eerkes-Medrano D, Thompson RC, Aldridge DC (2015) Microplastics in freshwater systems: a review of the emerging threats, identification of knowledge gaps and prioritisation of research needs. Water Res 75:63–82. https://doi.org/10.1016/j.watres.2015.02.012
Eskenazi B, Warner M, Samuels S, Young J, Gerthoux PM (2007) Serum dioxin concentrations and risk of uterine leiomyoma in the Seveso Women's health study. Am J Epidemiol 166:79–87. https://doi.org/10.1093/aje/kwm048
Fazey FMC, Ryan PG (2016) Biofouling on buoyant marine plastics: an experimental study into the effect of size on surface longevity. Environ Pollut 210:354–360. https://doi.org/10.1016/j.envpol.2016.01.026
Feil A, Pretz T (2020) Mechanical recycling of packaging waste. In: Letcher T (ed) Plastic waste and recycling. Academic Press, pp 283–319. https://doi.org/10.1016/b978-0-12-817880-5.00011-6
Forte M, Iachetta G, Tussellino M, Carotenuto R, Prisco M, De Falco M et al (2016) Polystyrene nanoparticles internalization in human gastric adenocarcinoma cells. Toxicol In Vitro 31:126–136. https://doi.org/10.1016/j.tiv.2015.11.006
Fry DM, Fefer SI, Sileo L (1987) Ingestion of plastic debris by Laysan albatrosses and wedge-tailed shearwaters in the Hawaiian islands. Mar Pollut Bull 18:339–343. https://doi.org/10.1016/S0025-326X(87)80022-X
Gall SC, Thompson RC (2015) The impact of debris on marine life. Mar Pollut Bull 92:170–179. https://doi.org/10.1016/j.marpolbul.2014.12.041
GEF (2012) Secretariat of the convention on biological diversity and the scientific and technical advisory panel. Impacts of marine debris on biodiversity: current status and potential solutions, Montreal, Technical Series No. 67, 61 pages
Geyer R, Jambeck JR, Law KL (2017) Production, use, and fate of all plastics ever made. Sci Adv 3: 25–29. https://doi.org/10.1126/sciadv.1700782
Grand Review Research (2020) Plastic market size, share & trends analysis report by product (PE, PP, PU, PVC, PET, Polystyrene, ABS, PBT, PPO, Epoxy Polymers, LCP, PC, Polyamide), By Application, By Region, And Segment Forecasts, 2020–2027. Available at: https://www.grandviewresearch.com/industryanalysis/global-plastics-market. Accessed 7 April 2021
Gregory MR (2009) Environmental implications of plastic debris in marine settings—entanglement, ingestion, smothering, hangers-on, hitch-hiking and alien invasions. Phil Trans R SocB: B3642013-2025. https://doi.org/10.1098/rstb.2008.0265

Hanni KD, Pyle P (2000) Entanglement of pinnipeds in synthetic materials at south-east Farallon Island, California, 1976-1998. Mar Pollut Bull 40:1076–1081. https://doi.org/10.1016/S0025-326X(00)00050-3

Henderson JR (2001) A pre- and post-MARPOL annex V summary of Hawaiian monk seal entanglements and marine debris accumulation in the northwestern Hawaiian islands, 1982–1998. Mar Pollut Bull 42:584–589. https://doi.org/10.1016/S0025-326X(00)00204-6

Hurley RR, Woodward JC, Rothwell JJ (2017) Ingestion of microplastics by freshwater tubifex worms. Environ Sci Technol 51(21):12844–12851. https://doi.org/10.1021/acs.est.7b03567

Huuskonen H, Subiron I, Folguera J, Kortet R, Akkanen J, Vainikka A, Janhunen M, Kekäläinen J (2020) Do whitefish (*Coregonus lavaretus*) larvae show adaptive variation in the avoidance of microplastic ingestion? Environ Pollut 262:114,353. https://doi.org/10.1016/j.envpol.2020.114353

Jambeck JR, Geyer R, Wilcox C, Siegler TR, Perryman M, Andrady A, Narayan R, Law KL (2015) Marine pollution. Plastic waste inputs from land into the ocean. Science 347:768–771. https://doi.org/10.1126/science.1260352

Jin FL, Li X, Park SJ (2015) Synthesis and application of epoxy resins: a review. J Ind Eng Chem 29:1–11. https://doi.org/10.1016/j.jiec.2015.03.026

Jin Y, Lu L, Tu W, Luo T, Fu Z (2019) Impacts of polystyrene microplastic on the gut barrier, microbiota and metabolism of mice. Sci Total Environ 649:308–317. https://doi.org/10.1016/j.scitotenv.2018.08.353

Kato T, Yashiro T, Murata Y, Herbert DC, Oshikawa K, Bando M et al (2003) Evidence that exogenous substances can be phagocytized by alveolar epithelial cells and transported into blood capillaries. Cell Tissue Res 311:47–51. https://doi.org/10.1007/s00441-002-0647-3

Kim HR, Lee HM, Yu HC, Jeon E, Lee S, Li J, Kim DH (2020) Biodegradation of polystyrene by pseudomonas sp. isolated from the gut of Superworms (larvae of Zophobas atratus). Environ Sci Technol 54:6987–6996. https://doi.org/10.1021/acs.est.0c01495

Kluser S, Neumann P, Chauzat MP, Pettis JS, Peduzzi P, Witt R, Fernandez N, Theuri M (2010) Global honey bee colony disorders and other threats to insect pollinators. Nairobi, Kenya, United Nations Environment Programme: UNON/Publishing Services

Kühn S, van Franeker JA (2020) Quantitative overview of marine debris ingested by marine megafauna. Mar Pollut Bull 151:110,858. https://doi.org/10.1016/j.marpolbul.2019.110858

Laist DW (1997) Impacts of marine debris: entanglement of marine life in marine debris including a comprehensive list of species with entanglement and ingestion records. In: Coe JM, Rogers DB (eds) Marine debris. Springer Series on Environmental Management. Springer, New York, NY, pp 99–139. https://doi.org/10.1007/978-1-4613-8486-1_10

Ballesteros LV, Matthews JL, Hoeksema BW (2018) Pollution and coral damage caused by derelict fishing gear on coral reefs around Koh Tao, Gulf of Thailand. Mar Pollut Bull 135:1107–1116. https://doi.org/10.1016/j.marpolbul.2018.08.033

Lebreton L, Slat B, Ferrari F, Sainte-Rose B, Aitken J, Marthouse R, Hajbane S, Cunsolo S, Schwarz A, Levivier A, Noble K, Debeljak P, Maral H, Schoeneich-Argent R, Brambini R, Reisser J (2018) Evidence that the great Pacific garbage patch is rapidly accumulating plastic. Sci Rep 8:4666. https://doi.org/10.1038/s41598-018-22939-w

López-Martínez S, Morales-Caselles C, Kadar J, Rivas ML (2021) Overview of global status of plastic presence in marine vertebrates. Glob Chang Biol 27:728–737. https://doi.org/10.1111/gcb.15416

Lu L, Wan Z, Luo T, Fu Z, Jin Y (2018) Polystyrene microplastics induce gut microbiota dysbiosis and hepatic lipid metabolism disorder in mice. Sci Total Environ 631–632:449–458. https://doi.org/10.1016/j.scitotenv.2018.03.051

Lusher AL, McHugh M, Thompson RC (2013) Occurrence of microplastics in the gastrointestinal tract of pelagic and demersal fish from the English Channel. Mar Pollut Bull 67:94–99. https://doi.org/10.1016/j.marpolbul.2012.11.028

Mahler GJ, Esch MB, Tako E, Southard TL, Archer SD, Glahn RP, Shuler ML (2012) Oral exposure to polystyrene nanoparticles affects iron absorption. Nat Nanotechnol 7:264-U1500

Maul J, Frushour BG, Kontoff JR, Eichenauer H, Ott KH, Schade C (2007) Polystyrene and styrene copolymers. In: Ullmann's encyclopedia of industrial chemistry. Wiley-VCH, GmbH & Co, Weinheim, Germany. https://doi.org/10.1002/14356007.a21_615.pub2

Moore JD (1973) Acrylonitrile-butadiene-styrene (ABS)—a review. Composites 4(3):118–130. https://doi.org/10.1016/0010-4361(73)90585-5

Murray F, Cowie PR (2011) Plastic contamination in the decapod crustacean *Nephrops norvegicus* (Linnaeus, 1758). Mar Pollut Bull 62:1207–1217. https://doi.org/10.1016/j.marpolbul.2011.03.032

Newell KJ, Taylor B, Walton JC, Tweedie EJ (2000) Plastic bread-bag clips in the gastrointestinal tract: report of 5 cases and review of the literature. CMAJ 162:527–529

Nunes B, Castro BB, Gomes J, Carvalho T, Gonçalves F (2018) Cholinesterases as environmental biomarkers to address the putative effects of low, realistic levels of waterborne uranium. Ecol Indic 95(Part 2):1019–1025. https://doi.org/10.1016/j.ecolind.2017.05.028

Plastics Europe (2018) Plastics-the Facts 2018 An analysis of European plastics production, demand and waste data. https://www.plasticseurope.org/application/files/6315/4510/9658/Plastics_the_facts_2018_AF_web.pdf. Accessed 10 April 2021

Poonian CNS, Lopez DD (2016) Small-scale mariculture: a potentially significant threat to dugongs (*Dugong dugon*) through incidental entanglement. Aquat Mamm 42:56–59. https://doi.org/10.1578/AM.42.1.2016.56

Priyanka M, Dey S (2018) Ruminal impaction due to plastic materials—an increasing threat to ruminants and its impact on human health in developing countries. Vet World 11(9):1307–1315. https://doi.org/10.14202/vetworld.2018.1307-1315

Procter J, Hopkins FE, Fileman ES, Lindeque PK (2019) Smells good enough to eat: dimethyl sulfide (DMS) enhances copepod ingestion of microplastics. Mar Pollut Bull 138:1–6. https://doi.org/10.1016/j.marpolbul.2018.11.014

Rayner JL, Wood C, Fenton SE (2004) Exposure parameters necessary for delayed puberty and mammary gland development in Long-Evans rats exposed in utero to atrazine. Toxicol Appl Pharmacol 195:23–34. https://doi.org/10.1016/j.taap.2003.11.005

Rosenfeld CS (2017) Neuroendocrine disruption in animal models due to exposure to bisphenol A analogues. Front Neuroendocrinol 47:123–133. https://doi.org/10.1016/j.yfrne.2017.08.001

Ryan PG (2018) Entanglement of birds in plastics and other synthetic materials. Mar Pollut Bull 135:159–164. https://doi.org/10.1016/j.marpolbul.2018.06.057

Savoca MS, Wohlfeil ME, Ebeler SE, Nevitt GA (2016) Marine plastic debris emits a keystone infochemical for olfactory foraging seabirds. Sci Adv 2:1–9. https://doi.org/10.1126/sciadv.1600395

Sazima I, Gadig OBF, Namora RC, Motta FS (2002) Plastic debris collars on juvenile carcharhinid sharks (*Rhizoprionodon lalandii*) in Southwest Atlantic. Mar Pollut Bull 44:1149–1151. https://doi.org/10.1016/S0025-326X(02)00141-8

Schuyler Q, Hardesty BD, Wilcox C, Townsend K (2014) Global analysis of anthropogenic debris ingestion by sea turtles. Conserv Biol 28:129–139. https://doi.org/10.1111/cobi.12126

Sheferaw D, Fikreysus GM, Tesfaye D, Etana D (2014) Ingestion of indigestible foreign materials by free grazing in Amhara region, Ethiopia. Trop Anim Health Prod 46:247–250

Stuart JN, Watson ML, Brown TL, Eustice C (2001) Plastic netting: an entanglement hazard to snakes and other wildlife. Herpetol Rev 32:162

Sussarellu R, Suquet M, Thomas Y et al (2016) Oyster reproduction is affected by exposure to polystyrene microplastics. Proc Natl Acad Sci U S A 113:2430–2435. https://doi.org/10.1073/pnas.1519019113

Warner M, Eskenazi B, Mocarelli P, Gerthoux PM, Samuels S (2002) Serum dioxin concentrations and breast cancer risk in the Seveso Women's health study. Environ Health Perspect 110:625–628. https://doi.org/10.1289/ehp.02110625

Welden NA (2020) The environmental impacts of plastic pollution. In: Letcher TM (ed) Plastic waste and recycling. Elsevier, London, pp 195–222

Wierckx N, Prieto MA, Pomposiello P, de Lorenzo V, O'Connor K, Blank LM (2015) Plastic waste as a novel substrate for industrial biotechnology. Microb Biotechnol 8(6):900–903. https://doi.org/10.1111/1751-7915.12312

Wilcox C, Van Sebille E, Hardesty BD (2015) Threat of plastic pollution to seabirds is global, pervasive, and increasing. Proc Natl Acad Sci U S A 112:11,899–11,904. https://doi.org/10.1073/pnas.1502108112

Wright SL, Rowe D, Thompson RC, Galloway TS (2013) Microplastic ingestion decreases energy reserves in marine worms. Curr Biol 23(23):R1031–R1033. https://doi.org/10.1016/j.cub.2013.10.068

Zhao S, Zhu L, Li D (2016) Microscopic anthropogenic litter in terrestrial birds from Shanghai, China: Not only plastics but also natural fibers. Sci Total Environ 550:1110–1115. https://doi.org/10.1016/j.scitotenv.2016.01.112

Potential of Plastic Waste in Enhancing the level of Pathogenicity of diverse Pathogens in the Marine Biota

16

Charles Oluwaseun Adetunji, Abel Inobeme, John Tadso, Olugbemi T. Olaniyan, Ogundolie Frank Abimbola, Mohd. Shahnawaz, and Osikemekha Anani

Abstract

It has been discovered that microplastics portend the capacity to enter the bodies of diverse living organisms through direct exposure. This might result into numerous health challenges such as oxidative stress, necrosis, inflammation, and genotoxicity which are connected to a wide range of deleterious health outcomes such as diabetes, cardiovascular diseases, bowel infections, chronic inflammation, and neurodegenerative diseases. This chapter intends to provide detailed information on the contaminants and toxins that are found in several

C. O. Adetunji (✉)
Applied Microbiology, Biotechnology and Nanotechnology Laboratory, Department of Microbiology, Edo State University Uzairue, PMB 04, Auchi, Edo State, Nigeria
e-mail: adetunji.charles@edouniversity.edu.ng

A. Inobeme
Department of Chemistry, Edo State University Uzairue, PMB 04, Auchi, Edo State, Nigeria

J. Tadso
Department of Chemistry, Ibrahim Badamasi Babangida University, Lapai, Niger State, Nigeria

O. T. Olaniyan
Laboratory for Reproductive Biology and Developmental Programming, Department of Physiology, Edo State University Uzairue, PMB 04, Auchi, Edo State, Nigeria

O. F. Abimbola
Department of Biotechnology, Baze University, Abuja, Nigeria

M. Shahnawaz
Biofuels Institute, School of the Environment and Safety Engineering, Jiangsu University, Zhenjiang, China

O. Anani
Laboratory for Ecotoxicology and Forensic Biology, Department of Biological Science, Edo State University Uzairue, PMB 04, Auchi, Edo State, Nigeria

M. Shahnawaz et al. (eds.), *Impact of Plastic Waste on the Marine Biota*,
https://doi.org/10.1007/978-981-16-5403-9_16

plastic wastes available in the sea water. Detailed information was also provided on different types of diseases that could be caused in marine biota as a result of plastic waste, while modes of action involved in the plastic degradation were also provided.

Keywords

Pathogenicity · Pathogens · Diseases · Marine biota · Microplastics · Health challenges

16.1 Introduction

Plastics are produced from organic polymeric materials of synthetic origin and have wide applications in various areas of human endeavors. They are useful in clothing, electronic materials, medical devices, packaging of food, and construction instruments among others (Proshad et al. 2018). It has been recognized as indispensable materials with unique properties and applications. Previously these materials were considered to be harmless due to their chemical inertness; however, it has been observed in recent times that the continual release of plastic materials into the environment has brought about serious problems in different aspects of the environment (Mülhaupt 2013). Plastics contain chemicals such as bisphenol, brominated flame retardants, phthalates, polyfluorinated agents, and antiminitroxide among others which are rated significant risk factors for health. Though they are incredibly relevant due to their broad applicability, the presence of toxic compounds in them has also been reported to be responsible for different diseases, besides they are durable and cannot be degraded biologically (Chandegara et al. 2015).

Recent findings have shown that over few decades ago, tonnes of plastic wastes have been dumped by humans into water bodies. Up to 580,000 pieces of plastic waste in each square kilometer were reported for Atlantic Ocean which is alarming. With continuity in this process, the ecosystem would be at great risk, and the food chain would be severely deteriorated. These plastic materials when accumulated in water bodies are capable of affecting water creatures causing harm and diseases. They also affect the health of eaters of seafoods thereby impacting the entire environment (Jambeck et al. 2015).

Since 1960, plastic wastes have gradually gathered in marine habitat up to the extent that huge masses of these materials are found floating in oceans and sea beaches that were once beautiful environments on sight. Globally, an approximate of 500 billion plastics is reported to be used on yearly basis, with about 8 million tonnes of plastics finding their ways into the oceans yearly. This is yet not surprising because of the estimation that 50% of plastics tend to be used for once prior to been thrown away (Abreu and Pedrotti 2019).

Since plastics are made of chemical components that are toxic in nature, they have the ability to cause serious harms to the environment-polluting air, land, and water. Environmental pollution by plastic wastes occurs when these materials have accumulated in an area to the extent of impacting the environment negatively, resulting to creation of problems for wildlife, plants, and also human population. Contamination of environment due to waste from plastic materials has been recognized as a global problem and has constituted significant environmental burden, most especially in the aquatic habitat where plastic materials are subjected to long-term breakdown through biophysical processes with alarming negative impact on aquatic lives (Wilcox et al. 2015).

Therefore, this chapter intends to provide detailed information on the plastic waste enhancing the pathogenicity of the pathogens to cause diseases in the marine biota.

16.2 Process on How Plastic Waste Could Generate Diseases in the Marine Biota

Generally, plastic wastes have been attributed to a lot of health concerns to living organisms. The pollution of our water bodies has led to the disorganization of the food pyramid flow of the aquatic habitat; it has also lead to several deaths of sea turtles, marine birds, fishes, seagulls, seabirds, and marine mammals such as whales, sea lions, dolphins and Burmeister's porpoise among others. These aquatic creatures are endangered through entanglement with various forms of plastics and ingesting plastic debris, causing suffocation or starvation due to overfilling of their stomach with plastics that makes them look filled even when there is no food in their stomach. The presence of these plastic wastes in the water bodies overtime also can encourage the growth of pathogens which in return causes harm to both sea creatures and humans.

A common product of plastic waste that is usually found in our water bodies is the plastic bottles which all over the world are used for handling different forms of liquids/solvents ranging from reagents, beverages, food, salts, and chemicals among others. These plastics can be natural, synthetic, or semisynthetic. Other types of plastic are is thermosetting plastics and thermoplastics (Kehinde et al. 2020). Depending on the composition of the plastic, they can take months and years to be degraded naturally in the water body. These materials can also be degraded through bacteria and fungi and then be taken by aquatic organisms as food which can result to blockage in their system especially gastrointestinal system and also cause death by suffocation (Desforges et al. 2015, Glaser 2019) (Fig. 16.1).

The most dangerous class of plastic polluting the aquatic habitat is the microplastics which can easily be swallowed by seabirds, sea turtles that feed on jelly fish and some fishes such as bigeye tuna and mammals such as fur seal and whale leading to their eventual deaths (Alabi et al. 2019). These marine organisms are source of food also to humans and will eventually bioaccumulate in them and end up in humans leading to several challenges. Plastics can be from polyester-styrene, polyethylene terephthalate, polyvinyl chloride polysulfone, polyesters, and high-density polyethylene (resin) and chemicals such as polyfluorinated, phthalates, and antimony trioxide. Also other chemicals such as bisphenol A (BPA) (an antioxidant

Fig. 16.1 A dead albatross chick found on Midway Atoll in the Pacific Ocean with plastic debris in its stomach (reprinted by permission from [Elsevier]: [Marine Pollution Bulletin] [Entanglement of birds in plastics and other synthetic materials, Peter G. Ryan © 2018]

or a stabilizer) and brominated flame retardants are present in plastics. The presence of these chemicals in the water bodies through the process of leaching has been reported by Alabi et al. (2019).

The adverse effects due to the exposure of humans and animals to these chemicals are fast and becoming an environmental and public health concern. Bisphenol A is an endocrine-disrupting chemical (Mathieu-Denoncourt et al. 2015; Vom and Vandenberg 2021) and has been linked to several diseases such as DNA damage (Lusher et al. 2017) prostate and ovarian cancer, erectile dysfunction, autism, (Hengstler et al. 2011; Rochester 2013; Alabi et al. 2019) liver diseases, heart diseases, and diabetes (Lang et al. 2008). Other diseases attributed to the consumption of plastics include oxidative stress and its various linked diseases, damage of the organs, hypercalcemia and hypocalcemia, bone osteolysis, changes in gene and protein expression, mortification, and caspase-mediated cell death (Lusher et al. 2017).

16.3 Different Types of Diseases That Could Be Caused in Marine Biota as a Result of Plastic Waste Deposition

The bioavailability of pollutants from microplastics is high for marine habitats and organisms due to their ubiquitous nature in the pelagic and benthic environment and their minute sizes (Barbosa and Gimenez 2015). In the marine ecological system, plastic wastes have been documented to accumulate and adsorb various contaminants and toxins that are found in sea water in different points. Commonly adsorbed contaminants such as phenanthrene, nonylphenol, and dichlorodiphenyldichloroethylene have been reported to have serious deleterious effect and have the unique tendencies of accumulating in several folds on debris of plastics when compared to the sea water within (Duis and Coors 2016).

Similar report has it that over 260 species of organisms in the marine habitat such as fish, mammals, turtles, sea birds, and invertebrate take in or are entangled with debris of plastics which brings about a decline in their ability to move, feeding potential, reproductive efficiency, laceration, ulcers, and finally death (Chandegara et al. 2015).

Plastic materials once in the environment as microplastics or in macroforms, they pollute and accumulate in the food chain and also affect the supply of water significantly which deplete the quality of water for aquatic lives. These wastes in the environment also leach out their toxic chemical components and additives or toxins adsorbed and concentrated in them which make them bioavailable to aquatic organisms and humans (Eriksen et al. 2018). Microplastics once in the bodies of organisms through direct exposure have serious health implication such as oxidative stress, necrosis, inflammation, and genotoxicity which are connected to a wide range of deleterious health outcomes such as diabetes, cardiovascular diseases, bowel infections, chronic inflammation, and neurodegenerative diseases among others (Bashir et al. 2020).

Up to a million of sea birds are killed yearly by plastic wastes. Similar to sea turtle, on ingestion of plastic materials by seabirds, a large portion of their stomach is taken up, which sometimes causes starvation in them as their ability to feed is remarkably affected. This has led to the dead of many sea birds with their stomachs usually filled with such wastes. It has been estimated by scientists that 60% of all species of sea birds have consumed pieces of plastic materials, with an alarming projection that this could increase to 99% by 2050 (Jambeck et al. 2015).

Recent report also has it that sea turtles on ingestion of 14 pieces of plastic materials have a higher risk of dying. The younger ones are more vulnerable due to their poorer selectivity and differentiating capacity on food and drift readily with current alongside plastics. In several beaches, contamination by plastic waste becomes so pervasive to the extent of altering the reproductive rates of turtles by affecting the temperatures within the sand where they carry out incubation (Bashir et al. 2020).

The entanglement of seabirds, fishes, marine mammals, and sea turtles with plastic debris also cause suffocation, drowning, and starvation; humans are also vulnerable since most of these microplastics are eventually passed across due to their nonbiodegradability.

16.4 Modes of Action and Chemistry of Plastic Degradation

It has been established that adequate understanding of how plastics interact with the environment in real-world situations is crucial. Because of this contact, the plastics deteriorate, and their surface properties often alter, leading to the formation of new functional groups. Plastics interact with the environment and organisms after this shift. It is a dynamic environment in which the parameters are always changing.

Polyethylene terephthalate (PET), polyethylene, and polypropylene degrade by the mechanisms of thermal, biodegradation, and photodegradation. Each of the three polymers degrades at a different rate and through different paths. Photo- and thermal deterioration is similar in normal situations. Photodegradation of polyethylene causes stronger peaks in the bands on their infrared spectra that represent esters, ketones, acids, as well as other compounds.

Polypropylene is similar to polyethylene, except it is more resistant to photodegradation. The photooxidation of polyethylene terephthalate comprises the generation of hydroperoxide classes via oxidation of the CH2 groups close to the ester linkages, as well as the creation of photoproducts from the hydroperoxide species via multiple routes. Interactions with bacteria and the production of biofilms varied between the three polymers. If the sample has been already photodegraded by UV exposure, biodegradation results in a drop in carbonyl indices. These findings are supported by studies using environmental samples; however, plastic breakdown is highly dependent on local environmental circumstances, which are often a mix of those reproduced in laboratories. Some studies have found that solar radiation can fracture plastic sheets on beaches in as little as a few months to a few years, whereas PET bottles can last up to 15 years on seabeds (Fotopoulou and Karapanagioti 2017).

Plastic garbage is currently being produced at a rate of around Mt. year^{-1}. Plastics are rapidly accumulating in the environment, but our understanding of their long-term persistence is limited. The available research on ecological degradation degrees as well as mechanisms for the main categories of thermoplastic polymers is summarized in this perspective (Chamas et al. 2020). The specific surface degradation rate, a metric for harmonizing diverse types of data, is implemented and utilized to extrapolate half-lives. The precise exterior degradation amount values span a wide range, with some of the variation owing to degradation tests carried out in various normal settings (Chamas et al. 2020). In the marine environment, specific surface degradation rate for high-density polyethylene (HDPE) ranges from practically 0 to around 11 Mt. year^{-1}. This method produces a lot of intriguing results. Linear extrapolation that used a mean specific surface degradation rate for high-density polyethylene in the maritime environment yields predicted half-lives varying from 58 years (bottles) to three centuries (pipes). In the maritime environment, for instance, specific surface degradation rates for HDPE are polylactic acid (PLA) are remarkably similar, despite the fact that PLA degrades 20 times quicker than HDPE on land (Chamas et al. 2020).

The accumulations of plastic materials in the surroundings are becoming more of a worry as worldwide usage grows with their normal opposition to degradation increases. This study seeks to provide a broad overview of current understanding in areas related to polymer biodegradation, particularly poly(ethylene terephthalate). An overview of the difficulties related through plastic contamination in the aquatic environs, an account of the materials, degradability of poly(ethylene terephthalate) along with commercial manufacturing, and a general idea of the prospective aimed at biodegradable polymers as well as biodegradation of conventional polymers at present in manufacture are all included (Webb et al. 2012).

Man-made plastics that are extensively used in daily goods are common and slow-degrading polymers in the environment. The ability of microbes to accelerate degradation is of particular importance. Because of their ability to digest and metabolize man-made polymers, some of the members of the metabolically assorted species *Pseudomonas* sp. are of unlimited notice. *Pseudomonas* sp. types inaccessible through ecological conditions have been found to efficiently degrade polypropylene, polyethylene, polyurethane, polystyrene, polyethylene terephthalate, polyvinyl chloride, polyethylene glycol, polyvinyl alcohol, and polyethylene succinate. The review of current understanding on the parameters influences *Pseudomonas* sp. abilities to metabolize these various polymeric polymers and by-products. It is also deliberate for future research avenues that must be pursued in order to fully utilize *Pseudomonas* sp. potential in bioremediation systems aimed at removing plastic trash (Wilkes and Aristilde 2017).

Plastic waste disposal and reusing have become a thoughtful worldwide concern because it touches all ecosystems and living beings. Researchers used practices which include FTIR, SEM, and others to investigate polyethylene through biodegradation by assessing decisions in different structural characteristics as well as physicochemical. These evidences, however, are inadequate to establish the precise biodegradation of polyethylene. Furthermore, the sophisticated skills can be used to accurately measure polyethylene degradation by means of isotope-labelled polyethylene to determine its absorption into end products such as $13CO_2$ (Ghatge et al. 2020).

Plastics are the most unavoidable substance in our daily lives, and they play an important role in human existence. Plastics are organic polymers made up of long chains of hydrocarbons and other compounds with a high molecular mass. Plastics are extremely stable compounds that are difficult to degrade or change. As a result, plastic contamination is regarded as part of the significant menace to the ecosystem along with other living things. To solve the problem, there are physical, chemical, and biological solutions. Due to environmental limits, traditional biological approaches such as the enzymatic process of plastic breakdown are ineffective. To dispose of plastic trash, landfills and incineration are commonly used, which results in a loss in water absorption in the soil, resulting in soil infertility. When plastic garbage is dumped into waterways, it accumulates in aquatic animals, posing a threat to aquatic biotic systems. Other methods of degradation take longer and are less dependable. Nanoparticles have a unique potential for polymer degradation, as they can speed up the process by affecting the proliferation of appropriate microorganisms and acting as powerful catalysts in the degradation or transformation process. Nanoparticles come in a variety of shapes and sizes to improve physicochemical stability and degradability. The purpose of this review is to go through the various nanoparticles for plastic degradation as well as the usefulness of nanoplastics (David et al. 2020).

Plastics are incredibly valuable materials that have had a wide range of effects on our civilization. The extensive usage of these materials, on the other hand, has resulted in massive amounts of plastic contamination in both natural along with man-made ecosystems. Plastic biodegradation is an important aspect in reducing the effects of plastic contamination. Furthermore, microbes which could digest even the most refractory polymers are appearing. Biodegradable plastics, however, are being produced which are fundamentally additional susceptible to bacteriological occurrence. The natural destinies of these two forms of plastics, their chemical linkages, and the enzyme activity are involved in their disintegration. Finally, a forecast is given regarding the biotechnological application of plastic waste as a substrate, either through the use of these enzymes or through thermochemical processing (Wierckx et al. 2018).

Plastic trash is inextricably incorporated into the environment, where it degrades physicochemical qualities along with mechanical which gives to the development of plastic particles that are classified as microplastics as soon as their extent is less than 5 mm. Microplastic contamination has been recorded in a variety of environments around the world in recent years, and it is now considered a possible hazard to ecosystems and human health. However, knowledge of how plastics degrade in the environment and also how microplastics develop is relatively limited. Potential hotspots for the buildup of plastic trash were recognized important devices in addition to characterization techniques for plastic degradation, and studies on plastic breakdown reviewed in the environment were assessed. Future research should focus on identifying the important environmental characteristics and plastic features that influence degradation in order to better anticipate the destiny of plastics in various contexts along with aid in the improvement of devices to reduce plastic contamination. More research on the degradation as well as formation of microplastics, particularly nanoplastics, is needed to evaluate their ecological concerns as well as fate in the environs (Zhang et al. 2021a; b).

One of the most pressing environmental issues confronting modern society is the enormous accumulation of plastic garbage. Because of their effects on all kinds of life, these wastes are considered a severe worldwide problem. As a result, there is a pressing need to demonstrate viable eco-environmental strategies that can mitigate the harmful environmental effects of existing disposal methods. Therefore, our current understanding of the predominant processes and biodegradation efficacy of synthetic plastics appears to be inadequate. Our review attempts to highlight the function of microbes, particularly algae, in the entire plastic biodegradation process, with a particular focus on the depolarization of modern organic plastic kinds. Furthermore, the potential of insects' gut microbial consortia to digest synthetic plastic wastes is highlighted in our review. The conceptual pathway of the decomposition processes of six types of synthetic plastics is discussed in this view. These discoveries could aid in the development of bio-upcycling methods for plastic trash in order to produce useful metabolic products (Sameh et al. 2021).

Plastics are organic raw materials with a high molecular weight. Plastic polymers disruptive impacts are putting the ecosystem at risk; hence solutions to degrade them are needed. Plastics are being converted into monomers using both biological and abiotic methods. Both approaches are aimed at reducing polymers to monomers. By releasing enzymes on polymers, microbes act on monomers to degrade them. The rate of microbial degradation is influenced by both the environment and the properties of the polymer.

The rate of biological deterioration is measured using a variety of approaches. Some plastics, on the other hand, are resistant to microbiological attack. The rate of deterioration is influenced by the environment and the properties of the polymer. The rate of biological deterioration is monitored using a variety of methods. The necessity of the hour is to develop bio-based plastics that can be decomposed effectively. These polymers can be degraded to monomers, which can subsequently be converted back into petrochemical goods. This will help meet the growing need for organic fuels and may even act as a next-generation fuel. There is currently no effective method for degrading plastics, so scientists are working to create methods that will not only decompose these polymers but also result in valuable compounds. This article aims to bring together some of the most prevalent methods for degrading various types of polymers, as well as a list of probable bacteria that can feed on them (Darakhshanda et al. 2019).

Plastic pollution and its consequences have engulfed the complete planet, placing all the branches of skill on the line to find solutions to this worldwide disaster. Because a single remediation strategy to address this crisis is frequently establish to be ineffective, considerable attention has been sparked in the improvement of multidisciplinary methods – comprising biological as well as physicochemical approaches to eliminate the consequences of plastic contamination in all likely ways. Despite the variety of treatments available, the employment of organic treatments to prevent plastic deterioration has extended traction. Some of the usually utilized plastics look as if in vast forms which includes polypropylene-based bottles, polyethylene plastic bags, polystyrene styrene cups, and polyvinyl chloride pipes. Plastic nicknamed as one of the hardest polymers such as polydicyclopentadiene, acrylonitrile butadiene styrene, and polycarbonate is commonly referred to as such because they are resistant to deterioration under normal ambient conditions. They are frequently damaged in harsh as well as nonhostile conditions characterized through radiation, temperature, pH, and other factors. On the other hand, creating such extreme settings for plastic deterioration is not always practicable. Gut bacteria that can tolerate the dangerous climate of the gut environment may be attractive candidates for plastic biodegradation in this scenario. The current research looks at the gut microorganisms of various biological agents and how they can help with plastic remediation. The current analysis summarizes the load applied for plastic cleanup, as well as the microbiological prospects, obstacles, and potential breakthroughs in effective plastic remediation (Sini et al. 2021).

Every year, enormous amounts of plastic are manufactured all around the world. Plastics accumulate in the environment after being discharged, where plastic has been trash in the oceans is a major environmental problem. Plastic makes up to about 60% among all the floating wreckage in the oceans, and the amount is growing every year. In the environment such as marine, plastic polymers are bare to sunshine, physical stress, and oxidants, as well as degrade over time. To detect along with evaluation of potential environmental dangers, it is necessary to understand the degradation processes and products. Although persistent organic pollutants as well as additives which sorb to the surface of plastic have received some attention, the compounds produced by the certain degradation of the polymers themselves have not been thoroughly explored through the ecological standpoint. In this work is the obtainable data on the degradation paths along with compounds produced by the six most commonly used plastic kinds in Europe. The data was extrapolated to potential degradation processes as well as produces under the ecological circumstances seen at the ocean's surface. The kind of polymer determines the possible degradation products and pathways. UV light and oxygen are the two most critical variables that cause polymers by means of a backbone carbon to carbon to degrade, resulting in chain scission.

Abiotic degradation is likely to occur before biodegradation because reduced polymer remains created by chain scission are more vulnerable to biodegradation. When heteroatoms are present in a polymer's main chain, hydrolysis, photooxidation, as well as biodegradation take place. Plastic polymer degradation can result in little-molecular-weight polymer fragments, like oligomers along with monomers, as well as the production of new end groups, most especially carboxylic acids (Gewert et al. 2015).

16.5 Conclusion and Future Recommendation to Knowledge

This chapter has provided detailed information on the different types of diseases that could be caused in marine biota as a result of plastic waste, while modes of action involved in the plastic degradation were also discussed. The application of next-generation sequencing techniques could be applied in the characterization of some microorganism that could induce diverse diseases that could affect the marine system and which could also affect the food chain through a long time accumulation of toxic substance in the ecosystem.

Acknowledgement COA, AI, JT, OTO, and OA are grateful to the management of Edo State University Uzairue, while OFA is grateful to the management of Federal University of Technology Akure, Nigeria, for providing a support for this research. MS is thankful to the authorities of the Jiangsu University for providing an opportunity to work as an Associate Professor under the Invited Talent Programme.

References

Alabi OA, Ologbonjaye KI, Awosolu O, Alalade OE (2019) Public and environmental health effects of plastic wastes disposal: a review. J Toxicol Risk Assessment 5(2):1–13

Abreu A, Pedrotti ML (2019) Microplastics in the oceans: the solutions lie on land. J Field Action 19:62–67

Barbosa LGA, Gimenez BCG (2015) Microplastics in the marine environment: current trends and future perspectives. Mar Pollut Bull 97:5–12

Bashir I, Lone FA, Bhat RA, Mir SA, Dar ZA, Dar SA (2020) Concerns and threats of contamination on aquatic ecosystems. In: Hakeem K, Bhat R, Qadri H (eds) Bioremediation and biotechnology: sustainable approaches to pollution degradation. Springer, Cham. https://doi.org/10.1007/978-3-030-35691-0_1

Chamas A, Moon H, Zheng J, Qiu Y, Tabassum T, Jang JH, Suh S (2020) Degradation rates of plastics in the environment. ACS Sustain Chem Eng. https://doi.org/10.1021/acssuschemeng.9b06635

Chandegara VK, Cholera SP, Nandasana JN, Kumpavat MT, Patel KC (2015) Plastic packaging waste impact on climate change and its mitigation. In: Subbaiah R, Prajapati GV (eds) Water management and climate smart agriculture. Adaptation of climatic resilient water management and agriculture, vol 3. Gyan Publishing House, New Delhi, pp 404–415

Darakhshanda I, Rafia R, Rana KI (2019) Usage of potential micro-organisms for degradation of plastics. J Environ Biol 34(65):23–76

David M, Prakash L, Sangeetha J, Naik J, Thangadurai D, Thimmappa SC (2020) Degradation of plastics using nanomaterials. In: Kharissova O, Martínez L, Kharisov B (eds) Handbook of nanomaterials and nanocomposites for energy and environmental applications. Springer, Cham, pp 1–13. https://doi.org/10.1007/978-3-030-11155-7_74-1

Desforges J-PW, Galbraith M, Ross PS (2015) Ingestion of microplastics by zooplankton in the Northeast Pacific Ocean. Arch Environ Contam Toxicol 69:320–330. https://doi.org/10.1007/s00244-015-0172-5

Duis K, Coors A (2016) Microplastics in the aquatic and terrestrial environment: sources (with a specific focus on personal care products), fate and effects. Environ Sci Eur 28:2. https://doi.org/10.1186/s12302-015-0069-y

Eriksen M, Thiel M, Prindiville M, Kiessling T (2018) Microplastic: what are the solutions? In: Wagner M, Lambert S (eds) Freshwater microplastics. The handbook of environmental chemistry, vol 58. Springer, Cham, pp 273–298. https://doi.org/10.1007/978-3-319-61615-5_13

Fotopoulou KN, Karapanagioti HK (2017) Degradation of various plastics in the environment. In: The handbook of environmental chemistry. Springer International Publishing. https://doi.org/10.1007/698_2017_11

Gewert B, Plassmann MM, MacLeod M (2015) Pathways for degradation of plastic polymers floating in the marine environment. Environ Sci: Processes Impacts 17(9):1513–1521. https://doi.org/10.1039/c5em00207a

Ghatge S, Yang Y, Ahn JH, Hur HG (2020) Biodegradation of polyethylene: a brief review. Appl Biol Chem 63(1):1–14. https://doi.org/10.1186/s13765-020-00511-3

John A. Glaser (2019). Biological degradation of polymers in the environment. In Alessio Gomiero (ed) Plastics in the environment. IntechOpen. doi: https://doi.org/10.5772/intechopen.85124. Available from: https://www.intechopen.com/books/plastics-in-the-environment/biological-degradation-of-polymers-in-the-environment

Hengstler JG, Foth H, Gebel T, Kramer PJ, Lilienblum W et al (2011) Critical evaluation of key evidence on the human health hazards of exposure to bisphenol A. Crit Rev Toxicol 41:263–291

Jambeck JR, Geyer R, Wilcox C, Siegler TR, Perryman M et al (2015) Plastic waste inputs from land into the ocean. Science 347:768–771

Kehinde O, Ramonuc OJ, Babaremua KO, Justin LD (2020) Plastic wastes: environmental hazard and instrument for wealth creation in Nigeria. Heliyon 6(10):e05131. https://doi.org/10.1016/j.heliyon.2020.e05131

Lang IA, Galloway TS, Scarlett A, Henley WE, Depledge M et al (2008) Association of urinary bisphenol A concentration with medical disorders and laboratory abnormalities in adults. J Am Med Assoc 300:1303–1310

Lusher A, Hollman P, Mendoza-Hill J (2017) Microplastics in fisheries and aquaculture. Status of knowledge on their occurrence and implications for aquatic organisms and food safety. FAO Fisheries and Aquaculture Technical Paper, p. 615

Mathieu-Denoncourt J, Wallace SJ, De Solla SR, Langlois VS (2015) Plasticizer endocrine disruption: highlighting developmental and reproductive effects in mammals and non-mammalian aquatic species. Gen Comp Endocrinol 219:74–88

Mülhaupt R (2013) Green polymer chemistry and bio-based plastics: dreams and reality. Macromol Chem Phys 214(2):159–174. https://doi.org/10.1002/macp.201200439
Proshad R, Islam MS, Kormoker T, Haque MA, Mahfuzur Rahman MD, et al. (2018) Toxic effects of plastic on human health and environment: a consequences of health risk assessment in Bangladesh, Int J Health, 6: 1–5
Rochester JR (2013) Bisphenol A and human health: a review of the literature. Reprod Toxicol 42: 132–155
Sameh SA, Tamer E, Rania A, Daochen Z, Yehia AGM, Eleni K, Metwally AM, Michael K, Jianzhong S (2021) Plastic wastes biodegradation: mechanisms, challenges and future prospects. Sci Total Environ 780:146,590. https://doi.org/10.1016/j.scitotenv.2021.146590
Sini F, Rebello S, Mathachan Aneesh E, Sindhu R, Binod P, Singh S, Pandey A (2021) Bioprospecting of gut microflora for plastic biodegradation. Bioengineered 12(1):1040–1053. https://doi.org/10.1080/21655979.2021.1902173
Vom SFS, Vandenberg LN (2021) Update on the health effects of bisphenol a: overwhelming evidence of harm. Endocrinology 162(3):bqaa171. https://doi.org/10.1210/endocr/bqaa171
Webb H, Arnott J, Crawford R, Ivanova E (2012) Plastic degradation and its environmental implications with special reference to poly(ethylene terephthalate). Polymers 5(1):1–18. https://doi.org/10.3390/polym5010001
Wierckx N, Narancic T, Eberlein C, Wei R, Drzyzga O, Magnin A et al (2018) Plastic biodegradation: challenges and opportunities. In: Steffan R (ed) Consequences of microbial interactions with hydrocarbons, oils, and lipids: biodegradation and bioremediation. Springer, Cham, pp 1–29. https://doi.org/10.1007/978-3-319-44535-9_23-1
Wilcox C, Van Sebille E, Hardesty BD (2015) Threat of plastic pollution to seabirds is global, seabirds, pervasive and increasing. Proc Natl Acad Sci U S A 38:11,899–11,904
Wilkes RA, Aristilde L (2017) Degradation and metabolism of synthetic plastics and associated products by *pseudomonas* sp.: capabilities and challenges. J Appl Microbiol 123(3):582–593. https://doi.org/10.1111/jam.13472
Zhang K, Amir HH, Aleksandra T, Zhang Y, James KHF, Chenxi W, Paul KSL (2021a) Understanding plastic degradation and microplastic formation in the environment: a review. Environ Pollut 274:116,554. https://doi.org/10.1016/j.envpol.2021.116554
Zhang N, Bin LY, He HR et al (2021b) You are what you eat: microplastics in the faeces of young men living in Beijing. Sci Total Environ 767:144,345. https://doi.org/10.1016/j.scitotenv.2020.144345

Correction to: Impact of Plastic Waste on the Marine Biota

Mohd. Shahnawaz, Manisha K. Sangale, Zhu Daochen, and Avinash B. Ade

Correction to:
M. Shahnawaz et al. (eds.),
Impact of Plastic Waste on the Marine Biota,
https://doi.org/10.1007/978-981-16-5403-9

The affiliation of the editors Mohd. Shahnawaz and Zhu Daochen in the front matter and Chapter 2 has been revised as Biofuels Institute, School of Environment and Safety Engineering, Jiangsu University, Zhenjiang, China.

The affiliation of the author Vasudeo P. Zambare in Chapter 14 has been revised as R&D Department, Om Biotechnologies, Nashik, Maharashtra, India and Aesthetika Eco Research Pvt Ltd, Nashik, Maharashtra, India.

The updated version of these chapters can be found at
https://doi.org/10.1007/978-981-16-5403-9
https://doi.org/10.1007/978-981-16-5403-9_2
https://doi.org/10.1007/978-981-16-5403-9_14

M. Shahnawaz et al. (eds.), *Impact of Plastic Waste on the Marine Biota*,
https://doi.org/10.1007/978-981-16-5403-9_17

MIX
Papier aus verantwortungsvollen Quellen
Paper from responsible sources
FSC® C105338

If you have any concerns about our products,
you can contact us on
ProductSafety@springernature.com

In case Publisher is established outside the EU,
the EU authorized representative is:
Springer Nature Customer Service Center GmbH
Europaplatz 3, 69115 Heidelberg, Germany

Printed by Libri Plureos GmbH
in Hamburg, Germany